NEUROMETHODS

AF786592

Series Editor
Wolfgang Walz
University of Saskatchewan
Saskatoon, SK, Canada

For other titles published in this series, go to
www.springer.com/series/7657

Animal Models for Retinal Diseases

Edited by

Iok-Hou Pang

Glaucoma Research, Alcon Research Ltd, Fort Worth, TX, USA

Abbot F. Clark

The North Texas Eye Research Institute and Department of Cell Biology and Anatomy, University of North Texas Health Science Center, Fort Worth, TX, USA

Humana Press

Editors
Iok-Hou Pang
Glaucoma Research
Alcon Research Ltd
Fort Worth, TX
USA
iok-hou.pang@alconlabs.com

Abbot F. Clark
The North Texas Eye Research Institute
and Department of Cell Biology and Anatomy
University of North Texas Health Science Center
Fort Worth, TX
USA
abclark@hsc.unt.edu

ISSN 0893-2336 e-ISSN 1940-6045
ISBN 978-1-4939-5750-7 ISBN 978-1-60761-541-5 (eBook)
DOI 10.1007/978-1-60761-541-5
Springer Dordrecht Heidelberg London New York

Softcover re-print of the Hardcover 1st edition 2010

Cover illustration: Inset: Cross section of the human eye (provided by National Eye Institute, National Institute of Health); Background: Cross section of the mouse retina (provided by Dr. Robert Mullins, University of lowa).

Printed on acid-free paper

Humana Press is part of Springer Science+Business Media (www.springer.com)

Preface to the Series

Under the guidance of its founders Alan Boulton and Glen Baker, the Neuromethods series by Humana Press has been very successful, since the first volume appeared in 1985. In about 17 years, 37 volumes have been published. In 2006, Springer Science + Business Media made a renewed commitment to this series. The new program will focus on methods that are either unique to the nervous system and excitable cells or which need special consideration to be applied to the neurosciences. The program will strike a balance between recent and exciting developments like those concerning new animal models of disease, imaging, in vivo methods, and more established techniques. These include immunocytochemistry and electrophysiological technologies. New trainees in neurosciences still need a sound footing in these older methods in order to apply a critical approach to their results. The careful application of methods is probably the most important step in the process of scientific inquiry. In the past, new methodologies led the way in developing new disciplines in the biological and medical sciences. For example, Physiology emerged out of Anatomy in the nineteenth century by harnessing new methods based on the newly discovered phenomenon of electricity. Nowadays, the relationships between disciplines and methods are more complex. Methods are now widely shared between disciplines and research areas. New developments in electronic publishing also make it possible for scientists to download chapters or protocols selectively within a very short time of encountering them. This new approach has been taken into account in the design of individual volumes and chapters in this series.

Wolfgang Walz

Preface

Retinal diseases are leading causes of irreversible visual impairment and blindness, affecting over 100 million individuals worldwide. Age-related macular degeneration and glaucoma are the leading causes of blindness in the elderly, while diabetic retinopathy is the leading cause of visual impairment in middle-aged individuals. The prevalence of all three of these retinal diseases will continue to increase as our world's populations continue to age and diabetes becomes endemic. There are a wide variety of additional important retinal diseases, including various acquired retinal degenerations (e.g., retinitis pigmentosa), maculopathies, retinal vascular disorders (e.g., ischemic retinopathies such as central retinal vein occlusion, sickle cell retinopathy, retinopathy of prematurity), and inflammatory retinopathy, each of which has devastating consequences on our most important sense perception.

Appropriate study models, especially animal models, are essential to the understanding of the etiology, pathology, and progression of these diseases. They are also critical to the evaluation, development, and improvement of therapeutic strategies for these diseases. The overall objective of this book is to provide a survey of valuable techniques as well as animal models for the prominent retinal diseases. The book starts with an overview of the morphology of the retina, visual behavior, and genetics and genomics approaches for retinal research, followed by animal models for the research of specific human retinal diseases, e.g., retinal degeneration, age-related macular degeneration, retinopathy of prematurity, diabetic retinopathy, glaucoma, retinal ischemia, and retinal inflammation. Each chapter was written by recognized experts in their respective fields. We hope that this book is valuable to ocular investigators and ophthalmologists currently working in the area of retinal diseases and ophthalmology. Its detailed and practical descriptions of the models should also appeal to those interested in entering this fascinating and important field of research.

Fort Worth, TX — *Iok-Hou Pang, Ph.D.*
Fort Worth, TX — *Abbot F. Clark, Ph.D.*

Contents

Contributors

JOSEPH CAPRIOLI • *Department of Ophthalmology, David Geffen School of Medicine, University of California Los Angeles, Los Angeles, CA, USA*

RACHEL R. CASPI • *Laboratory of Immunology, National Eye Institute, National Institutes of Health, Bethesda, MD, USA*

ABBOT F. CLARK • *The North Texas Eye Research Institute and Department of Cell Biology and Anatomy, University of North Texas Health Science Center, Fort Worth, TX, USA*

ROBERT M. DOUGLAS • *Department of Ophthalmology and Visual Sciences, University of British Columbia, Vancouver, BC, Canada*

CHRISTIAN GRIMM • *Lab for Retinal Cell Biology, Department of Ophthalmology and Center for Integrative Human Physiology, University of Zurich, Zurich, Switzerland*

RONALD S. HARWERTH • *College of Optometry, University of Houston, Houston, TX, USA*

JOE G. HOLLYFIELD • *Department of Ophthalmology, Cole Eye Institute, Cleveland Clinic Lerner College of Medicine, Cleveland Clinic, Cleveland, OH, USA*

REIKO HORAI • *Laboratory of Immunology, National Eye Institute, National Institutes of Health, Bethesda, MD, USA*

GARETH R. HOWELL • *The Jackson Laboratory, The Howard Hughes Medical Institute, Bar Harbor, ME, USA*

SIMON W.M. JOHN • *The Jackson Laboratory, The Howard Hughes Medical Institute, Bar Harbor, ME, USA*
Department of Ophthalmology, Tufts University of Medicine, Boston, MA, USA

SANDRINE JOLY • *Lab for Retinal Cell Biology, Department of Ophthalmology, University of Zurich, Zurich, Switzerland*

MALAIKA KURZ-LEVIN • *Department of Ophthalmology, University Eye Hospital, Zurich, Switzerland*

LISA KUTTNER-KONDO • *Department of Ophthalmology, Cole Eye Institute, Cleveland Clinic Lerner College of Medicine, Cleveland Clinic, Cleveland, OH, USA*

JACKY MAN KWONG KWONG • *Department of Ophthalmology, David Geffen School of Medicine, University of California Los Angeles, Los Angeles, CA, USA*

TREVOR J. MCGILL • *Casey Eye Institute, Oregon Health and Science University, Portland, OR, USA*

CASEY M. MILLER • *Alcon Research Ltd, Fort Worth, TX, USA*

ROBERT F. MULLINS • *Department of Ophthalmology and Visual Sciences, The University of Iowa, Iowa City, IA, USA*

STEPHAN C.F. NEUHAUSS • *Institute of Zoology, University of Zurich, Zurich, Switzerland*

IOK-HOU PANG • *Glaucoma Research, Alcon Research Ltd, Fort Worth, TX, USA*

JOHN S. PENN • *Vanderbilt University School of Medicine, Nashville, TN, USA*

GLEN T. PRUSKY • *Department of Physiology and Biophysics, Weill Cornell Medical College, New York, NY, USA*
Burke Medical Research Institute, White Plains, NY, USA
MARIJANA SAMARDZIJA • *Lab for Retinal Cell Biology, Department of Ophthalmology, University of Zurich, Zurich, Switzerland*
JESSICA M. SKEIE • *Department of Ophthalmology and Visual Sciences, The University of Iowa, Iowa City, IA, USA*
ADRIAN M. TIMMERS • *Alcon Research Ltd, Fort Worth, TX, USA*
SUSAN E. YANNI • *Vanderbilt University School of Medicine, Nashville, TN, USA*
LI ZHU • *Alcon Research Ltd, Fort Worth, TX, USA*

Chapter 1

Essentials of Retinal Morphology

Robert F. Mullins and Jessica M. Skeie

Abstract

The conversion of wavelengths of light into information useful to the brain requires a tremendous degree of anatomic and functional specialization. The mammalian retina is a remarkably refined and adapted tissue that is capable of light detection, processing, and transmission of information to other sites in the central nervous system. In this chapter, we provide a brief overview of the anatomical features of the mammalian retina and discuss regional variability observed in the eyes of humans and other species.

Key words: Retinal morphology, Retinal layers, Retinal cells

1. Introduction

The conversion of wavelengths of light into information useful to the brain is a daunting challenge. The mammalian retina is a remarkably refined and adapted tissue that is capable of light detection, for both forming images and for nonvisual light perception (for example, regulating circadian rhythms). These processes require an intricate anatomy and involve at least three different classes of light sensing cells, a set of neurons with long axons that project to the brain, and dozens of subtypes of intermediate neurons that modulate interactions between photoreceptor and ganglion cells. In this chapter, we provide a brief overview of the anatomical features of the mammalian retina.

The major axis that is useful in understanding ocular anatomy is the "inner-outer axis." The inner retina is the portion of the retina closest to the vitreous, whereas the outer retina is the portion of the retina closest to the choroid. Thus, one can remember the relative positions of the internal limiting membrane and external limiting membrane, inner and outer segments, and so on by

Iok-Hou Pang and Abbot F. Clark (eds.), *Animal Models for Retinal Diseases*, Neuromethods, vol. 46
DOI 10.1007/978-1-60761-541-5_1,

this axis. In addition, central and peripheral are terms that are also valuable in understanding how the retina varies regionally. We discuss this further below.

2. Layers of the Retina

The retina is generally considered to have two major components, the neurosensory retina and the pigment epithelium of the retina (RPE). Although the morphological features of the neural retina and RPE do not appear consistent with grouping these tissues together, during embryonic development the RPE and neural retina are derived from the same layer of neuroepithelium, and thus both are neuroectoderm and are considered "retina".

A casual examination of the histology of the neural retina (Fig. 1) reveals its characteristic lamination in which three principal layers of nuclei are ordered perpendicular to the inner–outer axis. These are, from internal to external, the relatively thin ganglion cell layer (the location of cell bodies of the various classes of ganglion cells and displaced amacrine cells), the inner nuclear

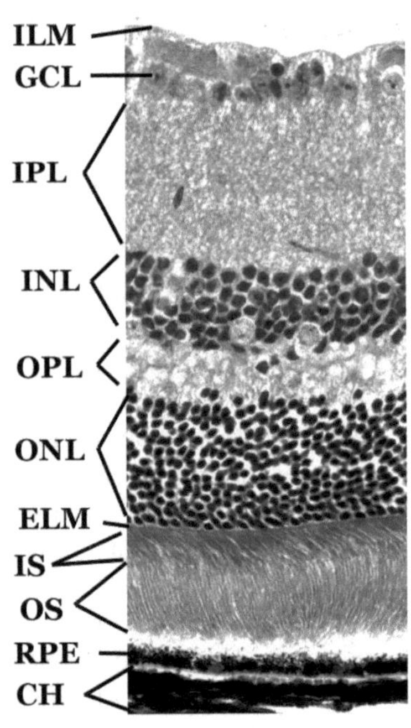

Fig. 1. Basic morphological features of the mouse retina. Hematoxylin–eosin-stained section (*Left*) shows the characteristic retinal lamination. Depicted are the internal limiting membrane (ILM), ganglion cell layer (GCL), inner plexiform layer (IPL), inner nuclear layer (INL), outer plexiform layer (OPL), outer nuclear layer (ONL), external limiting membrane (ELM), inner segments (IS), outer segments (OS), retinal pigment epithelium (RPE), and choroid (CH)

layer (the layer housing the nuclei of bipolar cells, amacrine cells, horizontal cells, and Muller cells), and the outer nuclear layer, in which the nuclei of rod and cone photoreceptors reside. These layers are depicted in Fig. 1.

Other compartments of the retina that are not occupied by cell nuclei include the two plexiform, or synaptic layers. The inner plexiform layer is located between the ganglion cell and inner nuclear layer, and the outer plexiform layer resides between the outer and inner nuclear layers. The retinal synaptic layers contain intricate arrangements of axons and dendrites. This circuitry supports early visual processing, allowing events such as adjustment to vastly different environmental levels of light, and detection of the movement of stimuli across the retina.

In addition, two compartments referred to as the internal and external limiting membranes (or inner and outer limiting membranes) are recognized. The nomenclature of "membrane" is a carryover from early histological observations in which the term membrane was used to describe any band-like structure observed on histology. The eye includes several "membranes," in the cornea, retina, and choroid. The external limiting membrane (or outer limiting membrane) lies between the outer nuclear layer and photoreceptor inner segments, and is comprised of intracellular junctions between photoreceptor cells and Muller cells. This layer can be variably observed in some preparations of retina, particularly using phase contrast or differential interference optics. Its molecular constituents include the adhesion molecules CD44 and ICAM-1 (1, 2). The internal limiting membrane, which is the interface between the retina and vitreous, is an attachment site for Muller cells; its composition includes extracellular matrix molecules such as collagen type IV and laminin typically found in basal laminae.

3. Cell Types of the Retina

The retina contains a rich diversity of cell types, including light-sensing neurons not found anywhere else in the mammalian body. This complexity was first noted by Santiago Ramon y Cajal, whose precise illustrations of silver-impregnated retinal neurons remain striking and accurate over a hundred years after they were created. This organization and complexity, and the fact that the retina is an accessible component of the central nervous system, is in part responsible for its popularity in research.

The RPE is the outermost layer of cells of the retina. It is a simple cuboidal epithelium, which contains large melanosomes that, along with those in the choroidal melanocytes, quench photons that are not absorbed during their first pass through the retina. Apart from this role in minimizing light scattering, the RPE displays a wide range of biological functions, including phagocytosis of outer segment discs, creation of the outer aspect of the blood–retinal barrier, maintenance of the choroidal vasculature, and recycling of vitamin A derivatives during the visual cycle. The RPE can be appreciated in unstained sections by its brown-black appearance due to the presence of melanosomes. The presence of melanin and, in older primate eyes, lipofuscin, can make detection of immunohistochemical signals challenging on these cells. In albino animals, species with a tapetum lucidum (in which the RPE is regionally unpigmented), or zebrafish raised in pigment inhibitor, structural features of the RPE can be more readily visualized and stained. Molecular markers for the RPE include anti-RPE65 antibodies (localized to the cytosol), anti-bestrophin antibodies (localized to the basolateral membrane), and anti-CRALBP antibodies, the last of which also recognize Muller cells. During some disease phenotypes, the RPE can become dedifferentiated and downregulate cell-type specific genes (3).

The retinal photoreceptor cells are the exquisitely specialized neurons in the outer retina. The light gathering component of these cells is called the outer segment, which is comprised of stacks of highly organized membranous discs enriched in an opsin protein, a heterotrimeric G protein (transducin) and other biochemical machinery necessary for phototransduction. On its inner aspect, the outer segment connects to the inner segment via a primary cilium with nine circumferential microtubule doublets and no core microtubules (i.e., the "9 + 0" arrangement observed ultrastructurally in primary, nonmotile cilia). The fact that phototransduction proteins are all delivered to the outer segment through this isthmus makes the connecting cilium a strategic site essential for the maintenance of photoreceptor cells, and it is notable that several cilium disrupting mutations result in photoreceptor degenerations (4). The inner segment of the photoreceptor cell contains a large number of organelles responsible for oxidative respiration and protein synthesis. Rods and cones can be readily distinguished by the appearance of their inner segments, which are narrow and cylindrical in rods and teardrop shaped in cones. Photoreceptor cells also have cell bodies with nuclei in the outer nuclear layer, axons that extend to the outer plexiform layer, and synaptic terminals (pedicles in cones and spherules in rods) that form intricate synaptic contacts with horizontal and bipolar cells. Useful markers for photoreceptor cells include anti-rhodopsin antibodies for rod outer segments, anti-cone opsin antibodies for cone outer segments, peanut agglutinin lectin for cone inner/outer segments, and various antibodies that recognize proteins involved in phototransduction.

The diversity of inner retinal neurons, based on profiles of shape, neurotransmitter receptor profile, sublayers in which their synapses form, and physiological responses to light is somewhat baffling to an outsider. In this brief discussion of structural features of the retina we make no attempt to demystify these complex patterns but refer interested readers to a few excellent review articles (5–7). Bipolar cells span from the outer plexiform layer to the inner plexiform layer, in which they form synaptic contacts with photoreceptor cells and ganglion cells, respectively. These contacts are further modulated by interdigitating horizontal and amacrine cells. The nuclei of bipolar cells are found in the inner nuclear layer. Horizontal cells are arranged in the outer aspect of the inner nuclear layer and, in contrast to bipolar cells, show an extended lateral morphology. In preparations of mouse eyes these cells frequently exhibit a large pale staining cytoplasm on the outer margin of the inner nuclear layer. Amacrine cells comprise a very large and diverse group, based on their structure, dendritic architecture and distribution of histochemical markers (8, 9). These cells are present in both the inner nuclear layer and the ganglion cell layer where they relay impulses and modify the functions of other retinal neurons. Markers for amacrine cells depend on cell class; antibodies directed against species of calcium binding proteins serve as markers for some amacrine cells.

A number of classes of ganglion cells have also been recognized, differing in their morphology, size, and the synaptic layer in which their dendrites ramify. Ganglion cells are distinguished in part by having long axons that run along the nerve fiber layer, perpendicular to the visual axis, and then turn and exit the eye through the optic nerve on their path the lateral geniculate nucleus and other cells in the central nervous system. Recently, a class of photosensitive ganglion cells containing a novel form of opsin has been discovered. These cells detect light but do not participate in the formation of images; instead they sense overall illumination. For a more detailed discussion of melanopsin and other ganglion cells, a few recent reviews are suggested (10, 11). Ganglion cells can be labeled with antibodies directed against synuclein, neurofilament heavy chains, and Thy1/CD90.

In addition to its neuronal populations, the retina also contains glial cells that support the retinal microenvironment through contributing to the inner aspect of the blood–retinal barrier, trafficking vitamin A derivatives, removing wastes and playing a space filling, structural role. These include Muller cells, also called radial glia, that span from the internal limiting membrane to the external limiting membrane, with nuclei in the inner nuclear layer. These cells possess pyramid-shaped endfeet along the internal limiting membrane and microvilli at the external limiting membrane, and show an electron dense cytosol on transmission electron microscopy. They also exhibit numerous smaller projections that extend into the spaces between retinal neurons. Markers for Muller cells include antibodies directed against cellular retinaldehyde-binding protein (also expressed in the RPE); glial fibrillar acidic protein is expressed as well but may be detectable only when Muller cells become activated, observed in some diseases. Astrocytes, the other type of true glia in the retina, are present in the ganglion cell layer and optic nerve head. These cells serve auxiliary functions in the retina, including contributing to the blood–retina barrier (12), and express glial fibrillar acidic protein and vimentin under normal conditions. Microglia are also found in the retina. These cells are not true neuroblast-derived glia, but are rather of hematopoietic origin, and can be detected in the human retina using antibodies directed against leukocytes (for example, MHC antigens, CD45). There has been considerable recent interest in these cells in age-related eye disease (for example (13–15)).

Examples of immunohistochemical markers expressed in retina are depicted in Fig. 2.

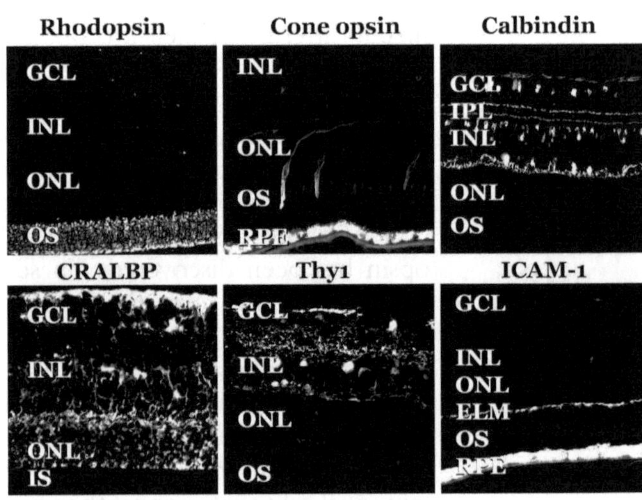

Fig. 2. Sections of mammalian eyes labeled with antibodies against different cell-type specific markers. Rhodopsin (*top left*) is present in rod outer segments on a normal mouse eye. Blue cone opsin (*top middle*) labeling is restricted to cone cell subsets in a human macula (note the autofluorescence of the RPE due to lipofuscin, commonly observed in normal aged human eyes). Calbindin reactivity in mouse (*top right panel*) is present in horizontal cells and other inner retinal neurons, including three sublayers of synapses of the inner plexiform layer. Human Muller cells are labeled with antibodies directed against cellular retinaldehyde-binding protein (CRALBP), whereas Thy1 reactivity in human eyes shows labeling predominantly of ganglion cells. ICAM-1 labeling in retina is present in the external limiting membrane. Abbreviations as given in Fig. 1

4. Vascular Supply

Cells of the inner retina are nourished by a set of continuous capillaries arising from the central retinal artery, while the outer retina, including the photoreceptor cells and RPE, is supplied by the choriocapillaris, arising from posterior ciliary arteries that breach the sclera at several positions (16). Structurally and functionally, the retinal capillaries and choroidal capillaries are quite distinct. The choriocapillaris endothelium is fenestrated and permeable to large macromolecules, presumable due to the actions of vascular endothelial growth factor (17). The diameters of choroidal capillaries are comparatively large. In contrast, the retinal capillary endothelium is of the nonfenestrated continuous type, with a large degree of cytosol and pericyte and glial concentric barriers around each capillary (18), which helps to preserve the highly regulated, defined milieu of the neural retina.

Endogenous alkaline phosphatase has been described on the endothelial cells of the arteries, arterioles, and capillaries (19, 20). Although the exact role of alkaline phosphatase in the retina and choroid is not fully understood, it can be used to visualize the vascular network of the retina and choroid in several species, as shown in Fig. 3.

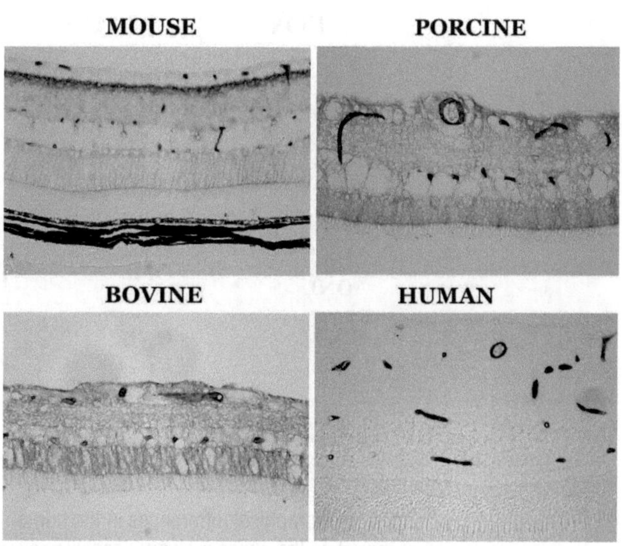

Fig. 3. Vascular endothelial cell labeling using alkaline phosphatase substrate in several species. The enzyme alkaline phosphatase is an enzyme present on the endothelium of the inner retinas of mouse, porcine, bovine, and human eyes. This labeling technique provides a reliable means of histochemically visualizing the vascular networks of the retina and can provide useful insights in terms of vascular growth or dropout, two common pathobiologies correlated with retinal disease

5. Topographic Variability in the Retina with Emphasis on Primate Eyes

Structural features of the retina vary in many species according to topographic position. Humans and nonhuman primates have a distinctive organization of the retina in which there are impressive central-peripheral differences in organization and cell density (Fig. 4). These species, in which rods vastly outnumber cones, possess a region called the fovea centralis, which has a very high density of photoreceptors. In humans and diurnal primates, the center of the fovea contains cones and no rods (21, 22), with variations in topography in nocturnal species (23). This high acuity, binocular vision is useful for tasks requiring depth perception such as seeing to leap from branch to branch. In humans, the density of cones is at its peak in the center of the fovea and drastically decreases to a consistent density in the periphery. These foveal cones are morphologically much more tapered than cones in the peripheral retina, although short wavelength cones (or S-cones), with a larger diameter than red and green cones, partially disrupt the packing of cones at the foveal center due to their larger dimensions.

The concentration of rods increases concentrically with increasing distance from the center of the fovea, with the highest packing of rods at 4.5 mm from the center of the fovea. The density of

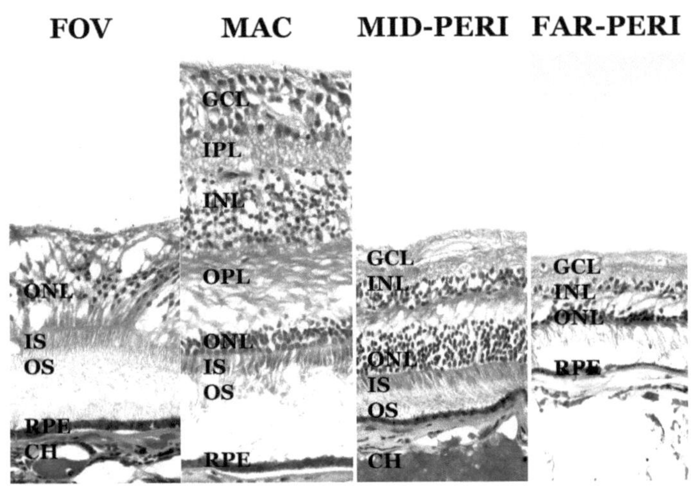

Fig. 4. Topographical differences in the human retina. In the fovea centralis (FOV) region of the retina the cones are very densely packed, having the highest density of cones per mm squared. There are no rods, ganglion cells, or bipolar cells in this region. Surrounding the fovea, the macular region just outside of the fovea is depicted (MAC). In this region, the thickness of each of the retinal layers is quite high. The macular ganglion cell layer is more than one cell thick, and there is a prominent Henle's fiber layer, where the OPL contains numerous cone axons. The retinal layers are progressively thinner in the mid-periphery and far periphery

cones falls dramatically at this distance from the fovea, displaying a consistent density in the peripheral retina, only slightly higher in the nasal hemisphere than in the temporal hemisphere. The optic nerve head, the site at which ganglion cell axons exit the retina, is free of both rods and cones.

The spatial organization of ganglion cells is very similar to that of the cones. In the macula, the ganglion cell layer is more than one nucleus in thickness. This is the only region where this is true. The high density of cones and associated ganglion cells in the macula are required in this region for high spatial acuity. Peripheral to the fovea, ganglion cells have higher densities in the nasal and superior hemispheres in contrast to the temporal and inferior hemispheres (24).

Histological and biochemical studies in human have found some molecular heterogeneity in macular and peripheral regions (see, for example (25–31)). Understanding the molecular profiles of macular and extramacular areas may be helpful in understanding a number of human diseases, in which there is sparing of either macular cells (e.g., some forms of retinitis pigmentosa) or peripheral cells (e.g., age-related macular degeneration).

Apart from primates, other orders of mammals show distinct distributions of photoreceptor cells and other elements (32, 33). One common variation is the presence of a visual streak, a horizontal band with enriched photoreceptor density similar in form but different in geographic distribution to the area centralis. This morphological adaptation permits a wide field of view useful for monitoring the horizon for approaching predators. In addition, superior–inferior gradients of photoreceptor cell densities and, an asymmetric, superior tapetum lucidum (responsible for reflective eyes of nocturnal animals) with nonpigmented RPE cells, are all indicators of regional specificity in mammals lacking an anatomic macula.

All of these regional anatomic specializations require that care be taken in evaluating mammalian eyes when comparing the effects of an intervention or a mutation on retinal morphology. For example, it is easy to be misled – and infer a pathologic phenotype – by comparing features like outer nuclear layer thickness collected from nonequivalent regions of the retina of treated vs. untreated animals. Thus the scientist interested in animal models of retinal disease needs to take into account the regional specializations of the model s/he is studying.

6. Closing Remarks

The retina and its supporting tissues are complex. Specializations in the retina include: migration of developing neurons and glia to distinct positions during development in an orderly array;

intricate details of synaptic wiring; dependence of retinal physiology on blood–retinal barriers and appropriate ionic conditions; removal of cellular debris without stimulating inflammation; axonal transport of macromolecules from ganglion cell bodies to distant synapses in the brain; and high demand for oxygen and other nutrients by photoreceptor cells.

We wish to make two points about these functional specializations. First, these functions are possible only because of a tremendous level of interdependency and interaction of cell types. In principal, the number of cells required for vision is a light-sensing photoreceptor cell and a cell to carry the signal to the higher processing centers of the brain. In contrast, the practical requirements of this process in vertebrates necessitate the profound number of supporting and modulating cell types found in the retina. Interactions between neuronal, glial, epithelial, endothelial, and myeloid cells create the exquisite process of vision. Understanding these interactions at a molecular level is a major goal of vision science.

Secondly, each of these specializations can, and does, malfunction in humans with different eye diseases. For this reason, the study of animal models of retinal diseases in which genetic, anatomic, and physiologic features can be manipulated continues to greatly improve our understanding of human disease. Current and future treatments for blinding diseases depend on a deeper understanding of the biology of the retina, which can, thankfully, be appreciated through the study of animal eyes.

Acknowledgments

The authors wish to thank the Iowa Lions Eye Bank for procuring the human tissue described in this report, as well as the technical assistance of Jasmine Hernandez, Elizabeth Faidley, Marissa Olvera, and Megan Riker in preparing the histology samples. We also thank Dr. Stewart Thompson for his helpful comments on the manuscript.

References

1. Duguid I, Boyd A, Mandel T (1992) Adhesion molecules are expressed in the human retina and choroid. Curr Eye Res 11:153–159
2. Chaitin MH, Wortham HS, Brun-Zinkernagel AM (1994) Immunocytochemical localization of CD44 in the mouse retina. Exp Eye Res 58:359–365
3. Guidry C, Medeiros NE, Curcio CA (2002) Phenotypic variation of retinal pigment epithelium in age-related macular degeneration. Invest Ophthalmol Vis Sci 43:267–273
4. Luby-Phelps K, Fogerty J, Baker SA, Pazour GJ, Besharse JC (2008) Spatial distribution of intraflagellar transport proteins in vertebrate photoreceptors. Vision Res 48:413–423
5. Luo DG, Xue T, Yau KW (2008) How vision begins: an odyssey. Proc Natl Acad Sci USA 105:9855–9862

6. Demb JB (2008) Functional circuitry of visual adaptation in the retina. J Physiol 586: 4377–4384
7. Masland RH (2005) The many roles of starburst amacrine cells. Trends Neurosci 28: 395–396
8. Masland RH (1988) Amacrine cells. Trends Neurosci 11:405–410
9. Kolb H (1997) Amacrine cells of the mammalian retina: neurocircuitry and functional roles. Eye 11(Pt 6):904–923
10. Berson DM (2007) Phototransduction in ganglion-cell photoreceptors. Pflugers Arch 454:849–855
11. Hankins MW, Peirson SN, Foster RG (2008) Melanopsin: an exciting photopigment. Trends Neurosci 31:27–36
12. Kaur C, Foulds WS, Ling EA (2008) Blood-retinal barrier in hypoxic ischaemic conditions: basic concepts, clinical features and management. Prog Retin Eye Res 27: 622–647
13. Penfold PL, Madigan MC, Gillies MC, Provis JM (2001) Immunological and aetiological aspects of macular degeneration. Prog Retin Eye Res 20:385–414
14. Combadiere C, Feumi C, Raoul W et al (2007) CX3CR1-dependent subretinal microglia cell accumulation is associated with cardinal features of age-related macular degeneration. J Clin Invest 117:2920–2928
15. Caicedo A, Espinosa-Heidmann D, Piña Y, Hernandez E, Cousins S (2005) Blood-derived macrophages infiltrate the retina and activate Muller glial cells under experimental choroidal neovascularization. Exp Eye Res 81:38–47
16. Hayreh S (1975) Segmental nature of the choroidal vasculature. Br J Ophthalmol 59:631–648
17. Saint-Geniez M, D'Amore PA (2004) Development and pathology of the hyaloid, choroidal and retinal vasculature. Int J Dev Biol 48: 1045–1058
18. Frank RN, Turczyn TJ, Das A (1990) Pericyte coverage of retinal and cerebral capillaries. Invest Ophthalmol Vis Sci 31:999–1007
19. Lutty GA, McLeod DS (2005) Phosphatase enzyme histochemistry for studying vascular hierarchy, pathology, and endothelial cell dysfunction in retina and choroid. Vision Res 45:3504–3511
20. Lessell S, Kuwabara T (1964) Phosphatase histochemistry of the eye. Arch Ophthalmol 71:851–860
21. Osterberg G (1935) Topography of the layer of rods and cones in the human retina. Acta Ophthalmol 6:1–103
22. Curcio CA, Sloan KR, Kalina RE, Hendrickson AE (1990) Human photoreceptor topography. J Comp Neurol 292:497–523
23. Finlay BL, Franco EC, Yamada ES et al (2008) Number and topography of cones, rods and optic nerve axons in new and old world primates. Vis Neurosci 25:289–299
24. Curcio CA, Allen KA (1990) Topography of ganglion cells in human retina. J Comp Neurol 300:5–25
25. Ethen CM, Reilly C, Feng X, Olsen TW, Ferrington DA (2006) The proteome of central and peripheral retina with progression of age-related macular degeneration. Invest Ophthalmol Vis Sci 47:2280–2290
26. Gulcan HG, Alvarez RA, Maude MB, Anderson RE (1993) Lipids of human retina, retinal pigment epithelium, and Bruch's membrane/choroid: comparison of macular and peripheral regions. Invest Ophthalmol Vis Sci 34:3187–3193
27. Ishibashi K, Tian J, Handa JT (2004) Similarity of mRNA phenotypes of morphologically normal macular and peripheral retinal pigment epithelial cells in older human eyes. Invest Ophthalmol Vis Sci 45:3291–3301
28. van Soest SS, de Wit GM, Essing AH et al (2007) Comparison of human retinal pigment epithelium gene expression in macula and periphery highlights potential topographic differences in Bruch's membrane. Mol Vis 13:1608–1617
29. Bowes Rickman C, Ebright JN, Zavodni ZJ et al (2006) Defining the human macula transcriptome and candidate retinal disease genes using EyeSAGE. Invest Ophthalmol Vis Sci 47:2305–2316
30. Mullins R, Skeie J, Malone E, Kuehn M (2006) Macular and peripheral distribution of ICAM-1 in the human choriocapillaris and retina. Mol Vis 12:224–235
31. Mullins R, Kuehn M, Faidley E, Syed N, Stone E (2007) Differential macular and peripheral expression of bestrophin in human eyes and its implication for Best disease. Invest Ophthalmol Vis Sci 48:3372–3380
32. Mowat FM, Petersen-Jones SM, Williamson H et al (2008) Topographical characterization of cone photoreceptors and the area centralis of the canine retina. Mol Vis 14: 2518–2527
33. Goodchild AK, Ghosh KK, Martin PR (1996) Comparison of photoreceptor spatial density and ganglion cell morphology in the retina of human, macaque monkey, cat, and the marmoset *Callithrix jacchus*. J Comp Neurol 366:55–75

Chapter 2

Visual Behavior

Robert M. Douglas, Trevor J. McGill, and Glen T. Prusky

Abstract

Because repairing visual dysfunction is the primary goal of therapy for retinal disease, a quantification of visual function is imperative for the evaluation of potential treatments for these diseases. The Visual Water Task and the Virtual Optokinetic System have been developed to conduct behavioral tests of vision in rodent models of retinal disease. These tests are less invasive and often more sensitive than physiological or anatomical measures of retinal function. Moreover, discrepancies between different measures of retinal function suggest that central and retinal adaptations may complicate the disease process.

Key words: Vision, Retinal disease, Behavior, Acuity

1. Why Measure Vision?

Typically a disturbance in vision, or outright vision loss is what prompts an initial visit to the doctor's office. The task of the clinician then is to identify the underlying pathology and possible causes. The order is reversed for the research scientist: given a hypothesized disease cause, the questions are how do changes in physiological processes occur, and how do they lead to the clinical symptoms. Unfortunately, few basic science laboratories have had the ability to take the final step; to see whether the altered physiology or proposed treatment actually benefits to visual behavior. As such, many researchers believe that because a retinal disease causes the degeneration of retinal cells, the appropriate measurement of treatment success should be the restoration of those cells, not whether vision is improved. Not only is this flaw in the research process a problem in translating basic research into clinical practice, it may also mean that fundamental aspects of biology have been overlooked.

Iok-Hou Pang and Abbot F. Clark (eds.), *Animal Models for Retinal Diseases*, Neuromethods, vol. 46
DOI 10.1007/978-1-60761-541-5_2, © Springer Science+Business Media, LLC 2010

There are also practical advantages to a greater use of behavioral measures. For example, time series studies requiring repeated measures are feasible, whereas other methodologies are often limited to a few sessions (i.e., unit recording), or just one endpoint (i.e., histology). Anesthetics are needed for most electrophysiological recordings but the anesthetic itself can alter cellular responses (1) and produce detrimental cumulative effects when used repeatedly.

A disease can give rise to more than one symptom and ideally an animal model should capture as many features as possible. Unfortunately, this has not been easy as no single behavioral task can measure all relevant aspects of vision in all species, and thus, multiple tests may be necessary. In addition, the behavioral task has to be matched to the species as each has its own evolutionary history and behavioral repertoire. To be efficient a visual task must be made to conform as closely as possible to what the animal would do naturally. While some species have certain advantages, such as large eyes (e.g. primates, dogs, pigs, and rabbits), a fovea (primates), or specific genetic mutations (RPE65 dog), behavioral testing of vision is not straightforward in many species. Rodents are an important exception. Learning in rats has been a common subject of psychological research over the last century, and there is now considerable experience with mice as well. Fortunately, this matches up well with the use of these rodents in retinal research where there are a large number of inbred and transgenic rodent models with various genetic mutations, each of which represent a specific analog of a human retinal disease. For example, the Royal College of Surgeons (RCS) rat has a mutation within the RPE, which indirectly causes progressive death of photoreceptors and deterioration of visual function similar to what is typically seen in patients with AMD. In addition, numerous mouse models also exist with intraphotoreceptor mutations that cause the progressive death of photoreceptors (rd1, rd2, $rho^{-/-}$, etc), lipofuscin (Elovl4 mice; (2)) as well as drusen accumulation and neovasularization, two classic characteristics of AMD (ccl2; (3)).

The biggest choice an experimenter must make is whether they need a task in which each animal is required to make a perceptual choice about whether it sees a pattern, or they may rely on simple visual reflexes to gage whether a visual signal is effective. We have used both approaches, and in this chapter we will describe our efforts in testing visual function in rats and mice, and why this is important for evaluating the progression of retinal degenerative diseases and potential therapies.

2. Choice Tasks

In choice tasks, the animal views several stimuli, either side-by-side or sequentially, and then indicates which one they can see. The stimuli are then changed until the animal can no longer tell the difference and, after several repetitions, a threshold value is established. The big difficulty is of course, getting the animal to tell the experimenter what it is seeing, and a wide variety of approaches and reinforcement regimes have been tried. In the 1930s Lashley had rats jump toward a door covered with the "positive" stimulus (Fig. 1a). If they made an error and jumped the wrong way to the "negative" stimulus, the door was locked

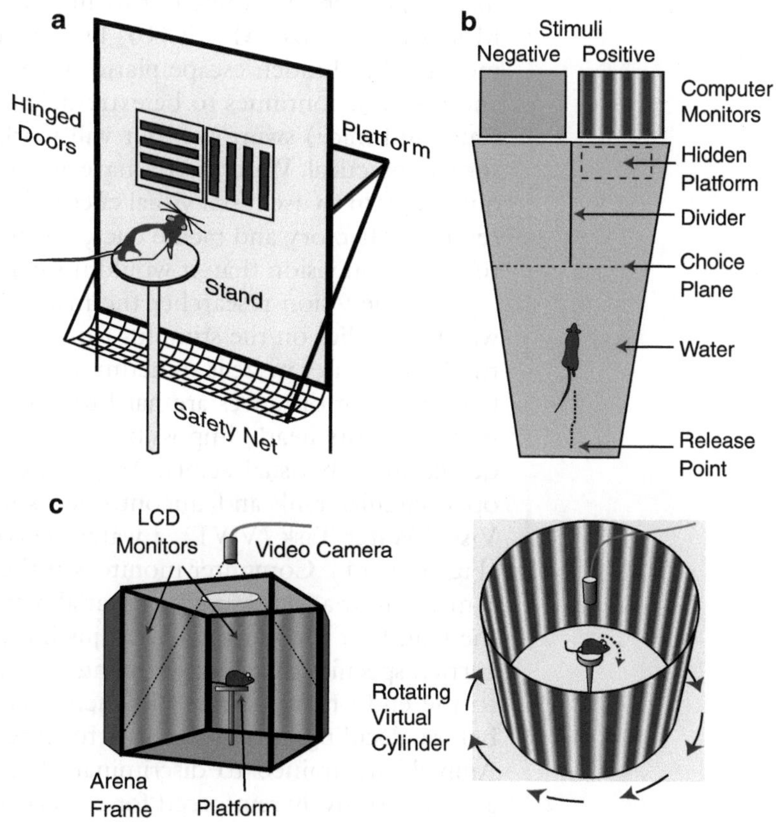

Fig. 1. Schematic of three systems used to study vision in rodents. (**a**) Lashley's jumping stand (4) in which a rat had to choose one of two doors to which it had to jump. The correct choice allowed the animal to pass through the door to safety, while a wrong resulted in the rat dropping to the net below. (**b**) In a more modern system, the Visual Water Task (11), the choices are displayed on computer monitors and the animal's task is to use the visual pattern as cue to find a platform that is just below the surface of the water. (**c**) In the Virtual Optokinetic System (22), the animal is in a virtual reality chamber (*left panel*) and sees grating pattern on cylinder (*right panel*). Rotation of the cylinder evokes a tracking only if the visual system detects the motion

and they fell to catch net (4). The jumping stand is still used to a limited extent (5) but has been superceded largely by operant techniques that are less aversive and easier to learn. For example, to study color vision Williams and Jacobs (6–8) have trained their animals to push illuminated panels to obtain soy milk. Using this approach they have characterized murine spectral sensitivity behaviorally and demonstrated color vision in normal mice (6), changes with age (7), and trichromatic color in mice with human photopigments (8). A similar approach has been used to measure acuity and contrast sensitivity in rats (9).

2.1. Visual Water Task

In our own laboratories we have used escape from water as a reinforcer, thus avoiding having to use the food or water deprivation that has been needed in the other operant tasks to motivate animals to perform. We came to this after having experience with the Morris water maze. Morris (10) built a circular water tank with a movable but hidden escape platform to study navigation in rats, and this task continues to be extremely popular for this purpose. Rats (and mice) swim well but will work to get out of water as soon as practical. Water mazes have an advantage that the rodents quickly learn to use distal visual cues to locate the platform. Water removes olfactory and tactile cues, possibly forcing the animal to rely more on vision than it would normally be inclined to do.

For the vision researcher the main problem with the Morris water maze lies on the stimulus front. While a visible platform in the Morris water maze is sometimes used to check for visual problems, viewing distances are hard to control, making quantitative measurements nearly impossible and the task can confuse visual detection with visual acuity. To get around the problems of an open circular tank and uncontrolled stimuli, we developed the Visual Water Task (VWT), a water version of the classic Y-maze (Fig. 1b) (11). Computer monitors at the ends of the arms of the Y maze allow a wide variety of spatial patterns to be displayed and the length of the arms, set by adjusting the length of an opaque barrier, specifies the closest viewing distances. Rats naturally swim to the end of the barrier and then peer around at each display before deciding on which arm to enter, making scoring easy. Animals are trained to discriminate between two visual stimuli and are positively reinforced for each correct response.

The most common measure of vision is acuity, the ability to resolve two high-contrast items as being distinct. For humans the Snellen chart provides a convenient series of tests as two parts of the same letter. For animals, and in many human studies, an equivalent task is to discriminate between a sine-wave grating and a gray screen of the same mean luminance. The spatial frequency is adjusted until a threshold established and the resultant acuity estimate is expressed in cycles per degree (c/d). Normal human acuity is 30 c/d while typical Long Evans rat and C57BL6 mouse thresholds are near 1.0 and 0.5 c/d, respectively (11). Albino rat strains have about half the acuity of pigmented ones (12). Computer monitors allow other parameters like contrast, orientation (13), and motion sensitivity to be measured (14). On the other hand, color perception in animals is almost impossible to study with current computer-driven monitors.

2.2. Example of Use

McGill et al. (15) characterized the visual acuity of RCS rats from 30 days of age until blindness at 11 months. This was the first longitudinal quantification of spatial vision in a model of retinal disease. There are two noteworthy aspects to the data shown in Fig. 2a. First, visual acuity measured at P28 is already impaired when compared with nondystrophic animals. Second, we obtained low but still measurable acuities long after ERGs had disappeared. These two results suggest that a behavioral assay may be more sensitive than the standard electrophysiological test. In a subsequent study we have used this task to show cellular therapies with both a human-derived RPE cell line (ARPE19) and human Schwann cells significantly limit the progression of visual deterioration (16).

2.3. Limitations

A significant limitation of the VWT is the time invested in training and testing animals is on the order of weeks, which limits longitudinal studies to a maximum of once per month measurement. Although, when the animals are well-trained, they do not require a full retraining before subsequently obtaining another threshold, 1–2 weeks are still needed. Thus daily measurements of vision are not possible in the VWT. Another consequence is that the earliest age at which visual thresholds can be measured from rodents in the VWT is approximately P30, about 2 weeks after the day of eye-opening.

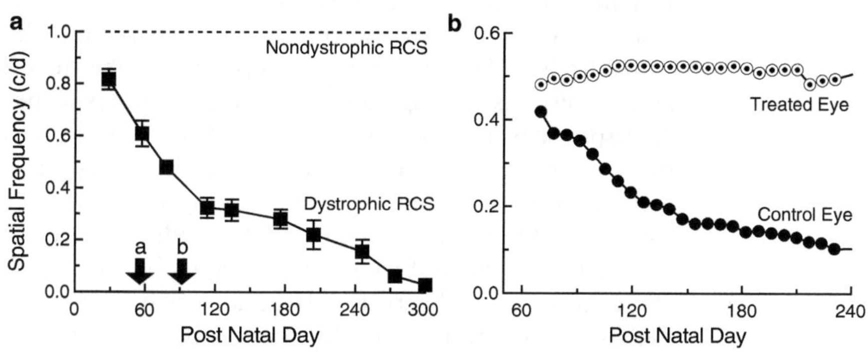

Fig. 2. Examples of data behavioral measures of retinal degeneration in RSC rats. (**a**) Acuity in dystrophic RCS rats as measured in the VWT declined as the animals aged (15). *Arrows* mark where the (**a**) and (**b**) wave of the ERG are lost. (**b**) Using the VOS there was no decrease in the spatial frequency threshold in one eye of dystrophic RCS rats treated with syngeneic Schwann cell transplants, while vision was gradually lost in the untreated fellow eye (17)

Typically behavioral testing is done in a photopic or mesopic levels during the light part of the day. Since a human experimenter must insert and remove the animal from the apparatus, testing at scotopic levels may also be difficult or impossible.

3. Visual Reflex Tasks

All choice tasks require time to train the animals and it can take days or weeks to establish a single threshold. Thus, one cannot follow the rapid changes early in development, or immediately after an intervention. But by using automatic, untrained responses, it is possible to devise visual tests that require no training. Even more importantly, because reflex responses are rapid and stereotyped, changes that happen over a few days can be followed.

Pupilloconstriction happens automatically when light is shone on the retinae. Lund and colleagues (18) have used this pupillary reflex to follow the gradual loss of visual function with age in Royal College of Surgeons (RCS) rats and to demonstrate restoration of function after retinal transplants (19). However, the presence of a pupillary response is a long way from the desired clinical result, and tests of spatial pattern vision are needed. The optokinetic system provides a near ideal way of evaluating vision in untrained animals. When large parts of the visual field are stimulated with a moving pattern, a feeling of self-motion or circular vection is induced. The tendency then is to rotate the direction of gaze (eyes, head, and body) in the same direction so as to reduce image motion on the retina.

Traditionally, an optokinetic test system consisted of a large mechanical drum driven by a servo-motor (20). Changing stimulus patterns on the drum was difficult and typically consisted of high-contrast square-wave patterns of checkerboards. The easiest stimulus to deliver is a prolonged rotation in one direction. Scoring the response is also not as simple as seeing whether the animal finds a platform or presses a panel. Unlike the digital-like discrete responses in operant choice tasks, the optokinetic tracking response is more analog. If the head is held stationary the eye shows a nystagmus with slow compensatory eye movements interrupted periodically by brief resetting quick phases (OKN). The gain of the system is calculated by dividing the velocity of the slow phase, compensatory eye movements by the velocity of the image movement. There is a large and rich literature studying OKN in many species, and this topic is beyond the scope of this chapter (see a recent review by Stahl (21)). However, we have a general concern that the head-fixed condition is artificial, and the use either of an anesthetic or restraint of an awake, conscious animal may have their own difficulties. Moreover, eye movement recording techniques (electrooculography or magnetic search coil) require surgical preparation that makes them a poor choice when dealing with a large number of or fragile animals.

3.1. Virtual Optokinetic System

Recently, we have created a new version of the optokinetic task to address these stimulus and response issues (22). Figure 1c shows our Virtual Optokinetic System (VOS). Rats or mice are placed on a platform at the epicenter of a testing arena surrounded by four monitors. The monitors display four views from inside a virtual optokinetic drum on which we have displayed sine-wave gratings. The virtual drum can be rotated clockwise or counter-clockwise, stimulating an optokinetic tracking (OKT) response. Furthermore, as we are using computer-generated patterns we can change spatial frequency, contrast, and direction of movement instantaneously. This has allowed us to measure thresholds for stimulus parameters, which formerly were largely unobtainable with a mechanical system.

The other key aspect of our approach has been to let the animals move freely and, by using a video camera mounted above the platform, track them as they follow the stimulus. The observed head position is used to keep the computed optokinetic drum centered on the animal's head to ensure that spatial frequency remains fixed as the animal moves around. We also use the camera to watch head and body movements for an indication of whether the animal can follow the grating. We can ignore the eye movements in this situation as the average eye over several seconds will be very close to center. In addition, it is not easy to calculate head and body velocities from a video image with the precision needed to compute a gain. Instead, we have adopted a simpler approach: an experimenter watches the video from the camera and makes a decision about the presence or direction of the tracking. The sine-wave grating is then increased in spatial frequency (making it more difficult to resolve) until the animal no longer responds. This process is repeated in a staircase procedure to establish a threshold. Experimentally "blind" testing procedures are also possible with this technology.

3.2. Example of Uses

The VOS has been used to quantify visual thresholds in developing mice (22) and rats (23) and in mice monocularly deprived in adulthood (24). The VOS has also been used to examine the visual function in mice without functional rods or cones (25), noerg-1 (26), and nob4 mice (27), and used to evaluate the role of ON bipolar cells that were engineered to be photosensitive (28). Spatial frequency thresholds and contrast sensitivity functions have also been characterized in the RCS rat as the retina degenerate as well as after one potential treatment, syngeneic Schwann cell transplants (17). Example data are shown in Fig. 2b.

3.3. Advantages

One of the major advantages of using the VOS is that animals require no reinforcement training before being tested. This greatly decreases the time required to generate a visual threshold, and therefore allows for much larger group sizes to be examined. In addition, the ability to generate thresholds quickly, in some cases as fast as 5–10 min, allows for spatial frequency and contrast sensitivity thresholds to be assessed on a daily basis. This is of particular importance in developmental studies as animals can be tested from the day of eye-opening, and for very fast retinal degenerations where visual decline can occur sooner than it can be measured with other tasks.

Another advantage is that the two eyes can be tested independently (29). In rodents with only one open we see tracking only when the stimulus moves in the temporal to nasal direction. When both eyes are open, when the virtual cylinder is rotated in the clockwise direction, only the left eye responds to the movement resulting in an elicited optokinetic behavior moving in the same direction as the grating movement. Conversely, if the cylinder is rotated in the counter-clockwise direction, only the right eye (again temporal to nasal) responds to the stimuli. This phenomenon is particularly advantageous as therapies for retinal disease are often performed monocularly, maintaining the contralateral eye as a control (See Fig. 2.2b as an example). Therefore, employing the VOS allows for within-animal controls to be used in the experimental design, a powerful method in research.

3.4. Limitations

The main limitation is that OKT "acuity" is less than the best the eye provides and thus it may miss small effects. The spatial frequency threshold is ~0.38 c/d in the VOS while in the VWT the threshold is greater than 0.5 c/d. This presumably reflects the lower resolution information carried by subcortical visual pathways. On the other hand, measuring contrast sensitivity at lower frequencies (<0.1 c/d) is problematic in the VWT task due to the smaller field of view. The VOS can go down to one cycle in 360 degrees.

The two approaches may also have differences in how much of the retina they need. Even in normal rodents the retina is not completely homogenous and different parts may have different thresholds. With the VOS we have been able to reduce the area of stimulation to the nasal (binocular) fields without any decrement in performance while the thresholds of the temporal, monocular fields are significantly worse. Spatial frequency sensitivity increases as the stimulation area falls below 50 degrees, and the thresholds with full-field stimulation match those of the most sensitive fields (23, 24).

One cannot be sure what threshold one is measuring. In the VOS, the tracking movements are subtle near threshold and require extensive experience to monitor; consequently the limits in part may be human threshold for seeing motion. Automating the tracking with a machine vision system does change this fundamentally as a computer algorithm will also have a threshold; indeed our experience so far is that it will be hard to beat a trained human observer. Moreover, even in the VWT is the threshold really a true sensory threshold, or some point above threshold at which the animal stops relying on it and adopts some other strategy (such as always going left)?

4. Future Directions

There is now abundant evidence that electrophysiological or anatomical measures of retinal disease cannot be equated with visual performance. For example, in Royal College of Surgeons (RCS) rats, photoreceptor degeneration is evident at ~3 weeks of age, and by 90 days of age the 12 photoreceptor cell body thick layer has declined to one cell body thick. Previously published electrophysiological data in the RCS rat suggests that the electroretinogram (ERG) is no longer measurable by ~4 months, and luminance thresholds measured from the brain are lost by ~7 months (30). Using the VWT and VOS visual thresholds can be measured reliably with each behavioral method up to ~10 months of age (15–17), significantly longer than expected given the degree of photoreceptor death and the earlier loss of electrophysiological responses. The differences are substantial. For example, at 4 months of age, a time when nearly all photoreceptors have been lost, and the ERG is no longer measurable, the RCS rat still retains approximately 40–50% of their visual acuity. Likewise, using a transgenic rat model with a mutation in the rhodopsin gene, Sagdullaev et al. (31) reported that by P60-100, photoreceptor cell death is near complete and multiunit recordings from the superior colliculus are nonresponsive; however, Thomas et al. (32) report measurable OKRs at both P135 and P205 using the same S334ter-3 rats.

Similar discrepancies between measures of the visual system can also be found in the clinic. For example, patients with advanced Retinitis Pigmentosa (RP) usually have a significantly reduced ERG while still having near normal acuity. Another example has been found in patients with Duchenne's Muscular Dystrophy where an absent or severely reduced ERG seems to have little effect on vision (33, 34). Clearly, the use of anatomical or electrophysiological measure to infer vision rescue following therapeutic intervention can be misleading, and because vision rescue is the ultimate clinical goal of treatment, vision needs to be measured in some form.

It is important to remember that behavioral tests are sensitive to changes all along the visual pathways. For example, induced strokes and aspiration lesions of the visual cortex result in a significant drop in VWT acuity from 1 c/d down to ~0.7 c/d (unpublished observations). Monocular deprivation early in life also decreases VWT acuity (35). Central adaptations can also improve performance. Early experience with VOS improves visual thresholds in rats (23), as can monocularly deprivation in both adult rats and mice (24). These enhancements, which we are currently studying in detail, offer additional capabilities as well as potential problems for experimenters more concerned with retinal function.

Finally, multiple measures of visual function may be needed. For example, we have reported that following subretinal transplantation of syngeneic Schwann cells without the aid of immune-suppression, spatial frequency thresholds in our VOS were preserved near normal levels up to 35 weeks of age (Fig. 2b) (17). However, that was not the only visual test conducted. In particular, whereas VOS contrast sensitivity was significantly preserved over controls, it was not maintained near normal levels. This suggested incomplete rescue of visual function, which was confirmed through evaluation of the visual acuity in the VWT task. The visual acuity showed a rescue of approximately 60% of normal vision, again better than controls but much lower than the VOS spatial frequency thresholds suggested.

5. Summary

Retinal disease causes irreversible vision loss for which there is currently no effective treatment. Many animal models exist for studying the mechanisms and potential treatments for retinal disease; rodent models appear to be the most advantageous. Historically, measuring vision in rodents has been difficult and constrained by the lack of effective testing devices. Recently, the Visual Water Task and the Virtual Optokinetic System have been developed, which have significantly advanced the field of vision science in rodents. This chapter describes the history of measuring vision, the development of the two tasks, and how the development of these tasks has enabled researchers to address specific queries regarding rodent models of retinal disease.

References

1. Pham TA, Graham SJ, Suzuki S et al (2004) A semi-persistent adult ocular dominance plasticity in visual cortex is stabilized by activated CREB. Learn Mem 11:738–747
2. Karan G, Lillo C, Yang Z et al (2005) Lipofuscin accumulation, abnormal electrophysiology, and photoreceptor degeneration in mutant ELOVL4 transgenic mice: a model

for macular degeneration. Proc Natl Acad Sci USA 102:4164–4169
3. Tuo J, Bojanowski CM, Zhou M et al (2007) Murine ccl2/cx3cr1 deficiency results in retinal lesions mimicking human age-related macular degeneration. Invest Ophthalmol Vis Sci 48:3827–3836
4. Lashley KS (1930) The mechanism of vision: I. A method for rapid analysis of pattern vision in the rat. J Gen Psych 37:453–460
5. Seymoure P, Juraska JM (1997) Vernier and grating acuity in adult hooded rats: the influence of sex. Behav Neurosci 111: 792–800
6. Jacobs GH, Williams GA, Fenwick JA (2004) Influence of cone pigment coexpression on spectral sensitivity and color vision in the mouse. Vis Res 44:1615–1622
7. Williams GA, Jacobs GH (2007) Cone-based vision in the aging mouse. Vis Res 47: 2037–2046
8. Jacobs GH, Williams GA, Cahill H, Nathans J (2007) Emergence of novel color vision in mice engineered to express a human cone photopigment. Science 315(5819):1723–1725
9. Keller J, Strasburger H, Cerutti DT, Sabel BA (2000) Assessing spatial vision - automated measurement of the contrast-sensitivity function in the hooded rat. J Neurosci Methods 97:103–110
10. Morris R (1984) Developments of a water-maze procedure for studying spatial learning in the rat. J Neurosci Methods 11:47–60
11. Prusky GT, West PW, Douglas RM (2000) Behavioral assessment of visual acuity in mice and rats. Vis Res 40:2201–2209
12. Prusky GT, Harker KT, Douglas RM, Whishaw IQ (2002) Variation in visual acuity within pigmented, and between pigmented and albino rat strains. Behav Brain Res 136:339–348
13. Bowden F, Douglas RM, Prusky GT. Horizontal bias in rat visual acuity. Program No 26018 Abstract Viewer/Itinerary Planner Washington, DC: Society for Neuroscience, 2002 Online 2002.
14. Douglas RM, Neve A, Quittenbaum JP, Alam NM, Prusky GT (2006) Perception of visual motion coherence by rats and mice. Vis Res 46:2842–2847
15. McGill TJ, Douglas RM, Lund RD, Prusky GT (2004) Quantification of spatial vision in the Royal College of Surgeons rat. Invest Ophthalmol Vis Sci 45:932–936
16. McGill TJ, Lund RD, Douglas RM, Wang S, Lu B, Prusky GT (2004) Preservation of vision following cell-based therapies in a model of retinal degenerative disease. Vision Res 44:2559–2566
17. McGill TJ, Lund RD, Douglas RM et al (2007) Syngeneic Schwann cell transplantation preserves vision in RCS rat without immunosuppression. Invest Ophthalmol Vis Sci 48:1906–1912
18. Whiteley SJ, Young MJ, Litchfield TM, Coffey PJ, Lund RD (1998) Changes in the pupillary light reflex of pigmented royal college of surgeons rats with Age. Exp Eye Res 66: 719–730
19. Whiteley SJ, Litchfield TM, Coffey PJ, Lund RD (1996) Improvement of the pupillary light reflex of Royal College of Surgeons rats following RPE cell grafts. Exp Neurol 140:100–104
20. Coffey PJ, Girman S, Wang SM, Hetherington L, Keegan DJ, Adamson P, Greenwood J, Lund RD (2002) Long-term preservation of cortically dependent visual function in RCS rats by transplantation. Nat Neurosci 5:53–56
21. Stahl JS (2008) Characteristics and applications of mouse eye movements. In: Chalupa LM, Williams DL (eds) Eye, Retina, and Visual System of the Mouse. MIT Press, Cambridge, pp 87–105
22. Prusky GT, Alam NM, Beekman S, Douglas RM (2004) Rapid quantification of adult and developing mouse spatial vision using a virtual optomotor system. Invest Ophthalmol Vis Sci 45:4611–4616
23. Prusky GT, Silver BD, Tschetter WW, Alam NM, Douglas RM (2008) Experience-dependent plasticity from eye opening enables lasting, visual cortex-dependent enhancement of motion vision. J Neurosci 28:9817–9827
24. Prusky GT, Alam NM, Douglas RM (2006) Enhancement of vision by monocular deprivation in adult mice. J Neurosci 26: 11554–11561
25. Schmucker C, Seeliger M, Humphries P, Biel M, Schaeffel F (2005) Grating acuity at different luminances in wild-type mice and in mice lacking rod or cone function. Invest Ophthalmol Vis Sci 46:398–407
26. Pinto LH, Vitaterna MH, Shimomura K et al (2005) Generation, characterization, and molecular cloning of the Noerg-1 mutation of rhodopsin in the mouse. Vis Neurosci 22: 619–629
27. Pinto LH, Vitaterna MH, Shimomura K et al (2007) Generation, identification and functional characterization of the nob4 mutation of Grm6 in the mouse. Vis Neurosci 24:111–123

28. Lagali PS, Balya D, Awatramani GB et al (2008) Light-activated channels targeted to ON bipolar cells restore visual function in retinal degeneration. Nat Neurosci 11:667–675
29. Douglas RM, Alam NM, Silver BD, McGill TJ, Tschetter WW, Prusky GT (2005) Independent visual threshold measurements in the two eyes of freely moving rats and mice using a virtual-reality optokinetic system. Vis Neurosci 22:677–684
30. Girman SV, Wang S, Lund RD (2005) Time course of deterioration of rod and cone function in RCS rat and the effects of subretinal cell grafting: a light- and dark-adaptation study. Vis Res 45:343–354
31. Sagdullaev BT, Aramant RB, Seiler MJ, Woch G, McCall MA (2003) Retinal transplantation-induced recovery of retinotectal visual function in a rodent model of retinitis pigmentosa. Invest Ophthalmol Vis Sci 44:1686–1695
32. Thomas BB, Seiler MJ, Sadda SR, Aramant RB (2004) Superior colliculus responses to light – preserved by transplantation in a slow degeneration rat model. Exp Eye Res 79:29–39
33. Cibis GW, Fitzgerald KM (2001) The negative ERG is not synonymous with nightblindness. Trans Am Ophthalmol Soc 99:171–175 discussion 5–6
34. Cibis GW, Fitzgerald KM, Harris DJ, Rothberg PG, Rupani M (1993) The effects of dystrophin gene mutations on the ERG in mice and humans. Invest Ophthalmol Vis Sci 34:3646–3652
35. Prusky GT, West PW, Douglas RM (2000) Experience-dependent plasticity of visual acuity in rats. Eur J NeuroSci 12:3781–3786

Chapter 3

Genetic and Genomic Approaches for Understanding Retinal Diseases

Gareth R. Howell and Simon W.M. John

Abstract

Here we review both established and emerging approaches for studying retinal diseases. We primarily focus on the use of the mouse as a genetic model, as it is a mammalian model with many resources and is amenable to a variety of genetic manipulations. Additionally, we highlight two other organisms, zebrafish and fruit fly that are emerging as valuable genetic tools to study retinal disease. We discuss the ways in which near-complete genome sequences of these three organisms are revolutionizing our ability to investigate the complex mechanisms involved in retinal diseases.

Key words: Genetics, Genomics, Mouse, Zebrafish, Fruit fly, Genome manipulation, Genetic screen

1. Introduction

The neural retina is the transparent tissue lining the interior of the eye. It is responsible for detecting and processing visual stimuli. The many different neurons in the retina include the photoreceptors that receive the initial light stimuli, amacrine, horizontal, and bi-polar cells that process the visual signals and the retinal ganglion cells (RGCs), the output neurons that translate the processed signals to the brain. The retinal neurons are supported by a variety of cell types including glia, immune cells, and vascular-associated cells. There are numerous diseases that affect the different populations (and subpopulations) of retinal neurons, including various forms of retinal degeneration (defined as diseases primarily affecting the photoreceptors) and glaucoma (primarily affecting the RGCs). Animal models for these and other retinal diseases are reviewed in detail in other chapters. In this chapter, we focus on modeling genetic disorders of the retina.

Iok-Hou Pang and Abbot F. Clark (eds.), *Animal Models for Retinal Diseases*, Neuromethods, vol. 46
DOI 10.1007/978-1-60761-541-5_3, © Springer Science+Business Media, LLC 2010

Table 1
Review of useful genetic and genomic resources

Database	Web address	Database type	Summary
Mouse Mutant Resource (MMR)	http://mousemutant.jax.org	Organism specific	Catalogs spontaneous mutations that arise in the breeding colonies at The Jackson Laboratory
The Knockout Mouse Project (KOMP)	http://www.knockout-mouse.org	Organism specific	Project to generate a knockout in every mouse gene
European Conditional Mouse Mutagenesis Program (EUCOMM)	http://www.eucomm.org	Organism specific	Project to generate a knockout in every mouse gene
Mouse Genome Informatics (MGI)	http://informatics.jax.org	Organism specific	Catalogs all pertinent data for the mouse including a catalog of genes and mutant strains available
The Zebrafish Model Organism database (ZFin)	http://zfin.org	Organism specific	Catalogs all pertinent data for the zebrafish including a catalog of genes and mutant lines available
Flybase	http://flybase.org	Organism specific	Catalogs all pertinent data for fruit fly including a catalog of genes and mutant lines available
Ensembl	http://www.ensembl.org	Genome browser and catalog of orthologs	Genome sequence and annotation for vertebrates and other eukaryotes. Includes the most comprehensive catalog of orthologous genes
UCSC Genome Bioinformatics	http://genome.ucsc.edu	Genome browser	Reference sequence and annotation for a large collection of genomes
National Center for Biotechnology Information (NCBI)	http://www.ncbi.nlm.nih.gov	Molecular biology resource	Wide range of integrated databases encompassing Pubmed, nucleotide and protein sequences and gene catalogs
EntrezGene@NCBI	http://www.ncbi.nlm.nih.gov/gene	Gene catalog	Most comprehensive catalog of genes for completed genomes
Homologene@NCBI	http://www.ncbi.nlm.nih.gov/homologene	Ortholog catalog	Groups together orthologous genes for completed genomes

We also highlight databases that contain valuable genetic and genomic information that aides in investigating complex mechanisms involved in retinal diseases. These databases are summarized in Table 1.

2. Generating Animal Models of Human Retinal Diseases

In this section, we discuss the techniques available for both inducing and identifying retinal disease-causing mutations in animal models, information that can then be used to identify mutations in the equivalent retinal diseases in humans as well as to aid in the understanding of the pathogenesis of the disease.

Recent advances have greatly facilitated the identification of disease genes in the human population. These advances provide great promise for understanding complex human diseases that have been notoriously difficult to understand at the genetic and molecular levels (1). For example, common high-risk and protective alleles for age-related macular degeneration (AMD) have been identified in genes such as complement factor H, complement components 2 and 3 (reviewed in (2)). Recently, a 412Phe variant (encoded by the T allele of a single nucleotide polymorphism) of the Toll-like receptor 3 gene (*TLR3*) was associated with protection against geographic atrophy, or "dry" AMD (3). This association was replicated in two independent case-control series. Despite these advances, many challenges remain. This is especially true for complex diseases such as glaucoma that are common but do not appear to be affected by high frequency alleles. Instead, it is likely that low-frequency alleles in many different genes confer risk. The low frequency of these alleles added to the confounding effects of other genetic factors and environment will continue to make the identification of the genes determining disease risk challenging.

To aid in the identification of specific retinal disease-causing mutations it can be advantageous to turn to animal models where well-defined large pedigrees are easy to produce and gene mapping is possible at very high resolution. In cases where the causative mouse gene is located in a chromosomal region that matches a region of the human genome that is implicated in the same disease (concordance), the mouse information can help to identify the human gene even if the implicated region of human chromosome is large (4). Identifying a disease-causing mutation in a gene within a small critical region of the genome in an animal model provides candidate gene(s) to assess in patients. In cases where the associated genetic interval contains multiple genes that are all sequenced, but causative mutations are not obvious, information from multiple species may be combined to identify the specific gene that is important. The ability to manipulate genes and determine the effects of specific mutations in experimental organisms will be very important for determining the important genes and how they interact. It will also provide disease models to investigate molecular mechanisms and to test new treatment ideas.

A number of features influence the usefulness of a particular animal as a genetic model for human retinal diseases. These include: (1) the similarity of the retina of the animal model to the human retina, (2) the availability of resources such as well-characterized genomes and genetically defined strains, (3) the ability to manipulate the genome to generate new disease models, and (4) the ease and cost of housing and breeding the animal model. In this review, we highlight three animal models (mouse, zebrafish, and fruit fly) that are being used as genetic tools to study human retinal disorders. Other animal models are emerging, such as the chicken and the frog, and many of the points discussed below will be applicable to these emerging models as the respective resources are developed.

2.1. The Mouse as a Model for Retinal Diseases

The mouse (*Mus musculus*) is the most widely used animal model for genetic studies of human retinal diseases. Several factors contribute to this. First, mice have essentially the same, layered structure of the retina as humans. The biggest differences are the absence of a macula and fovea in mice (5). In addition to directly studying the retina, the mouse provides a powerful system for studying mechanisms in other tissues that affect retinal disease. For example, the aqueous humor drainage structures of mice and people have key similarities. Therefore, the mouse is a powerful model to study mechanisms of IOP elevation, a strong risk factor for glaucoma (6). Second, there are many genetically defined and uniform strains of mice. It can be important to study any disease on controlled genetic backgrounds, as genotype/phenotype interactions are not confounded by differing and uncontrolled genetic interactions that may modify the phenotype (discussed below). Third, the mouse is the mammal most amenable to genetic manipulation, including the ability to carry out large-scale genetic screens. Last, in practical terms, mice are relatively inexpensive to house (compared particularly to larger mammals), and breed quickly (a gestation time of around 20 days).

2.1.1. Mouse Model Resources of Retinal Diseases: Inbred Strains and Spontaneous Mutations

There are approximately 100 "classical" common inbred mouse strains, including the popular C57BL/6J mouse – the strain used in the initial sequencing of the mouse genome (7). Each distinct inbred mouse strain has a constant genetic background where each mouse is genetically identical to every other mouse of the same strain. This uniformity is the result of successive filial matings designed to render the mice homozygous at every locus in the genome (reviewed in (8)). More recently, the collaborative cross was initiated to generate a large panel of recombinant inbred strains derived from a genetically diverse set of founder strains. The collaborative cross strains will provide a common reference panel designed to assess complex traits and interactions between genes, environments, pathogens, and other factors (9).

The classical inbred strains have unique sets of mutations that give each strain distinctive characteristics, such as coat color, weight, and lifespan. In some cases, spontaneous mutations that are fixed in these strains model human retinal diseases. The C3H/HeJ strain is homozygous for the retinal degeneration allele *Pde6b*^{rd1}, and has been widely used to study retinal degeneration in mice (10). Similarly, the DBA/2J strain is now widely used to study glaucoma. DBA/2J mice have acquired mutations in two genes, *Tyrp1* and *Gpnmb*, which cause iris pigment dispersion (*ipd*) and iris stromal atrophy (*isa*) leading to elevation in intraocular pressure (IOP) (11, 12). The elevation in IOP is necessary to cause glaucomatous RGC degeneration and associated optic nerve degeneration.

2.1.2. Spontaneous Mutations Model Human Disease Mutations

Spontaneous mutations can be loss of function, gain of function, and dominant-negative mutations. The larger the collection of mouse strains the more likely that spontaneous mutations will arise. The Jackson Laboratory, Bar Harbor, is one of the largest repositories for mouse strains. The Mouse Mutant Resource (MMR – see http://mousemutant.jax.org) at The Jackson Laboratory is the primary repository of strains and stocks carrying spontaneous genetic mutations. The MMR performs basic characterization (both genetic and phenotypic) of mice with spontaneous mutations that arise in the large breeding colonies, and they maintain and distribute established stocks. This collection includes a significant number of models of retinal diseases (highlighted in Table 2 and reviewed in (13)). For example, mutations in *CRB1* have been shown to cause retinal diseases and Leber congenital amaurosis in humans (14, 15). A mouse model (retinal degeneration 8, *rd8*) distributed by The Mouse Mutant Resource was discovered at The Jackson Laboratory and shown to be caused by a single base deletion in the *Crb1* gene (16)

2.1.3. Identifying Spontaneous Disease-Causing Mutations by Genetic Mapping and Sequencing

Although the affected gene(s) involved in a new spontaneous disease model may not necessarily be known, researchers are able to study aspects of the disease without understanding the underlying genetic change. However, it is important to identify the genetic alteration that led to the observed phenotype. In this section, we briefly describe methods for mutation detection in mice that are also applicable for other animal models including zebrafish and fruit fly.

Table 2
Key mouse models of retinal diseases (adapted from (13))

Gene/Strain	Location	Phenotype description
Pde6b^rd1	Chr 5	Retinal degeneration 1
Agtpbp1^pcd	Chr 13	Purkinje cell degeneration, retinal degeneration
Rds^Rd2	Chr 17	Retinal degeneration 2
Rd3	Chr 1	Retinal degeneration 3
Tub	Chr 7	Retinal degeneration 5
Mfrp^rd6	Chr 9	Retinal degeneration 6
Nr2e3^rd7	Chr 9	Retinal degeneration 7
Crb1^rd8	Chr 1	Retinal degeneration 8
Pde6b^rd10	Chr 5	Retinal degeneration 10
Rpe65^rd12	Chr 3	Retinal degeneration 12
Cep290^rd16	Chr 10	Retinal degeneration 16
Col2a1^sedc	Chr 15	Retinoschisis
nob2	Chr X	Missing outer plexiform layer and poor ERG
Pde6c^cpfl1	Chr 19	Cone photoreceptor function loss 1
Gnat^cpfl3	Chr 3	Cone photoreceptor function loss 3
Vldlr^tm1Her	Chr 19	Retinal spots and subretinal neovascularization
Rpl24^Bst	Chr 16	Age-related, subretinal neovascularization
DBA/2 J	Chr4/ Chr6	Pigmentary glaucoma caused by mutations in *Tyrp1* and *Gpnmb*

As genetic maps for mouse were made available, many spontaneous mutations were identified. Detailed descriptions of mutation mapping and mutant gene identification are given in reference (8). To map a given mutation that exists in an inbred strain, the original mutation-carrying strain is "outcrossed" to a second inbred line (the "mapping" strain) that is genetically distinct (genome has a large number of informative genetic markers) to the original strain. The resulting progeny (known as F1s) are now heterozygous for the original two-parent strains at every locus that differed between the parental strains. F1 progeny can then be either intercrossed (bred among themselves) or backcrossed (bred with one or both of the parental strains) to generate F2 or N2 progeny, respectively. The genomes of progeny will consist of random combinations of the original and mapping strain (both heterozygous and homozygous). Determining how the phenotype of the progeny corresponds to inheritance of alleles from each parental strain localizes the mutation of interest to a particular region of the genome, which is commonly called the critical region.

Following gene mapping, the most common mutation detection strategy is to amplify by the polymerase chain reaction (PCR) and sequence all functional elements (coding regions, untranslated regions (UTRs), and promoters) for genes contained within the critical region. This relatively simple PCR amplification and Sanger sequencing strategy has been made possible by the availability of a well-annotated mouse genome sequence. By comparing affected and nonaffected individuals, potential disease-causing mutations are identified. Mutation identification is not always simple. A mutation may not be identified even after screening all known functional elements, in the critical region. Not all functional elements in the mouse genome have been identified. Mutation detection will be more successful with further improvements in the annotation of the mouse genome as well as with the evolution of next generation sequencing enabling larger regions of the genome to be sequenced more efficiently (see Sect. 4). After identifying potential mutations, their ability to cause disease is further evaluated. Candidate mutations are assessed by screening a number of affected and nonaffected mice to determine if the mutation segregates with the phenotype of interest. Follow up experiments include generating a targeted mutation in the gene harboring the mutation to test if it is actually disease causing. Rescuing the mutant phenotype by transgenesis also tests causation. Similar strategies were used successfully to identify and prove that the *Gpnmb*^{R150X} and *Tyrp1*^{b} mutations cause glaucoma in DBA/2J mice (11, 12).

2.1.4. Manipulation of the Mouse Genome to Generate Targeted Models

A major advantage of using mouse as a model system is the depth of well-developed techniques available for manipulating the genome. In this section, we highlight some of the more common methods used as more detailed descriptions are provided elsewhere (17, 18). The ability to target and alter a specific gene(s) is an important and necessary tool. For instance, the function of a gene can be completely ablated (loss of function) in all cell types, or conditionally in specific cell types. Conversely, a gene can be mutated in such a way to generate a novel function (a gain of function) such as engineering a known human mutation into a mouse gene. Inducing mutations in genes of choice and assessing phenotype is commonly termed "reverse genetics," as it is the opposite of "forward genetic" approaches whereby spontaneous/induced mutations are discovered as a result of an overt phenotype.

Here, we briefly summarize the key features of traditional gene targeting technology. A targeting construct is designed to mutate a functionally important part of the gene of interest or to replicate a human mutation. The construct is integrated into the genome of embryonic stem (ES) cells by homologous recombination, thereby removing the functionally important region in the gene of interest. The targeted ES cells are then injected into blastocysts that are then implanted into surrogate females. The resulting pups will be chimeras made up of cells derived from both the targeted ES cell line and the host blastocyst. Commonly, coat color mutations are used to distinguish between ES cells and host and to determine the degree of chimerism. Chimeras derived largely from the targeted ES cell line are most likely to transmit the targeted mutation to offspring and are used to establish a mouse line with the desired targeted allele in every cell.

In addition to "knocking out" the function of a gene in all cell types as described above, genes can also be ablated or altered "conditionally". Conditional knockouts are vital to study genes that have multiple functions, particularly if one function is necessary for embryonic or early development. In these cases, simply removing the function of the gene in all cells is likely to cause embryonic or premature lethality, preventing assessment of gene function in adults. In conditional knockouts, the function of a gene is altered in either a spatial and/or temporal manner. Conditional knockouts are often generated using recombination sequences known as *loxP* sites that are recognized and recombined by Cre recombinase from bacteriophage P1. Another system utilizes FRT sites and FLP recombinase from yeast. *loxP* sites are incorporated into a targeting construct flanking the functionally important sequence(s) to be deleted from the gene of interest. Commonly, after targeting in the ES cells and generation of the mouse line, the gene of interest is still functional, as the *loxP* sites do not disrupt the normal transcription and translation of the gene. However, in the presence of Cre recombinase, the portion of DNA flanked by the two *loxP* sites is removed. Cre recombinase is not natural in mice, but can be introduced into specific cells or tissues by crossing the line with the conditional allele to a second mouse strain that has been engineered to express Cre recombinase (a Cre line) in particular cell types using cell-specific promoters.

An additional advantage of the Cre recombinase system is the ability to induce Cre recombinase expression using tamoxifen. In this case, the Cre recombinase gene is fused with a Tamoxifen-binding domain (TBD) and recombinase activity is induced only in the presence of Tamoxifen. This fusion gene is known as CreERT (see (17)). This system has been used to study choroideremia (CHM), an X-linked degeneration of the retinal pigment epithelium, photoreceptors, and choroid caused by loss of

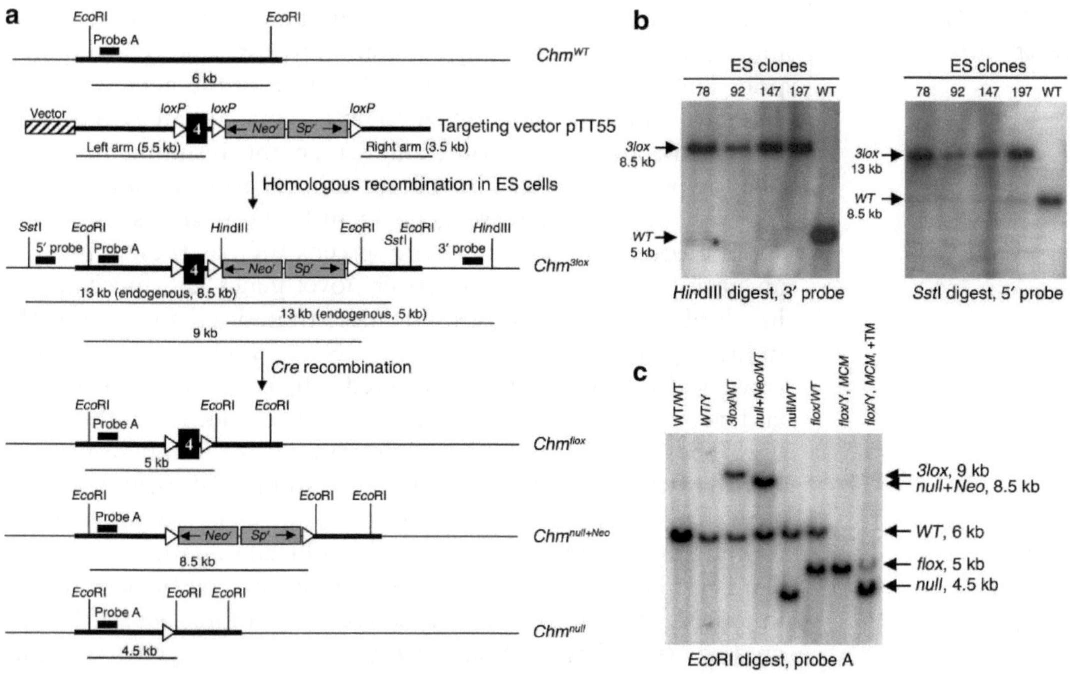

Fig. 1. Generating conditional alleles using Cre recombinase. In this example, a targeting vector was designed carrying 3 *loxP* sites, a neomycin and spectinomycin resistance cassette (Neo^r, Sp^r), and two homology arms for the *Chm* gene in mice. The *Chm*^{3lox} allele was generated by homologous recombination in ES cell clones. Cre-mediated recombination between the 3 *loxP* sites within the *Chm*^{3lox} allele resulted in three possible alleles: *Chm*^{flox}, *Chm*$^{null+neo}$, and *Chm*^{null}, which were distinguished by Southern blot analysis using *Eco*R1 digestion and probe A. Reproduced from (19)

function of the *CHM* gene (Fig. 1) (19). Complete knockouts of the *Chm* in mice are embryonic lethal. Therefore, a conditional knockout of the *Chm* gene was created using a tamoxifen-inducible and tissue-specific Cre recombinase. These experiments show that the degenerative process likely proceeds independently in the RPE and the photoreceptors, an important consideration when developing therapeutic strategies for human CHM patients.

2.1.5. Generating a Knockout for Every Gene in the Mouse Genome

Generating a targeted knockout can be expensive and time consuming for individual research labs and so efforts are underway to generate a knockout for every gene in the mouse genome. The Knockout Mouse Project (KOMP – http://www.knockoutmouse.org) is initially targeting approximately 10,000 genes, either by completely removing the functional protein, or generating a conditional allele. Similarly, the European Conditional Mouse Mutagenesis Program (EUCOMM- http://www.eucomm.org) is generating up to 13,000 conditional mutations across the mouse genome. The selected genes are intended to be a different subset to those chosen by KOMP. These resources will be important for both studying mechanisms involved in retinal disease as well as generating new retinal disease models.

2.1.6. High Throughput Genetic Screens

The probability that a spontaneous mutation arises in a gene of interest is low, and targeting a mutation to a specific gene requires prior knowledge that ablating the function of the gene will have an effect on retinal function. A complementary strategy is to randomly introduce mutations, and screen the resulting mice for phenotypes of interest. The mutant animals provide valuable new models and allow disease gene identification, as discussed for spontaneous mutations. Genetic screens such as these represent an unbiased approach to uncover novel genes and mechanisms involved in retinal diseases. In mice, mutations are commonly induced using ethylnitrosourea (ENU), a chemical mutagen (reviewed in (20)). ENU is injected into male mice (G0 males) and causes mutations to be introduced at random into the germ line. The dose of ENU injected determines the number of mutations created and varies with different strains. G0 male mice are then mated to females and the resulting progeny (G1s) will each carry a unique complement of ENU-induced mutations. G1 mice can be screened for phenotypes of interest caused by dominantly inherited mutations. To assess recessive mutations, G1s are crossed to wild-type females to generate G2 progeny. G2s can then be either intercrossed or backcrossed with the parental G1 male to produce G3s that are assessed for phenotypes of interest. Once phenotypes of interest are identified, the mutation can be identified using genetic mapping and mutation detection strategies described above.

A novel ENU-induced mutation in the mouse Optic atrophy 1 (*Opa1*) gene has provided useful insights into autosomal dominant optic atrophy (ADOA) (21). Mutations in the human *OPA1* gene cause the most common form of optic neuropathy in which RGCs are lost and visual acuity is impaired from an early age (22, 23). OPA1 is involved in mitochondrial homeostasis, resulting mitochondrial fusion and sequestration of cytochrome C to the mitochondria. An ENU mutagenized DNA archive from 10,000 C3HeB/FeJ male mice (24) was screened for point mutations in *Opa1* exons and a Q285X protein-truncation mutation was identified. The *Opa1*$^{+/-}$ mice exhibited abnormal mitochondrial morphology, optic nerve myelination anomalies and visual deficits (Fig. 2). This mouse is a valuable tool to directly investigate the pathophysiology of ADOA. *OPA1* has now been suggested to be elevated in human glaucoma and suggests mitochondrial fission may play an important role in the early pathogenesis of glaucoma (25).

2.1.7. Assessing Mutations on Multiple Genetic Backgrounds can Identify Quantitative Trait Loci and Genetic Modifiers

An important component of studying the consequences of genetic mutations is the assessment of the phenotype on different genetic backgrounds. In human diseases, a mutation in one individual can result in severe symptoms, whereas that same mutation can cause little or no disease in others. This is often due to differences in their genetic background. Therefore, consideration of genetic background is important both when generating a novel mouse

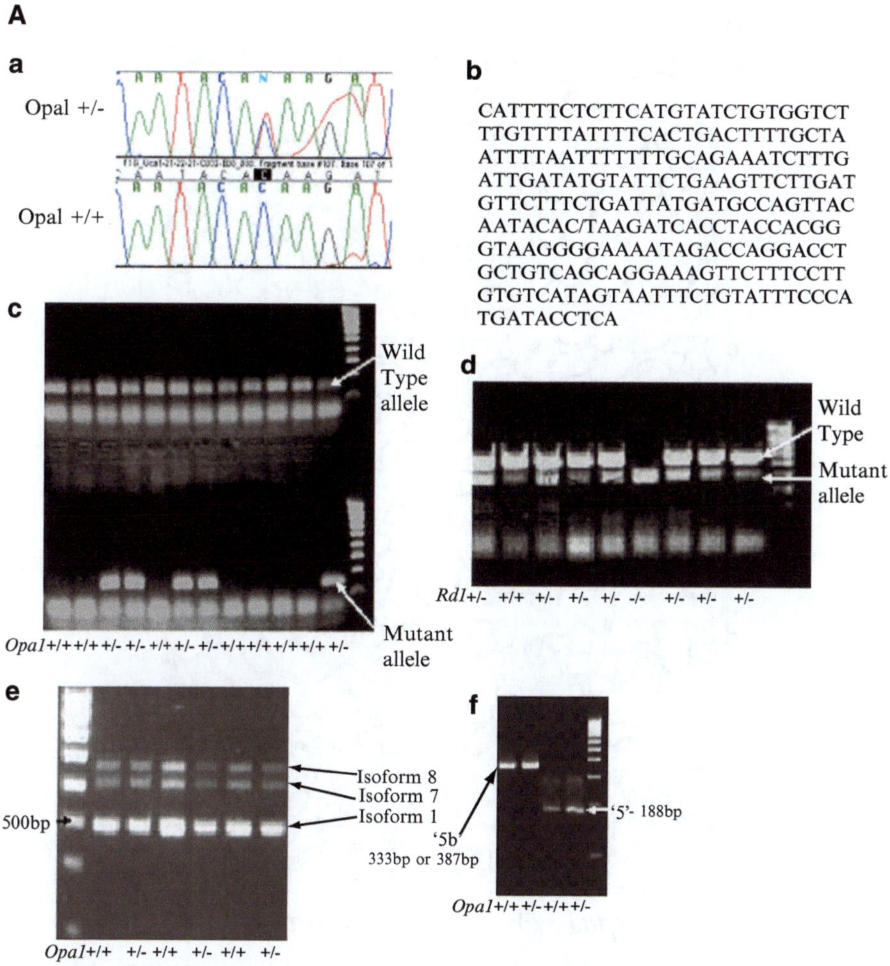

Fig. 2. An ENU-induced mutation causes autosomal dominant optic atrophy in mice. (**A**) DNA sequence chromatogram of B6;C3-*Opa1*^{Q285STOP} *Opa1*$^{+/-}$ mutant compared with *Opa1*$^{+/+}$ control, illustrating the exon 8 C–T transition at 1051 bp. (**b**) Mouse genomic sequence showing C–T transition at 1051 bp, which leads to Q285 STOP. (**c**) PCR allele-specific genotyping for Opa1 for a selection of *Opa1*$^{+/-}$ and *Opa1*$^{+/+}$ mice. Both wild-type and mutant fragments are 160 bp. (**d**) Mouse genotyping for a selection of *Rd1*$^{+/+}$ and *Rd1*$^{+/-}$ mice by multiplex PCR for Rd1 alleles. (**e**) RT–PCR from retinal cDNA extracted from a selection of *Opa1*$^{+/+}$ and *Opa1*$^{+/-}$ mice and resolved on 3% agarose using primers F3 and R8/9. Three isoforms are shown. (**f**) RT–PCR on retinal cDNA using primers F3 and R5 and R5b

model of a human disease and also in understanding the pathogenesis of disease. Identifying genetic modifiers can gain insight into important pathways involved in disease as well as identifying molecular targets for potential new treatments. A major advantage of the mouse as a model organism is the ability to assess causative mutations on different controlled genetic backgrounds and identify quantitative trait loci (QTLs) and genetic modifiers (reviewed in (26)).

In mice, QTLs have been identified that modify the retinal degenerations *rd3* (27) and *rd7* (28), as well as susceptibility of RGCs to death after optic nerve crush (29). Mutations in the human RD3 gene has been shown to cause Leber Congenital

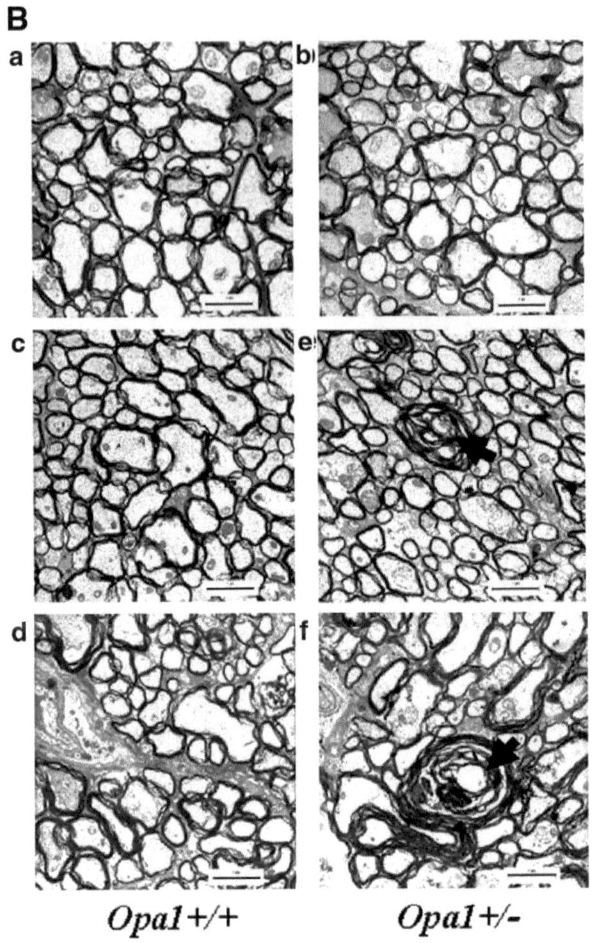

Fig. 2. (continued) (**B**) Transmission electron micrographs of optic nerve taken from 5–6 months old *Opa1*$^{+/+}$ (**Ba**) and *Opa1*$^{+/-}$ (**Bb**), 9 months old *Opa1*$^{+/+}$ (**Bc**) and *Opa1*$^{+/-}$ (**Bd**), 15 months old C57BL/6J^{cr1} (**Be**), and 18 months old *Opa1*$^{+/-}$ (**Bf**) mice. The optic nerves appear comparable at 6 months of age between genotypes; however, by 9 months of age anomalies, shown by *black arrows*, identified as whirls of myelin, were observed in the optic nerves of the Opa1$^{+/-}$ mice, which continued to 18 months of age. Bar = 2 μm. Reproduced from (21)

Amaurosis, type XII (30). Mutations in the mouse ortholog, *Rd3* cause a similar retinal degeneration. *RD3* has no known molecular function in either humans or mice. However, the degeneration phenotype is significantly faster on a BALB/cByJ genetic background (BALB-*rd3/rd3*) compared to an albino C57BL/6 J strain (B6a-*rd3/rd3*) (27). Potential modifiers were identified using a reciprocal (BALB-*rd3/rd3* × B6a-*rd3/rd3*) F1 intercross and six highly significant QTLs were identified (31). In a similar study, Li and colleagues observed varying levels of susceptibility to optic nerve crush in 15 inbred strains of mice (32). Optic nerve crush causes optic nerve degeneration and RGC death, important components of glaucoma. They went on to map a dominant modifying QTL, RGC susceptible 1 (*Rgcs1*) to a 58 Mb region on mouse chromosome 5 (29). Further refinement of QTLs by additional mapping and testing of genes within the critical region is required to identify individual genetic modifiers.

The identification of genetic modifiers can suggest new avenues for improved therapy in human disease. In humans, mutations in the cytochrome P450 family 1, subfamily B, polypeptide 1 (*CYP1B1*) gene cause primary congenital glaucoma (PCG) but striking phenotypic differences occur between individuals. *Cyp1b1*-deficient mice have focal angle abnormalities similar to that observed in PCG but they do not develop high IOP or glaucoma. However, *Cyp1b1*-deficient mice that are also deficient for the Tyrosinase (*Tyr*) have more severe angle malformations than do mice carrying the *Cyp1b1* mutation alone (33). *Tyr* also modified the phenotype of *Foxc1*-deficient mice, another gene whose ortholog causes human glaucoma. Importantly for potential treatments in humans, Tyrosinase produces L-DOPA and administration of L-DOPA in the drinking water of pregnant mice deficient in both CYP1B1 and tyrosinase substantially alleviated the developmental abnormalities (33).

2.2. The Zebrafish as a Model for Retinal Diseases

Over the last decade, the zebrafish (*Danio rerio*) has become a powerful model system for studying vertebrate retinal development and emerges as a model organism for retinal disorders. The zebrafish retina has the same basic morphology as the mammalian retina (reviewed in (34)). Even for glaucoma risk factors, the zebrafish appears more relevant than was initially believed. The aqueous humor outflow pathway has recently been shown to have some similarities to that of mammals, suggesting molecular conservation of some functions, and strengthening the potential for zebrafish as a model for studying IOP elevation and glaucoma (Fig. 3) (35). Zebrafish are easy to maintain in the laboratory setting and they breed easily. A single breeding pair can produce more than 200 offspring from a single mating. The eggs are fertilized externally and are transparent, as are the developing embryos. Within 24 h after fertilization all major organ systems are formed and by 3 days there are free swimming and feeding larvae. Disadvantages of using zebrafish include the lack of genetically inbred strains and the greater evolutionary distance from humans compared to mammals. Despite this, the great power and speed of genetic analyses in zebrafish make it a very important and complementary model system.

2.2.1. The Zebrafish is Amenable to Large-Scale Chemical and Genetic Mutagenesis

The success of the zebrafish as a model organism has largely been due to its amenability for large-scale genetic and chemical screens to study vertebrate development. However, these tools are now being applied to the study of retinal diseases including photoreceptor degeneration and glaucoma. Multiple laboratories have developed methods to carry out large ENU-based mutagenesis screens to uncover genes involved in development (36–38).

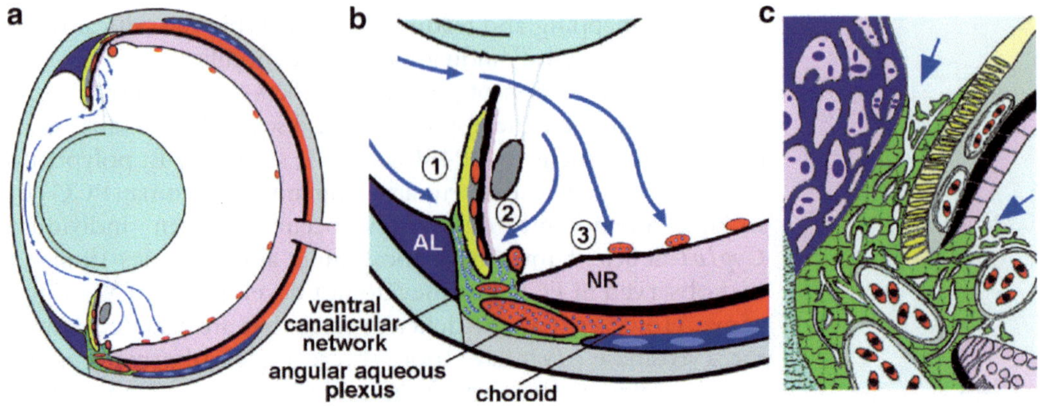

Fig. 3. Model of aqueous humor dynamics in the zebrafish eye. (**a**) Overview showing the vectorial flow of aqueous humor (*blue arrows*) from the dorsal ciliary epithelium to the ventral canalicular network and ventral vitreal-retinal vessels. (**b**) Higher magnification of aqueous humor outflow indicating absorption into (1) the iridocorneal and (2) ciliary openings of the ventral canalicular network and (3) ventral vitreal-retinal vessels. (**c**) Higher detail of the outflow pathway showing juxtacanalicular connective tissue cells at the iridocorneal and ciliary openings (indicated by *arrows*) and the tortuous lacunae created by endothelial cells (*green*) lining the canalicular network. Lens and cornea, *light blue*; annular ligament (AL), *purple*; blood-filled vessels and sinuses, *red*; iris argentea, *yellow*; iris stroma, lentis retractor, and sclera, *gray*; neural retina (NR), *pink*; scleral ossicle, *dark blue*; aqueous humor in outflow tissues, *blue–white* dots in (**a**) and (**b**) and pale blue in (**c**). Reproduced from (35)

Recessive mutations were recovered using the same three-generation mating scheme as described for mouse. Many more lines could be analyzed efficiently and at a cheaper cost than equivalent screens in mice. Mutations induced phenotypes relevant to the eye and, specifically, retinal development. Subsequently, more focused screens targeting development and function of the retina have been carried out. Although many mutations affect photoreceptor morphogenesis, a subset shows striking similarities to human retinal disorders such as retinitis pigmentosa, cone dystrophies or macular degenerations (reviewed in (39)). Zebrafish larvae are cone rich and it has been suggested that they may be of particular importance for studying macular degenerations (39), a disease that causes 50% of blindness in the western world. Although few studies have evaluated disease of later age of onset in zebrafish, preliminary studies have identified elevated IOP and enlarged eyes, providing models of potential importance for glaucoma (40).

An additional major advantage of the zebrafish is the ability with which insertional mutagenesis can be carried out. A pseudotype retrovirus that was first developed for gene therapy and genetic studies can infect many organisms including zebrafish where the transformation rate approaches 100% (34). Most founders transmit on an average ten proviral inserts to their progeny. The retrovirus inserts randomly into the zebrafish genome, and phenotypes are observed when the retroviral insertion alters the function of a gene (such as disrupting transcription or translation). The proviral insert acts as a marker and the affected gene can be readily identified. A study by Gross and colleagues identified 40 genes that when mutated by insertional mutagenesis cause defects in visual system development and function, the majority of which showed photoreceptor degeneration (41).

2.2.2. Knocking Down Genes in Zebrafish Using Morpholinos

As discussed earlier, the ability to specifically knockout the function, a gene of interest, is a powerful strategy in elucidating both its primary function as well as possible roles in disease pathogenesis. In the mouse, targeted gene knockout procedures are well-established. In zebrafish, the necessary resources, such as ES cells, and targeted mutagenesis by homologous recombination are not yet developed. An alternative strategy that has been used to elucidate the function of known genes is to carry out posttranscriptional gene silencing or gene "knockdown." The process by which translation of an mRNA transcript is prevented by RNA interference (RNAi) was first developed in the nematode, *Caenorhabditis elegans*. In zebrafish, gene knockdown is achieved by the injection of modified antisense oligonucleotides (morpholinos) that block translation of the desired mRNA. Morpholinos are commonly injected at the 1–2 cell stage and knockdown persists for up to 6 days. These experiments also allow the function of groups of genes to be assessed at one time as morpholinos targeting multiple genes can be injected at the same time. Morpholino-based knockdown has been used to study the pathogenic mechanisms of Oculo-Auricular Syndrome (OAS) (42). Mutations in *HMX1* (also known as *NKX5-3*) cause a variety of ophthalmic anomalies including microphthalmia, abnormalities of the RPE and rod/cone dystrophy. Morpholinos designed to knockdown the zebrafish ortholog *nkx5-3* were injected into zebrafish embryos at the 1–2 cell stage and by 5 days after fertilization the eye showed microphthalmia and an absence or delay of stratification of the retina. Although injection of the *nkx5-3* morpholino does not completely recapitulate human OAS, these experiments were able to implicate *nkx5-3* in axial patterning of the retina (42).

2.3. The Fruit Fly as a Model for Retinal Diseases

The compound eye of the fruit fly (*Drosophila melanogaster*), is very different to a vertebrate eye. The compound eye is comprised of approximately 800 ommatidia (simple eyes), each of which contains 20 cells including eight photoreceptor cells (PRCs) (reviewed in (43)). The fruit fly retina has no equivalent structure to vertebrate RGCs and optic nerve. Instead, the axons of the PRCs project into the lamina and medulla, and synapse with the lamina and medulla neurons. Despite these differences the fruit fly is still a useful genetic model for retinal diseases, but primarily those affecting photoreceptors (43). The PRCs of the fruit fly have similar molecular functions to the photoreceptors in vertebrates. For various mutant genes that cause PRC degeneration in the fruit fly, the orthologous genes cause photoreceptor degeneration in humans. The first fly mutations linked to retinal degeneration were in *ninaE*, which encodes the major rhodopsin, *Rh1*. This observation had human relevance as mutations in human rhodopsins were subsequently shown to account for a large percentage of autosomal dominant retinitis pigmentosa (44). Mutations in almost any gene that functions in the phototransduction pathway cause retinal degeneration in fruit fly; the molecular processes underlying these degenerations are diverse (reviewed in (45)).

2.3.1. Forward and Reverse Genetic Screens for Genes Required in Photoreceptor Development and Function

A major advantage of the fruit fly is the ability to carry out large-scale genetic screens efficiently and in a cost effective manner. The majority of genes involved in phototransduction are not required for fruit fly viability and so genetic screens for mutations affecting these pathways have been successful in *Drosophila*. Initial studies screened for visual cues but more recently, many studies use electroretinograms (ERGs). ERGs are extracellular recordings that measure the summed responses of all retinal cells to light. ERGs are generally carried out using white light and by placing an electrode on the surface of the compound eye (46). Mutations in genes involved in axonal and synaptic function can be essential for embryo or larval development. However, unique to *Drosophila*, it is possible to generate homozygous ENU-induced mutations in just the PRC cells of the eye, while the rest of the fly remains heterozygous. The system utilizes the GAL4/UAS and FLP/FRT, using an eye-specific GAL4 driver, in combination with a photoreceptor-specific, cell lethal transgene GMR-*hid* (47). This method was used to identify *milton*, a kinesin-associated protein required to localize mitochondria to nerve terminals. Photoreceptors, mutant for *milton*, show aberrant synaptic transmission despite normal phototransduction (48, 49).

Forward genetic screens in the fruit fly commonly utilize transposable elements. Transposons cause mutations by "hopping" into or close to functional genetic elements causing disruption of the function of a gene. The P element was initially discovered in a wild-type strain and so has been the most commonly used transposable element. Transposon-based forward genetic screens use two strains, the "jump starter" strain, that carries the P element transposase that when intercrossed efficiently mobilizes the transposon present in the genome of the "mutator" strain. P elements have been adapted to "trap" different genomic elements of transcription units including exons, enhancers and poly-adenylation signals (46).

3. Sequencing and Accessing Annotated Genomes

The availability of well-annotated genome sequence has revolutionized studies of human diseases, including those of the retina. The efficiency of finding disease-causing mutations within a given region of the chromosome is greatly improved with a detailed map of the functional elements contained within. Also, a well-annotated genome is necessary for efficiently manipulating genomes to generate new animal models of retinal diseases. The human genome was sequenced using two different approaches. The publicly funded approach adopted a large insert clone mapping strategy (50). Briefly, human genome DNA (derived from many different individuals) was fragmented and subcloned into large insert clones (primarily bacterial artificial chromosomes – BACs). A physical map consisting of many overlapping BAC clones was generated for each human chromosome using genetic markers and DNA fingerprinting (51, 52). A minimum set of overlapping clones (a tile path) was subcloned into sequencing vectors and sequenced to generate contiguous portions of sequence (contigs) covering each human chromosome. The second approach by Celera (a privately funded enterprise) used a whole genome shotgun (WGS) strategy which very quickly generated sequence covering large amounts of the human genome (53). WGS sequencing of large genomes was developed to sequence the 168-Mb fruit fly genome (54). The WGS approach generates sequence faster as it bypasses the need for physical mapping. DNA is sheared into smaller fragments that are cloned into sequencing vectors and sequenced at random. Ultimately this version of the human genome sequence was less accurate and covered the human genome with more discontiguous expanses of sequence. The physical mapping approach allows for sequence contigs to be placed more accurately onto chromosomes and gaps between contigs to be closed using additional mapping and sequencing.

Sequencing of the mouse genome used a combination of the strategies developed during the sequencing of the fruit fly and human genomes. This hybrid approach combines the advantages of both the WGS and physical mapping strategies. A WGS step efficiently and quickly generated sequence covering large regions of the mouse genome, and in parallel a physical map was generated and tile path clones sequenced. Sequence from both approaches was combined to generate an accurate sequence assembly of the mouse genome. The mouse genome sequence was initially generated from DNA isolated from C57BL/6J, a commonly used inbred mouse strain and is considered to be the "reference" mouse genome sequence. Subsequently, resequencing using high-density oligonucleotide arrays of 4 wild-derived and 11 classical strains of mice has generated a dense map of genetic variation in the laboratory mouse genome (55). The zebrafish genome is 1.7 Gb, two-thirds the size of the human and mouse genomes and was sequenced using the hybrid WGS and physical mapping strategy. As there is no inbred strain of zebrafish, the reference sequence was generated from DNA isolated from a number of different individual zebrafish.

3.1. Identifying Genes and Other Functional Elements

A near-complete genome sequence is only useful if the genes and other functional elements are identified and well-annotated. In this section, we summarize the current status of the annotation of the genomes of human, mouse, zebrafish, and fruit fly (data taken from Ensembl as of January 2009). After human, the mouse is the next best vertebrate genome annotated, and only the worm and fruit fly have similar levels of annotation among all other eukaryotes. The mouse genome sequence is 2.7 Gb and contains 23,435 known protein-coding genes and 3,164 noncoding RNA genes. This compares to 3.03 Gb of human genome sequence containing 21,649 protein-coding genes and 4,810 noncoding RNA genes. More than 80% of human protein-coding genes are predicted to have orthologs in the mouse genome (and this expected to get higher as gene annotation continues to improve). Over 90% of disease-causing human genes have a counterpart in mouse. The 1.7-Gb zebrafish genome contains 17,330 protein-coding genes. Sixty percent of human protein-coding genes are predicted to have at least one ortholog in the zebrafish genome. However, there is good evidence to suggest that the zebrafish genome has undergone at least one genome duplication making it more difficult to identify the ortholog that has the same function as the equivalent gene in humans. In approximately 20% of cases, a single human gene is predicted to have at least two zebrafish orthologs (reviewed in (56)). The fruit fly genome is predicted to encode 14,141 protein-coding genes and 949 RNA genes. Approximately 20% of the predicted protein-coding genes in human are expected to have an ortholog in the fruit fly.

Although the annotation of these genomes is advanced, there are many biological features that are still not well-represented in the genome browsers. True transcription start sites are not well-annotated for many genes, as these are difficult to identify. Annotation of transcription start sites relies almost exclusively on aligning the sequence of a full-length cDNA clone to the genomic sequence. Unfortunately, many cDNA clones are not always full length (commonly missing 5′ sequence). Therefore, the true transcription start site of a given gene may be upstream of the annotated start site. In addition, although a significant number of alternatively spliced transcripts are annotated, many of these are a "best guess" and may not represent real biologically relevant transcripts. It is also likely that many alternatively spliced transcripts are not annotated at all. Even less well-annotated are other functional elements such as promoter regions and regulatory elements. These tend to be short elements with low primary sequence conservation and so are difficult to identify in genomic sequence. To address this, the ENCODE (Encyclopedia of DNA Elements) consortium was established by the National Human Genome Research Institute (NHGRI) in 2003 to identify and characterize functional elements in the human genome. These elements include 5′ end of transcripts, chromatin structure, and replication features. In the pilot phase over 200 studies identified functional elements in 1% of the human genome (made up of 44 regions) using a broad range of assays (57). The results showed a number of interesting findings including a large number of intercalated transcripts and transcription factors that were previously thought to bind only promoters actually bind more widely. These studies have been essential in defining the direction for identification and annotation of all functional elements in complex genomes.

3.1.1. An Interlinked Network of Online Databases Provides Access to Genome Sequences, Gene Catalogs, and Mutants

The current versions of the annotated genome sequences for a wide variety of eukaryotic organisms including human, mouse, zebrafish, and fruit fly can be viewed at Ensembl (http://www.ensembl.org) or at UCSC Genome Browser (http://genome.ucsc.edu). In addition, databases are available that provide comprehensive resources specific for each organism. These databases include Mouse Genome Informatics (MGI – http://mgi.jax.org), the zebrafish-specific database ZFIN (http://zfin.org) and Flybase (http://flybase.org), the fruit fly-specific database. Although each database is designed and maintained by different groups, they are similarly structured. These databases provide a wealth of information for each organism including an up-to-date catalog of every known or predicted gene in the genome, genetic mapping information where available, associated phenotype information, and links to the genome browsers and other important databases. The National Center for Biotechnology Information (NCBI – http://www.ncbi.nlm.nih.gov) is a collection of a wide variety of databases that includes Pubmed and Entrez Gene, a database that catalogs known and predicted genes from organisms with completed genome sequences. All pertinent information for a given gene, including nucleotide and protein sequence, known alternative splice forms and salient references are included in each Entrez Gene entry.

An important feature of all databases mentioned is their interlinked nature. This allows for easy navigation between databases to access functional information for homologous genes from different organisms (orthologs). For example, as discussed earlier *OPA1* causes ADOA in humans and has been suggested to play a role in glaucoma. The orthologs of *OPA1* can be identified using either the Homologene database at the NCBI (http://www.ncbi.nlm.nih.gov/homologene) or the Predicted Ortholog section in the Human Ensembl genome browser. Homologene utilizes similarity searches and taxonomy information to group potential orthologs together. Ortholog prediction in Ensembl uses similarity searches combined with a phylogenetic tree strategy to determine the most likely orthologs. The homologene entry for *OPA1* (ID:14618) includes the orthologs in mouse, zebrafish, and fruit fly. Due to the increased sensitivity of the ortholog prediction in Ensembl, *OPA1* orthologs can be identified for more distant organisms including yeast.

A summary of the database highlighted in this section can be found in Table 1.

4. Genome-Wide Gene Expression Profiling

The availability of improving gene catalogs for organisms such as human, mouse, zebrafish, and fruit fly enable the roles of all genes in a given genome to be interrogated at one time. In this section, we briefly highlight important features regarding genome-wide expression profiling. Most commonly, microarrays have been used where probe sets for as many genes as possible are designed and manufactured onto glass slides. Fluorescently labeled RNAs derived from tissues of interest are hybridized to glass slides and the number of copies of a gene can be determined based on the intensity of fluorescent signal. Genome-wide gene expression profiling is used extensively as a tool to gain insight into molecular changes occurring in disease. RNA from diseased tissue is compared to healthy control tissue. Recently, Punzo and colleagues have used gene expression profiling to identify important pathways in retinitis pigmentosa (RP) (58). RP is commonly caused by mutations in rod-photoreceptor specific genes and yet as the disease progresses, cone-photoreceptors also die. To understand this progression, expression profiles of four mouse models of RP, harboring mutations in rod-specific genes, were analyzed by Affymetrix microarray analysis. Changes were found in the insulin/mammalian target of rapamycin pathway. Mice treated systemically with insulin, prolonged cone survival, whereas depletion of insulin had the opposite effect (58).

Microarray analysis can also be used to determine cell-specific gene expression profiles. Individual cell types are isolated from tissue using, for instance, immunopanning where antigenically unique cell types can be purified (>95%) using specific antibodies. Examples include gene expression profiles that have been carried out on purified astrocytes, neurons, and oligodendrocytes (59).

These data created the foundation for determining that astrocytes can induce RGCs to upregulate complement receptor C1q, a necessary step for RGC synapse elimination during development and possibly glaucoma (60). In a separate study, the transcriptome of Muller glia, the major glial type in the mammalian retina identified core Muller glia cell genes as well as growth factors and chemokines that may allow communication between Muller glia and surrounding retinal neurons (61). Gene expression profiles of cell types are a necessary resource in understanding normal functions of cells and then importantly how these become perturbed in disease.

5. The Sequencing Revolution: Massively Parallel Sequencing

As discussed in earlier sections of this review, a major limitation of genetic studies in both humans and animal models has been the ability to identify causative mutations within large regions of genomic sequence. Disease-causing mutations in humans can often only be localized to large regions, sometimes many megabases in size. Also, although the mapping of spontaneous or induced mutations in animal models is not as restrictive as in humans, mutations may still remain unidentified if the mutations are not easily ascertained using traditional sequencing methods within known functional elements. Historically, sequencing has been carried out using Sanger sequencing where short fragments of DNA sequence are generated (500 bp–1 Kb per lane or capillary).

Massively parallel sequencing (MPS) allows for hundreds of megabases of sequence to be produced in a single lane of a sequencing cell (reviewed in (62)). Three systems are currently available that use different chemistries, but all are able to produce large amounts of sequencing data in an efficient manner. The major systems are the Solexa Genome Analyzer from Illumina, the 454 Genome Sequencer from Roche and the SOLiD system from Applied Biosystems. A major driving force for improving MPS is to sequence 1,000 human genomes to create the most detailed and medically useful picture of human genetic variation (see http://www.1000genomes.org/). The sequencing is being carried out for a fraction of the cost and in a fraction of the time taken for the reference human genome sequence. The ultimate target is to sequence whole genomes for $1,000.

In conjunction with MPS, DNA enrichment technologies are being developed to enable specific regions of genomes (for example, those harboring potential disease-causing mutations) to be isolated and sequenced. DNA enrichment relies either on solution based capture (63) or hybridization-based methods where tiled probes are used to capture the regions of interest (64). A few megabases to as much as 30 Mb of sequence can be captured in a single experiment and can be either a contiguous stretch of genomic sequence or a selected set of functional elements dispersed throughout the genome. Ultimately it will be possible to enrich for the entire exome (all known exons) or all functional elements within the genomes of human or animal models.

Currently, MPS is being used in a number of different areas, for example, to better understand cancer biology by assessing tumor-specific transcriptomes and identifying somatic mutations in tumors. In one study, microRNAs, a class of nonprotein-coding genes, are being shown to be important in the progression of disease in a mouse model of leukemia (65). A second study used copy number variations in tumor genomes to identify potential cancer-causing genes (66). Next generation sequencing will enable the identification of mutations relevant to retinal diseases, in particular identifying DNA variations that contribute to complex diseases such as glaucoma and age-related macular degeneration.

6. Summary

In this review, we have highlighted a variety of genetic and genomic approaches to study mechanisms of disease in animal models. The continued development of genetic resources, in combination with rapid development of genomic technologies such as MPS are greatly increasing our ability to identify many new genes and pathways involved in the pathogenesis of complex diseases. As is being suggested for many diseases (67), systems genetics, where mutation detection, DNA variation, transcriptome analysis, and phenotyping in both human and animal studies are combined, will be the most powerful approach for developing improved human therapies for retinal diseases.

References

1. Libby RT, Gould DB, Anderson MG, John SW (2005) Complex genetics of glaucoma susceptibility. Annu Rev Genomics Hum Genet 6:15–44
2. Ding X, Patel M, Chan CC (2009) Molecular pathology of age-related macular degeneration. Prog Retin Eye Res 28(1):1–18
3. Yang Z, Stratton C, Francis PJ, Kleinman ME, Tan PL, Gibbs D, Tong Z, Chen H, Constantine R, Yang X et al (2008) Toll-like receptor 3 and geographic atrophy in age-related macular degeneration. N Engl J Med 359(14):1456–1463
4. Wang X, Ishimori N, Korstanje R, Rollins J, Paigen B (2005) Identifying novel genes for atherosclerosis through mouse-human comparative genetics. Am J Hum Genet 77(1):1–15
5. Smith RS, John SWM, Nishina PM, Sundberg JP (eds) (2002) Systematic evaluation of the mouse eye: anatomy, pathology and biomethods. CRC Press, Baco Raton

6. Gould DB, Smith RS, John SW (2004) Anterior segment development relevant to glaucoma. Int J Dev Biol 48(8–9):1015–1029
7. Waterston RH, Lindblad-Toh K, Birney E, Rogers J, Abril JF, Agarwal P, Agarwala R, Ainscough R, Alexandersson M, An P et al (2002) Initial sequencing and comparative analysis of the mouse genome. Nature 420(6915):520–562
8. Silver L (1995) Mouse genetics. Oxford University Press, Oxford
9. Churchill GA, Airey DC, Allayee H, Angel JM, Attie AD, Beatty J, Beavis WD, Belknap JK, Bennett B, Berrettini W et al (2004) The collaborative cross, a community resource for the genetic analysis of complex traits. Nat Genet 36(11):1133–1137
10. Bowes C, Li T, Danciger M, Baxter LC, Applebury ML, Farber DB (1990) Retinal degeneration in the rd mouse is caused by a defect in the beta subunit of rod cGMP-phosphodiesterase. Nature 347(6294):677–680
11. Anderson MG, Smith RS, Hawes NL, Zabaleta A, Chang B, Wiggs JL, John SW (2002) Mutations in genes encoding melanosomal proteins cause pigmentary glaucoma in DBA/2 J mice. Nat Genet 30(1):81–85
12. Chang B, Smith RS, Hawes NL, Anderson MG, Zabaleta A, Savinova O, Roderick TH, Heckenlively JR, Davisson MT, John SW (1999) Interacting loci cause severe iris atrophy and glaucoma in DBA/2 J mice. Nat Genet 21(4):405–409
13. Chang B, Hawes NL, Hurd RE, Wang J, Howell D, Davisson MT, Roderick TH, Nusinowitz S, Heckenlively JR (2005) Mouse models of ocular diseases. Vis Neurosci 22(5):587–593
14. den Hollander AI, Heckenlively JR, van den Born LI, de Kok YJ, van der Velde-Visser SD, Kellner U, Jurklies B, van Schooneveld MJ, Blankenagel A, Rohrschneider K et al (2001) Leber congenital amaurosis and retinitis pigmentosa with Coats-like exudative vasculopathy are associated with mutations in the crumbs homologue 1 (CRB1) gene. Am J Hum Genet 69(1):198–203
15. den Hollander AI, ten Brink JB, de Kok YJ, van Soest S, van den Born LI, van Driel MA, van de Pol DJ, Payne AM, Bhattacharya SS, Kellner U et al (1999) Mutations in a human homologue of Drosophila crumbs cause retinitis pigmentosa (RP12). Nat Genet 23(2): 217–221
16. Mehalow AK, Kameya S, Smith RS, Hawes NL, Denegre JM, Young JA, Bechtold L, Haider NB, Tepass U, Heckenlively JR et al (2003) CRB1 is essential for external limiting membrane integrity and photoreceptor morphogenesis in the mammalian retina. Hum Mol Genet 12(17):2179–2189
17. Adams DJ, van der Weyden L (2008) Contemporary approaches for modifying the mouse genome. Physiol Genomics 34(3): 225–238
18. Nagy A (2003) Manipulating the mouse embryo, 3rd edn. Cold Spring Harbor Laboratory, New York
19. Tolmachova T, Anders R, Abrink M, Bugeon L, Dallman MJ, Futter CE, Ramalho JS, Tonagel F, Tanimoto N, Seeliger MW et al (2006) Independent degeneration of photoreceptors and retinal pigment epithelium in conditional knockout mouse models of choroideremia. J Clin Invest 116(2):386–394
20. Justice MJ, Noveroske JK, Weber JS, Zheng B, Bradley A (1999) Mouse ENU mutagenesis. Hum Mol Genet 8(10):1955–1963
21. Davies VJ, Hollins AJ, Piechota MJ, Yip W, Davies JR, White KE, Nicols PP, Boulton ME, Votruba M (2007) Opa1 deficiency in a mouse model of autosomal dominant optic atrophy impairs mitochondrial morphology, optic nerve structure and visual function. Hum Mol Genet 16(11):1307–1318
22. Alexander C, Votruba M, Pesch UE, Thiselton DL, Mayer S, Moore A, Rodriguez M, Kellner U, Leo-Kottler B, Auburger G et al (2000) OPA1, encoding a dynamin-related GTPase, is mutated in autosomal dominant optic atrophy linked to chromosome 3q28. Nat Genet 26(2):211–215
23. Delettre C, Lenaers G, Griffoin JM, Gigarel N, Lorenzo C, Belenguer P, Pelloquin L, Grosgeorge J, Turc-Carel C, Perret E et al (2000) Nuclear gene OPA1, encoding a mitochondrial dynamin-related protein, is mutated in dominant optic atrophy. Nat Genet 26(2):207–210
24. de Hrabe Angelis MH, Flaswinkel H, Fuchs H, Rathkolb B, Soewarto D, Marschall S, Heffner S, Pargent W, Wuensch K, Jung M et al (2000) Genome-wide, large-scale production of mutant mice by ENU mutagenesis. Nat Genet 25(4):444–447
25. Ju WK, Kim KY, Lindsey JD, Angert M, Duong-Polk KX, Scott RT, Kim JJ, Kukhmazov I, Ellisman MH, Perkins GA et al (2008) Intraocular pressure elevation induces mitochondrial fission and triggers OPA1 release in glaucomatous optic nerve. Invest Ophthalmol Vis Sci 49(11):4903–4911
26. Hunter KW, Crawford NP (2008) The future of mouse QTL mapping to diagnose disease in mice in the age of whole-genome association studies. Annu Rev Genet 42:131–141

27. Linberg KA, Fariss RN, Heckenlively JR, Farber DB, Fisher SK (2005) Morphological characterization of the retinal degeneration in three strains of mice carrying the rd-3 mutation. Vis Neurosci 22(6):721–734
28. Haider NB, Zhang W, Hurd R, Ikeda A, Nystuen AM, Naggert JK, Nishina PM (2008) Mapping of genetic modifiers of Nr2e3 rd7/rd7 that suppress retinal degeneration and restore blue cone cells to normal quantity. Mamm Genome 19(3):145–154
29. Dietz JA, Li Y, Chung LM, Yandell BS, Schlamp CL, Nickells RW (2008) Rgcs1, a dominant QTL that affects retinal ganglion cell death after optic nerve crush in mice. BMC Neurosci 9:74
30. Friedman JS, Chang B, Kannabiran C, Chakarova C, Singh HP, Jalali S, Hawes NL, Branham K, Othman M, Filippova E et al (2006) Premature truncation of a novel protein, RD3, exhibiting subnuclear localization is associated with retinal degeneration. Am J Hum Genet 79(6):1059–1070
31. Danciger M, Ogando D, Yang H, Matthes MT, Yu N, Ahern K, Yasumura D, Williams RW, Lavail MM (2008) Genetic modifiers of retinal degeneration in the rd3 mouse. Investig Ophthalmol Vis Sci 49(7): 2863–2869
32. Li Y, Semaan SJ, Schlamp CL, Nickells RW (2007) Dominant inheritance of retinal ganglion cell resistance to optic nerve crush in mice. BMC Neurosci 8:19
33. Libby RT, Smith RS, Savinova OV, Zabaleta A, Martin JE, Gonzalez FJ, John SW (2003) Modification of ocular defects in mouse developmental glaucoma models by tyrosinase. Science 299(5612):1578–1581
34. Fadool JM, Dowling JE (2008) Zebrafish: a model system for the study of eye genetics. Prog Retin Eye Res 27(1):89–110
35. Gray MP, Smith RS, Soules KA, John SW, Link B (2008) The aqueous humor outflow pathway of zebrafish. Invest Ophthalmol Vis Sci 50:1515–1521
36. Driever W, Stemple D, Schier A, Solnica-Krezel L (1994) Zebrafish: genetic tools for studying vertebrate development. Trends Genet 10(5):152–159
37. Mullins MC, Nusslein-Volhard C (1993) Mutational approaches to studying embryonic pattern formation in the zebrafish. Curr Opin Genet Dev 3(4):648–654
38. Solnica-Krezel L, Schier AF, Driever W (1994) Efficient recovery of ENU-induced mutations from the zebrafish germline. Genetics 136(4):1401–1420
39. Malicki JJ, Pujic Z, Thisse C, Thisse B, Wei X (2002) Forward and reverse genetic approaches to the analysis of eye development in zebrafish. Vision Res 42(4):527–533
40. Link BA, Gray MP, Smith RS, John SW (2004) Intraocular pressure in zebrafish: comparison of inbred strains and identification of a reduced melanin mutant with raised IOP. Invest Ophthalmol Vis Sci 45(12): 4415–4422
41. Gross JM, Perkins BD, Amsterdam A, Egana A, Darland T, Matsui JI, Sciascia S, Hopkins N, Dowling JE (2005) Identification of zebrafish insertional mutants with defects in visual system development and function. Genetics 170(1):245–261
42. Schorderet DF, Nichini O, Boisset G, Polok B, Tiab L, Mayeur H, Raji B, de la Houssaye G, Abitbol MM, Munier FL (2008) Mutation in the human homeobox gene NKX5-3 causes an oculo-auricular syndrome. Am J Hum Genet 82(5):1178–1184
43. Ting CY, Lee CH (2007) Visual circuit development in Drosophila. Curr Opin Neurobiol 17(1):65–72
44. Iakhine R, Chorna-Ornan I, Zars T, Elia N, Cheng Y, Selinger Z, Minke B, Hyde DR (2004) Novel dominant rhodopsin mutation triggers two mechanisms of retinal degeneration and photoreceptor desensitization. J Neurosci 24(10):2516–2526
45. Wang T, Montell C (2007) Phototransduction and retinal degeneration in Drosophila. Pflugers Arch 454(5):821–847
46. Mates L, Izsvak Z, Ivics Z (2007) Technology transfer from worms and flies to vertebrates: transposition-based genome manipulations and their future perspectives. Genome Biol 8(Suppl 1):S1
47. Stowers RS, Schwarz TL (1999) A genetic method for generating Drosophila eyes composed exclusively of mitotic clones of a single genotype. Genetics 152(4):1631–1639
48. Gorska-Andrzejak J, Stowers RS, Borycz J, Kostyleva R, Schwarz TL, Meinertzhagen IA (2003) Mitochondria are redistributed in Drosophila photoreceptors lacking milton, a kinesin-associated protein. J Comp Neurol 463(4):372–388
49. Stowers RS, Megeath LJ, Gorska-Andrzejak J, Meinertzhagen IA, Schwarz TL (2002) Axonal transport of mitochondria to synapses depends on milton, a novel Drosophila protein. Neuron 36(6):1063–1077
50. Lander ES, Linton LM, Birren B, Nusbaum C, Zody MC, Baldwin J, Devon K, Dewar K, Doyle M, FitzHugh W et al (2001) Initial

sequencing and analysis of the human genome. Nature 409(6822):860–921

51. Bentley DR, Deloukas P, Dunham A, French L, Gregory SG, Humphray SJ, Mungall AJ, Ross MT, Carter NP, Dunham I et al (2001) The physical maps for sequencing human chromosomes 1, 6, 9, 10, 13, 20 and X. Nature 409(6822):942–943
52. McPherson JD, Marra M, Hillier L, Waterston RH, Chinwalla A, Wallis J, Sekhon M, Wylie K, Mardis ER, Wilson RK et al (2001) A physical map of the human genome. Nature 409(6822):934–941
53. Venter JC, Adams MD, Myers EW, Li PW, Mural RJ, Sutton GG, Smith HO, Yandell M, Evans CA, Holt RA et al (2001) The sequence of the human genome. Science 291(5507): 1304–1351
54. Adams MD, Celniker SE, Holt RA, Evans CA, Gocayne JD, Amanatides PG, Scherer SE, Li PW, Hoskins RA, Galle RF et al (2000) The genome sequence of *Drosophila melanogaster*. Science 287(5461):2185–2195
55. Frazer KA, Eskin E, Kang HM, Bogue MA, Hinds DA, Beilharz EJ, Gupta RV, Montgomery J, Morenzoni MM, Nilsen GB et al (2007) A sequence-based variation map of 8.27 million SNPs in inbred mouse strains. Nature 448(7157):1050–1053
56. Postlethwait JH (2007) The zebrafish genome in context: ohnologs gone missing. J Exp Zool B Mol Dev Evol 308(5):563–577
57. Birney E, Stamatoyannopoulos JA, Dutta A, Guigo R, Gingeras TR, Margulies EH, Weng Z, Snyder M, Dermitzakis ET, Thurman RE et al (2007) Identification and analysis of functional elements in 1% of the human genome by the ENCODE pilot project. Nature 447(7146):799–816
58. Punzo C, Kornacker K, Cepko CL (2009) Stimulation of the insulin/mTOR pathway delays cone death in a mouse model of retinitis pigmentosa. Nat Neurosci 12(1):44–52
59. Cahoy JD, Emery B, Kaushal A, Foo LC, Zamanian JL, Christopherson KS, Xing Y, Lubischer JL, Krieg PA, Krupenko SA et al (2008) A transcriptome database for astrocytes, neurons, and oligodendrocytes: a new resource for understanding brain development and function. J Neurosci 28(1): 264–278
60. Stevens B, Allen NJ, Vazquez LE, Howell GR, Christopherson KS, Nouri N, Micheva KD, Mehalow AK, Huberman AD, Stafford B et al (2007) The classical complement cascade mediates CNS synapse elimination. Cell 131(6):1164–1178
61. Roesch K, Jadhav AP, Trimarchi JM, Stadler MB, Roska B, Sun BB, Cepko CL (2008) The transcriptome of retinal Muller glial cells. J Comp Neurol 509(2):225–238
62. Shendure J, Ji H (2008) Next-generation DNA sequencing. Nat Biotechnol 26(10): 1135–1145
63. Gnirke A, Melnikov A, Maguire J, Rogov P, LeProust EM, Brockman W, Fennell T, Giannoukos G, Fisher S, Russ C et al (2009) Solution hybrid selection with ultra-long oligonucleotides for massively parallel targeted sequencing. Nat Biotechnol 27(2):182–189
64. Albert TJ, Molla MN, Muzny DM, Nazareth L, Wheeler D, Song X, Richmond TA, Middle CM, Rodesch MJ, Packard CJ et al (2007) Direct selection of human genomic loci by microarray hybridization. Nat Methods 4(11):903–905
65. Kuchenbauer F, Morin RD, Argiropoulos B, Petriv OI, Griffith M, Heuser M, Yung E, Piper J, Delaney A, Prabhu AL et al (2008) In-depth characterization of the microRNA transcriptome in a leukemia progression model. Genome Res 18(11):1787–1797
66. Chiang DY, Getz G, Jaffe DB, O'Kelly MJ, Zhao X, Carter SL, Russ C, Nusbaum C, Meyerson M, Lander ES (2009) High-resolution mapping of copy-number alterations with massively parallel sequencing. Nat Methods 6(1):99–103
67. Sieberts SK, Schadt EE (2007) Moving toward a system genetics view of disease. Mamm Genome 18(6–7):389–401

Chapter 4

Animal Models for Retinal Degeneration

Marijana Samardzija, Stephan C.F. Neuhauss, Sandrine Joly, Malaika Kurz-Levin, and Christian Grimm

Abstract

Retinal degeneration is often used to describe a category of human eye diseases, which are characterized by photoreceptor loss leading to severe visual impairment and blindness. An important, yet heterogeneous group of such diseases is called Retinitis Pigmentosa (RP). To understand the molecular mechanisms of disease induction and progression and to develop therapeutical strategies for the preservation of vision in RP patients, appropriate animal models are used in many research laboratories worldwide. The largest category of models consists of mutant (spontaneous and genetically engineered) mice. However, in recent years, zebrafish has been established as a highly valuable tool for the study of various biological problems, including retinal degeneration. In this review, we summarize the currently available mouse and zebrafish models to study retinal degeneration and give a short overview about recent developments in the field.

Key words: Mouse models, Zebrafish, Retinal degeneration, Retinitis Pigmentosa

1. Retinal Degeneration in Human Patients

Inherited retinal dystrophies are a heterogeneous group of diseases of which Retinitis Pigmentosa (RP) is the largest subgroup. RP is a genetically determined retinal degeneration affecting primarily the photoreceptor layer in both eyes.

With a prevalence of about 1:4,000 (1), more than 1 million people are affected worldwide.

The name "retinitis" is inaccurate, in that inflammation is not a prominent feature of the pathophysiology of this disease. Instead, progressive loss of retinal function and cell death leading to retinal atrophy are hallmarks of RP. The rod photoreceptors are first affected with a variable degree of subsequent degeneration of cones, therefore the disease is often also termed a rod–cone dystrophy.

Iok-Hou Pang and Abbot F. Clark (eds.), *Animal Models for Retinal Diseases*, Neuromethods, vol. 46
DOI 10.1007/978-1-60761-541-5_4, © Springer Science+Business Media, LLC 2010

Usually, visual impairment manifests itself initially by some degree of night blindness caused by rod photoreceptor dysfunction as well as visual field loss followed by the loss of central, cone-mediated vision. The age of disease onset depends mainly on the different types of RP and the underlying gene mutation and can vary from early childhood to late adulthood. The same is true for the severity of disease, ranging from mild unnoticed visual field loss with full vision to tunnel vision caused by profound field loss and markedly reduced central visual function. Complete blindness is often the final stage of the disease.

The most commonly used subclassification is based on the mode of inheritance. Typical RP can be inherited in an autosomal recessive, autosomal dominant, or X-linked recessive mode. Some cases occur in association with systemic disease such as Usher syndrome (autosomal recessive congenital deafness with RP) or Bardet–Biedl syndrome (RP, polydactyly, obesity, mental retardation, and hypogenitalism). Over the course of the last decade the knowledge about gene mutations associated with RP has quickly grown and many disease-associated genes and mutations have been described (http://www.sph.uth.tmc.edu/RetNet/sum-dis.htm). In a substantial proportion of patients, however, the causative mutation has not yet been identified, stressing the need for further research in this field. An accurate determination of the gene defect would facilitate an early and precise diagnosis as well as specific genetic counseling. Eventually, it might help to apply mutation-specific therapies in the future. Until now, no treatment is available for RP although there is some evidence that supplementation with oral vitamin A palmitate may slow the course of the disease (2). Fields of research aiming at the protection or restoration of useful vision for human patients include – but are not limited to – gene therapy, neuroprotection, stem cell transplantation or implantation of prosthetic devices.

2. Mouse Models to Study Retinal Degeneration and Photoreceptor Apoptosis

Mouse models used to study retinal degeneration can be classified into two basic categories: the induced and the inherited models. In the induced models, exposure of wild-type or genetically engineered animals (mostly mice and rats) to physical (e.g., light) or chemical (e.g., *N*-methyl-*N*-nitrosourea; MNU) treatments causes induction and progression of photoreceptor cell death leading to mild or severe retinal degeneration. The main advantage for using induced models lies in the controllable severity grade of the degeneration and the timing of disease onset.

The underlying cause of photoreceptor degeneration in the inherited group of animal models is a gene mutation or the expression of a transgene. Gene mutations can appear spontaneously as in the *rd1* (retinal degeneration 1) mouse which has a null mutation in the β subunit of the rod cGMP phosphodiesterase PDEβ or the RCS (Royal College of Surgeon) rat, which shows a functional deficiency in the MerTK gene (3). Larger animal models (dogs, cats) generally belong to this category. The vast majority of (mouse) models (Table 1), however, has been generated through various techniques of genetic modification. These models are highly useful to study the consequences of gene mutations on a molecular, cellular, tissue, or even systemic level.

2.1. Induced Models of Retinal Degeneration – Focus on Light Exposure

Most of these models mimic the human pathology in many but certainly not all aspects. A large variety of toxic factors or stimuli have been used to induce photoreceptor damage. Prominent examples are the DNA alkylating agent MNU (4–6), iron overload (7), autoantibody treatments (8), excitotoxicity (9), laser treatment (10), and others. The most frequently used model, however, applies either white light or visible light of different wavelengths and intensities to induce photoreceptor cell death. Light-induced photoreceptor degeneration is mediated by the visual pigment rhodopsin (6, 11). This model can easily be controlled and induction and extension of retinal damage can be adjusted according to the needs of the particular experiment. The observation that light-damaged photoreceptors die via an apoptotic process similar to the cells in RP or AMD (age-related macular degeneration) patients (12–15) and the suggestion that light may be an important co-factor for these diseases (16, 17) further increase the relevance of the light damage model for the study of disease mechanisms important for human pathology.

Two fundamentally different light exposure protocols are used to study the molecular processes leading to blindness. A long-term exposure (days to months) of low-level light (18, 19) induces a rather slowly progressing degeneration, whereas exposure to high levels of light for minutes to a few hours (14) generates a synchronous burst of photoreceptor apoptosis, which can lead to an almost complete loss of photoreceptors within 10 days after treatment. The two protocols not only induce different kinetics of disease progression but also different molecular pathways for induction and execution of cell death (20). Whereas c-Fos/AP-1 is important for execution of photoreceptor death in the short-term exposure model (21–23), c-Fos seems without effect in the long-term, low-light-level model (20). Transducin (or phototransduction) on the other hand is dispensable for the short-term but not for the long-term exposure model (20).

Members of the Jak/STAT signaling pathway are also critically involved in the execution of photoreceptor death after exposure to

Table 1
Animal models for retinal degeneration

Nr	Gene	Enzyme/protein	Disease	OMIM	Animal model	Ref
1	ABCA4 (ABCR)	ATP-binding cassette retina	arBBS	601691	KO	(104)
2	ABCC6	ATP-binding cassette, subfamily C (CFTR/ MRP), member 6	arBBS	603234	KO, KO	(105, 106)
3	AIPL1	Aryl hydrocarbon receptor interacting protein-like 1	arBBS	604392	KO, KO, KO	(107–109)
4	ALMS1	Alstrom syndrome 1	adMD, adOR, arOR	606844	SM, GTI, CI	(110–112)
5	ATXN7 (SCA7)	Ataxin 7	adRP	607640	KI(266Q/5Q)	(113)
6	BBS1	Bardet–Biedl syndrome 1	arCSNB	209901	KI	(114)
7	BBS2	Bardet–Biedl syndrome 2	xCSNB	606151	KO	(115)
8	BBS4	Bardet–Biedl syndrome 4	arCCRD	600374	KO, GTI	(116, 117)
9	BEST1 (VMD2)	Bestrophin 1	arUS	607854	KO, *dog C73T (cmr1) and G482A (cmr2)*	(118, 119)
10	CA4	Carbonic anhydrase IV	arOR	114760	KO	(120)
11	CABP4 (CSNB2B)	Calcium-binding protein 4	arLCA, arSSDR	608965	KO,	(121)
12	CACNA1F	Calcium channel, voltage-dependent, L type, alpha 1F subunit	arMD	300110	SM(nob2), KO	(122, 123)
13	CACNA2D4	Calcium channel, voltage-dependent, alpha 2/ delta subunit 4	xOR	608171	SM	(124)
14	CDH23 (USH1D)	Cadherin-like 23	arSSDR	605516	SM(waltzer, waltzer 2)	(125, 126)
15	CDH3	Cadherin 3, type 1, P-cadherin (placental)	arOR	114021	KO	(127)
16	CEP290	Centrosomal protein 290 kDa	arRP	610142	SM(rd16)	(128)

17	CFH	Complement factor H	arCCRD, arOR	134370	KO	(129, 130)
18	CHM	Choroideremia (Rab escort protein 1)	adSSDR	303100	KO, floxed	(131, 132)
19	CLN3	Ceroid-lipofuscinosis, neuronal 3	adSSDR	204200	KO, KO	(133, 134)
20	CNGA3	Cyclic nucleotide-gated channel, alpha-3	arSSDR	600053	KO	(135)
21	CNGB1	Cyclic nucleotide-gated channel, beta 1	arLCA, arRP, adOR	600724	KO	(136)
22	CNGB3	Cyclic nucleotide-gated channel, beta 3	adCCRD, adLCA, arLCA, adRP	605080	*Alaskan Malamute-derived dogs*	(137)
23	COL11A1	Collagen, type XI, alpha 1	arUS	120280	SM	(138)
24	COL2A1	Collagen, type II, alpha 1	xOR	120140	tg, KO, SM	(139–141)
25	COL9A1	Collagen, type IX, alpha 1	adMD	120210	KO	(142)
26	CRB1	Crumbs homolog 1 (Drosophila)	adMD	604210	SM(rd8), KO, KI(C249W)	(143–145)
27	CRX	Cone–rod homeobox-containing gene	adMD, adRP	602225	KO	(146)
28	DFNB31 (USH2D, whirlin)	Deafness, autosomal recessive 31	adOR	607928	SM	(147)
29	DMD	Dystrophin	adCSNB	300377	CI (mdx-)	(148)
30	EFEMP1	EGF-containing fibulin-like extracellular matrix protein 1	arOR	601548	KI(R345W), KO, KI(R345W)	(149–151)
31	ELOVL4	Elongation of very long chain fatty acids (FEN1/Elo2, SUR4/Elo3, yeast)-like 4	arUS	605512	KI, KO, KO	(152–154)
32	FSCN2	Ascin homolog 2, actin-bundling protein, retinal (*Strongylocentrotus purpuratus*)	arCSNB	607643	KO	(155)
33	FZD4	Frizzled homolog 4 (Drosophila)	arCSNB	604579	KO	(156)
34	GNAT1 (Tα)	Guanine nucleotide-binding protein, alpha transducing 1	adCCRD	139330	KO	(157)

(continued)

Table 1 (continued)

Nr	Gene	Enzyme/protein	Disease	OMIM	Animal model	Ref
35	GNAT2 (cTα)	Guanine nucleotide-binding protein, alpha transducing 2	adMD, adRP	139340	SM(cpfl3)	(158)
36	GPR98 (USH2C)	G-protein-coupled receptor 98	adCCRD, arLCA	602851	KO(VLGR/delTM7)	(159)
37	GRK1 (RHOK, RK)	G-protein-coupled receptor kinase 1	adLCA, adRP	180381	KO	(160)
38	GRM6 (mGluR6)	Glutamate receptor, metabotropic 6	arSSDR	604096	KO	(161)
39	GUCA1A (GCAP1)	Guanylate cyclase activator 1A (retina)	adSSDR	600364	KO, tg(G1T), tg(Y99C)	(162–164)
40	GUCA1B (GCAP2)	Guanylate cyclase activator 1B (retina)	arLCA, arRP	602275	KO	(162)
41	GUCY2D (RETGC)	Guanylate cyclase 2D, membrane (retina-specific)	arSSDR, adOR, arOR	600179	KO, KI(lacZ), KI(GFP), *chicken SM*	(165–168)
42	IMPDH1	IMP (inosine monophosphate) dehydrogenase 1	arRP	146690	KO	(169)
43	INVS	Inversin	arOR	243305	tg(gene disruption)	(170)
44	JAG1	Jagged 1 (Alagille syndrome)	arBBS	601920	KO	(171)
45	LRAT	Lecithin retinol acyl transferase	arSSDR	604863	KO, KO, KO	(172–174)
46	LRP5	Low-density lipoprotein receptor-related protein 5	arUS	603506	KO, CI	(175, 176)
47	MERTK	c-mer proto-oncogene tyrosine kinase	xOR	604705	*RCS rat*, KO, KO, KO	(3, 177–179)
48	MFRP	Membrane frizzled-related protein	arSSDR	606227	SM(rd6)	(180)
49	MKKS	McKusick–Kaufman syndrome	adRP, arRP, arOR	604896	KO, GTI	(181, 182)
50	MTTP	Microsomal triglyceride transfer protein	adRP, arRP	157147	KO	(183)
51	MYO7A (USH1B)	Myosin VIIA	xCSNB	276903	SM(shaker1)	(184)
52	NDP	Norrie disease (pseudoglioma)	arOR	300658	KO	(185)

53	NPHP3	Nephronophthisis 3 (adolescent)	adOA	608002	SM(pcy)	(186)
54	NR2E3	Nuclear receptor subfamily 2, group E, member 3	xOR	604485	SM(rd7), KO	(187, 188)
55	NRL	Neural retina leucine zipper	arSSDR	162080	KO	(189)
56	NYX	Nyctalopin	adSSDR	300278	SM(nob)	(190)
57	OAT	Ornithine aminotransferase (gyrate atrophy)	arUS	258870	KO	(191)
58	OPA1	Optic atrophy 1 (autosomal dominant)	adCSNB, arRP	605290	CI (opa+/−)	(192)
59	OPN1MW	Opsin 1 (cone pigments), medium-wave-sensitive	arSSDR	303800	KI	(193)
60	PANK2	Pantothenate kinase 2	arRP	606157	KO	(194)
61	pax2	Paired box2	adCCRD, adMD, arRP, arOR	167409	KO, CI	(195, 196)
62	PCDH15 (USH1F)	Protocadherin 15	adRP	605514	Ames waltzer (CI and SM)	(197)
63	PDE6B	Phosphodiesterase 6B, cGMP-specific, rod, beta	adMD, adRP	180072	SM(rd1, rd10)	(29, 198)
64	PEX7	Peroxisomal biogenesis factor 7	arSSDR	601757	KO	(199)
65	PRCD	Progressive rod–cone degeneration	adOR	610598	*dog SM*	(200)
66	PROM1	Prominin 1	arOR	604365	KO	(201)
67	PRPF3	PRP3 pre-mRNA processing factor 3 homolog (*Saccharomyces cerevisiae*)	arLCA	607301	KO	(202)
68	PRPH2 (rd2, RDS)	Peripherin 2 (retinal degeneration, slow)	arLCA, adRP	179605	KO, tg(L185P), KI(rds-307), tg(R172W), tg(hIRBP)	(203–207)
69	PXMP3 (Pex2)	Peroxisomal membrane protein 3, 35 kDa	arCCRD, arCSNB	170993	KO	(208)
70	RB1	Retinoblastoma 1	adCAD, arRP	180200	KO, KI(R654W)	(209, 210)
71	RBP4	Retinol-binding protein 4, plasma	arOR	180250	KO	(211)
72	RD3	Retinal degeneration 3	arOR	180040	SM(rd3)	(212, 213)

(continued)

Table 1
(continued)

Nr	Gene	Enzyme/protein	Disease	OMIM	Animal model	Ref
73	RDH12	Retinol dehydrogenase 12 (all-*trans*/9-*cis*/11-*cis*)	adCSNB, adRP, arRP	608830	KO, KO	(214, 215)
74	RDH5 (11cRDH)	Retinol dehydrogenase 5 (11-*cis*/9-*cis*)	adCCRD	601617	KO, KO	(216, 217)
75	RGR	Retinal G-protein-coupled receptor	arRP, arOR	600342	KO	(218)
76	RGS9	Regulator of G-protein signaling 9	adRP	604067	KO	(219)
77	RGS9BP	Regulator of G-protein signaling 9 binding protein	adRP, arRP	607814	KO	(220)
78	RHO	Rhodopsin	xRP	180380	tg(P23H), tgVPP, KO, KO, Palm(−/−), KI(E112Q), tg(P347S)	(221–227)
	RHO	Rhodopsin	xRP	180380	*tg pig (P347L), tg rat (P23H), tg rat (S334ter), tg frog (P23H)*	(228–231)
79	RIMS1	Regulating synaptic membrane exocytosis 1	arLCA, arRP	606629	KO	(232)
80	RLBP1 (CRALBP)	Cellular retinaldehyde-binding protein	xCCRD, xMD, xRP	180090	KO	(233)
81	ROM1	Rod outer segment membrane protein 1	arLCA	180721	KO	(234)
82	RP1	Retinitis pigmentosa 1 (autosomal dominant)	arSSDR	603937	KO, KO	(235, 236)
83	RP2	Retinitis pigmentosa 2 (X-linked recessive)	xOR	312600	*Siberian Husky dog*	(237)
84	RPE65	RPE65	arCSNB, arRP	180069	KO, SM(rd12), KI(R91W), *Briard dog*	(238)
85	RPGR	Retinitis pigmentosa GTPase regulator	adCCRD, adRP	312610	KO, tg	(239, 240)

86	RPGRIP1	Retinitis pigmentosa GTPase regulator-interacting protein	adCAD	605446	KO, *miniature longhaired dachshunds*	(241, 242)
87	RPGRIP1L	RPGRIP1L	adMD	610937	KO	(243)
88	RS1	Retinoschisin 1	arSSDR	312700	KO (lacZ), KO, CI	(244–246)
89	SAG (Arr)	S-antigen; retina and pineal gland (arrestin)	arLCA, arRP	181031	KO	(247)
90	SEMA4A	Semaphorin 4A	adCCRD	607292	GTI	(248)
91	TEAD1	TEA domain family member 1	arUS	189967	GTI, KO	(249, 250)
92	TIMP3	TIMP metallopeptidase inhibitor 3	arUS	188826	KO, KO	(251, 252)
93	TTPA	Tocopherol (alpha) transfer protein	arRP, arUS	600415	KO	(253)
94	TULP1	Tubby like protein 1	adORDD, adSSDR	602280	KO	(254)
95	UNC119 (HRG4)	unc-119 homolog (*Caenorhabditis elegans*)	arSSDR	604011	tg, KO	(255, 256)
96	USH1C (harmonin)	Usher syndrome 1C (autosomal recessive, severe)	arBBS	605242	SM, KO, KI(216G>A)	(257–259)
97	USH1G (sans)	Usher syndrome 1G (autosomal recessive)	arBBS	607696	SM(shaker and seal)	(260, 261)
98	USH2A	Usher syndrome 2A (autosomal recessive, mild)	arBBS	608400	KO	(262)
99	VCAN	Versican	adMD, adOR, arOR	118661	KO	(263)
100	WFS1	Wolfram syndrome 1 (wolframin)	adRP	606201	KO, floxed	(264, 265)

The table does not validate the listed models but merely gives an overview over the animal models which harbor a mutation in a gene identified to cause a particular retinal disease in human patients. Animal models other than mice (rat, dog, chicken, and frog) are indicated in italics. Zebrafish models are listed separately in Table 4.2. The search for animal models was done according to the 144 mapped and identified retinal disease genes published at RetNet (http://www.sph.uth.tmc.edu/RetNet/sum-dis.htm, last updated 20.11.08). Web pages used were: http://www.ncbi.nlm.nih.gov/sites/entrez?db=pubmed and http://www.informatics.jax.org/. Disease nomenclature was taken from RetNet

Disease: ***ad*** autosomal dominant, ***ar*** autosomal recessive, ***mt*** mitochondrial, ***x*** X-linked, *BBS* Bardet–Biedl syndrome, *CAD* chorioretinal atrophy or degeneration, *CCRD* cone or cone-rod dystrophy, *CSNB* congenital stationary night blindness, *LCA* leber congenital amaurosis, *MD* macular degeneration, *OA* optic atrophy, *OR* other retinopathy, *ORDD* ocular-retinal developmental disease, *RP* retinitis pigmentosa, *SSDR* syndromic/systemic diseases with retinopathy, *US* Usher syndrome

Animal models: *CI* chemically induced, *GTI* gene trap insertion, *KI* knock-in, *KO* knock-out, *SM* spontaneous mutation, *tg* transgenic high levels of light and in models of inherited retinal degeneration (see below) (24, 25). Rhodopsin regeneration in the visual cycle is an important factor for the determination of light damage susceptibility and caspase-1 may be involved in not yet clearly defined mechanisms (26–28).

Despite their advantages, the induced models can only partially represent the mechanisms and molecular processes responsible for retinal degeneration in human patients. For a complete understanding of the human disease and for the development of therapeutic strategies, it is of high importance that findings from the induced models are evaluated in animal models with an inherited retinal degeneration. Since cell death in most inherited models stretches over a long period of time, only few cells die at any given time making it difficult to identify the biochemical and molecular events which are causative for cell death. However, the knowledge gained from the induced models allows a targeted approach and candidate molecules and signaling systems can be tested specifically. This facilitates the investigations and may allow finding common retinal mechanisms which are either involved in retinal cell death or in endogenous rescue systems evolved to protect cell viability and visual function.

2.2. Inherited Models of Retinal Degeneration – Focus on rd1 and VPP as Models for RP

More than 190 genes are mapped and more than 140 genes have been identified which, when mutated, cause retinal degeneration, photoreceptor apoptosis and/or retinal dysfunction (http://www.sph.uth.tmc.edu/Retnet/home.htm). Of those, more than 40 are associated with RP. Animal models exist for most of the identified genes. A comprehensive overview of these models, along with their genes, associated human disease, and respective references are given in Table 1. Due to space restrictions, this review cannot describe each individual model in detail. Instead, we will focus on the three models (*rd1*, VPP, and $Rpe65_{R91W}$) mainly used in our own laboratory and summarize the findings from our work, which may help to understand the consequences of the particular genetic defect.

The *rd1* mouse is a model for autosomal recessive RP. Its nonsense mutation in the β-subunit of the phosphodiesterase causes a light-independent early onset rod–cone dystrophy (29) starting around postnatal day (PND) 10. At PND21, most rod photoreceptors have died and the debris has been cleared from the subretinal space with one row of cone cells remaining. With no support from rods, cones also die with almost no detectable photoreceptors left at 6 months of age.

VPP mice express a transgene in rods encoding a mutant rod opsin protein. The protein harbors three amino acid substitutions at the N-terminal end (V20G, P23H, and P27L). P23H accounts for approximately 10% of autosomal dominant RP in the USA. VPP-mediated retinal degeneration is influenced by light, has a later onset (around PND 15) and a slower progression (several weeks) than the degeneration caused by the *rd1* mutation (30, 31).

Both mouse lines have been studied to elucidate the molecular mechanisms for photoreceptor degeneration and to find therapeutical approaches to preserve vision. Although the *rd1* mouse is one of the most frequently used model for retinal degeneration, the mechanisms of cell death are still not completely clear. Due to the nonfunctional phosphodiesterase, *rd1* photoreceptors have elevated levels of cGMP and thus an increased influx of calcium (Ca^{2+}) leading to Ca^{2+} overload (32, 33). This may induce Ca^{2+}-activated cysteine proteases (calpains), which may be at least partially causative for photoreceptor degeneration in the *rd1* mouse (34). The reason for photoreceptor death in the VPP mouse is even less clear. Although most of the mutant protein may be correctly localized to rod outer segments and partially functional (35), some of the mutant rhodopsin molecules mislocalize, accumulate in intracellular inclusion bodies and form dimers (36, 37), which may lead to metabolic stress and death of photoreceptors.

Immune-related processes like the generation of auto-antibodies or the misregulation of the innate immune response have been implicated in various degenerative diseases of the neuronal system including RP and AMD (38–42). Similarly, in most of our models we found induced expression of several genes related to an immune response like the inflammatory protease caspase-1, interleukin-1β, monocyte chemoattractant protein 1 (MCP-1), and the complement components C1qα and CFH, among others (24, 28, 43, 44). However, we could not assign any specific role of these factors for the degenerative processes in photoreceptors after light damage or in the *rd1* and VPP mice. One important exception may be caspase-1. Genetic inactivation of the *Casp-1* gene increased survival of photoreceptors in the VPP mouse, but not in the rd1 mouse or the light-induced model of retinal degeneration (44). The significance of this observation is currently the focus of ongoing research.

Another molecular pathway which seems to be activated in many forms of retinal degeneration belongs to the Jak/STAT signaling system. In this signaling pathway, extracellular cytokines of the interleukin-6 (IL6) family of proteins interact with a bi- or tripartite transmembrane receptor causing the phosphorylation of intracellular members of the Janus kinase (Jak) family. Activated Jak kinases phosphorylate target proteins, among which signal transducer and activator of transcription (STAT) factors are prominent. Activation of STATs results in the differential regulation of target genes and the induction of a negative feedback loop involving the SOCS (suppressor of cytokine signaling) proteins (45). After light exposure as well as in the *rd1* and the VPP mouse, we observed a strongly elevated expression of leukemia inhibitory factor (LIF) and of cardiotrophin like cytokine (CLC) in the retina along with the activation of STAT3 and SOCS3 (25). Jak2 was prominently phosphorylated after light damage but not (*rd1*) or only weakly (VPP) in the inherited models (25). In addition, a strong activation of *Jak3* gene expression was observed in all models analyzed (unpublished observation). The detailed analysis of this signaling system in the VPP mouse revealed that LIF controls an elaborate signaling system, which is activated during retinal degeneration. In this system, photoreceptor injury induces expression of LIF in a subset of Müller glia cells. Increased expression of LIF is required to induce endothelin-2 (Edn2) expression (presumably in photoreceptors (46)), which, in turn, may activate Müller cells via the Ednrb receptor. Müller cells respond with a gliotic reaction and the upregulation of glial fibrillary acidic protein (GFAP). This may activate the expression of the growth and survival factor FGF2 (fibroblast growth factor 2), which seems to support survival of injured photoreceptors (24). In the absence of LIF, none of these downstream factors is activated and photoreceptor death in the VPP mouse is strongly accelerated. Since similar observations were made in the light damage model (unpublished data), this molecular response may be a common reaction of the retina to injury with the goal to protect viable cells from further damage and to rescue visual function. Since application of recombinant LIF already successfully protected photoreceptors against light damage (47), the LIF-Jak/STAT signaling system may be a promising target for the development of therapeutic strategies.

In a series of other experiments we addressed the role of the visual pigment and its regeneration in the visual cycle in induced or inherited retinal degeneration. RPE65 is the isomerohydrolase (48–50), which is required for the regeneration of 11-*cis*-retinal in the retinal pigment epithelium. Lack of functional RPE65 leads to a chromophore-deficient retina (48) and completely protects photoreceptor cells against light damage (6). This shows that light damage is mediated by the visual pigment as it has been suggested earlier (11). It may thus not be surprising that the kinetics of the regeneration of the visual pigment is a major determinant of the susceptibility to light-induced photoreceptor apoptosis and also influences the degenerative process in the VPP mouse (26, 27, 31, 51).

In addition to its role in modulating retinal degeneration, RPE65 is currently in the focus of the research community because of clinical trials aiming at the use of gene therapy to reconstitute vision in human RPE65 patients (52–55). Mutations in *RPE65* lead to Leber Congenital Amaurosis (LCA) or to Early Onset Severe Retinal Degeneration (EOSRD) in affected patients. For a better understanding of disease induction and progression, we generated a knock-in mouse expressing the most commonly found mutation in RPE65 ($Rpe65_{R91W}$). Analysis of the retinal pathology in this mouse model revealed that the R91W mutation reduces RPE65 protein levels and function, which leads to low levels of 11-*cis*-retinal and a progressive cone-rod dystrophy. In these conditions of limited chromophore supply, rods and cones compete for the available 11-*cis*-retinal. Since rods massively outnumber cones and because rhodopsin is thermo-dynamically more stable than cone opsin, rods incorporate most of the available chromophore leaving cones without sufficient 11-*cis*-retinal for proper opsin localization and function. This leads to cone opsin mislocalization and cone cell death (56, 57). These findings are not only relevant for the elucidation of the molecular cell death mechanisms but also for the design of treatment strategies using gene therapy to repair RPE65 deficiency. It seems important that such a treatment aims at the cone-rich region of the macula to increase local concentrations of 11-*cis*-retinal. Sufficiently high levels of 11-*cis*-retinal are required for cones to correctly localize opsin as a prerequisite to increase cell survival and prolong retinal function.

2.3. Strategies to Protect Photoreceptor Cells and to Preserve Vision

Neuroprotection is one possible approach to rescue photoreceptor cells from death to prolong the period of useful vision in patients. This approach is based on the observation that many patients have useful vision at younger ages, despite the presence of the disease causing gene mutation. This suggests that photoreceptor cells can function in such conditions but that a metabolic, cellular, or tissue-based stress induces the degeneration of the retina. Since it may not always be possible to prevent or reduce the underlying stress factor (e.g., the mutation) directly, increasing the resistance of photoreceptors is a valid approach to preserve vision. Recently, a phase I clinical trial using encapsulated cell therapy delivering ciliary neurotrophic factor (CNTF) to support survival of photoreceptor cells gave promising results (58). To establish a variety of effective therapies, however, it is essential to find and test additional neuroprotective treatments in animal models of retinal degeneration. Of particular interest are neuroprotective factors like lens epithelium-derived growth factor (LEDGF) (59), pigment epithelium-derived factor (PEDF) (60), glial cell-derived neurotrophic factor (GDNF) (61), rod-derived cone viability factor (RdCVF) (62), basic fibroblast growth factor (bFGF) (63), CNTF (64), and LIF (47) among others. Equally important are approaches to find and establish new factors which might be used to protect visual cells from degeneration. Recently, we showed that a short period of hypoxic preconditioning completely protected photoreceptors against light-induced degeneration (65). Part of the protective hypoxic response was mediated by erythropoietin (EPO) (65, 66) but additional factors are suspected to participate in the protection. The analysis of the retinal transcriptome after hypoxic preconditioning revealed that many genes involved in an oxidative stress response, apoptosis (pro or anti) or cell cycle control were differentially regulated by the short exposure period to low oxygen (67). The analysis of these candidate genes may lead to the identification of additional factors, which may be used to protect visual cells in a therapeutic approach.

3. Zebrafish as a New and Powerful Model to Study Retinal Physiology and Retinal Degeneration

While the mouse is a well-established model for ocular disease, the zebrafish is a more recent and increasingly popular addition to the animal models used for research on heritable eye diseases (see Table 2 for a collection of available Zebrafish models for the study of photoreceptor physiology and pathophysiology). Due to the rapid and external embryogenesis and high fecundity of the zebrafish, it is currently one of the best studied models for vertebrate development. Its superb genetics, in particular the feasibility to perform chemical induced mutagenesis screens, has led to an impressive number of mutations affecting all aspects of vertebrate development. More recently favorable features of visual system development have endeared the zebrafish to a growing number of vision researchers. The visual system of the zebrafish develops precociously, so that at 5 days postfertilization (dpf), a well-established retina that supports a number of visual behaviors has formed.

Another favorable feature of the zebrafish retina is the cone dominance. The zebrafish retina contains four different cone types, which absorb light maximally at 570 nm (red), 480 nm (green), 415 nm (blue), and 363 nm (ultraviolet) (68, 69). Although some rods are already present at 5 dpf, their functional contribution only becomes apparent at about 15 dpf (70). Hence all visual responses at 5 dpf, the typical age of analysis, can be considered purely cone driven. Therefore, research on the zebrafish retina nicely complements studies on the rod dominant retina of the nocturnal mouse.

Over the last few years, both the genetic tool kit to generate models of ocular diseases and to analyze them has been expanded dramatically.

Zebrafish can be efficiently mutagenized by simply adding the alkylating agent ethyl nitrosourea (ENU) to the water. The offspring of heterozygous parents can be screened for a variety of phenotypes of interest. The first two large-scale screens have isolated more than 6,000 mutant lines that cover all aspects of embryogenesis by simple visual inspection of the developing embryos (71, 72). These initial screens revealed a number of mutants affecting the development of the retina (73), which were

Table 2
Zebrafish photoreceptor mutants

Mutant	Disrupted gene	Ocular phenotype	References
bleached (blc)	Unknown	Hypopigmentation, general retinal degeneration	(266)
brudas (bru)	Unknown	Lack of outer segment, photoreceptor degeneration	(73)
elipsa (eli)	Unknown	Lack of outer segment, photoreceptor degeneration	(73, 93, 94)
fade out (fad)	Unknown	Photoreceptor degeneration	(91)
fading vision (fdv)	Silva	Hypopigmentation, shorter and misaligned photoreceptor outer segments	(99)
fleer (flr)	Unknown	Lack of outer segment, photoreceptor degeneration	(73, 267)
gantenbein (gnn)	Unknown	Photoreceptor degeneration	(268)
grumpy, laminin b1 (lamab1)	lamab1	Shorter photoreceptor outer segments	(269, 270)
ift57	ift57	Lack of outer segment, photoreceptor degeneration	(83, 96)
ift172	ift172	Lack of outer segment, photoreceptor degeneration	(83)
mikre oko (mok)	Dynactin 1	Photoreceptor degeneration	(73, 271, 272)
niezerka (nie)	Unknown	Photoreceptor degeneration	(73)
Night blindness (nba, nbb, nbc, nbd, nbe, nbf, nbg)	Unknown	Late onset night blindness (dominant)	(273–276)
no optokinetic response f (nof)	gnat2	Cone visual transduction	(277)
oval (ovl)	ift88	Lack of outer segment, photoreceptor degeneration	(73, 93–95)
partial optokinetic response a (pob)	pob	Specific red cone degeneration	(278)
Rab escort protein 1 (rep1)	rep1	Photoreceptor degeneration	(96)
sleppy, laminin g1 (lamac1)	lamac1	Shorter photoreceptor outer segments	(269, 270)

complemented by a number of subsequent screens (74). The strength of the zebrafish is the ease in which functional alterations of the visual system can be revealed. The most common approach has been to test larvae for the optokinetic response, in which moving stripes in the surround – typically provided by a moving drum or a projection – elicit stereotyped eye movements (reviewed in (75)). Larvae that display spontaneous eye movements, but fail to elicit them in response to a moving grating in the surround, are likely defective in vision. The position of the defect in the visual pathway can subsequently be specified by histological analysis and electroretinography (ERG). The ERG, which records sum field potentials of the retina in response to light, can easily be measured in zebrafish larvae and is directly comparable to mouse and human ERG recordings (76–80).

Such chemically induced mutants may reveal interesting phenotypes, but the underlying molecular defect still has to be identified by genomic mapping techniques. An alternative is offered by viral insertion mutagenesis. Here the mutating agent is a pseudotyped virus that enables rapid cloning of the disrupted gene (81, 82). Although this treatment is, in orders of magnitude, less efficient in the generation of mutations, the effort is balanced by the ease of cloning any disrupted gene(s). A rescreen of such insertion mutagenized lines for defects in vision have yielded a collection of mutants with photoreceptor defects caused by known molecular defects (83). The rapid growth of transgenic technology in the fish may soon result in the efficient generation of mutations by transposon-mediated insertions (84).

Although the number of available zebrafish mutants with characterized gene defects is growing fast, it is still not possible to knock-out a gene of interest by homologous recombination. One alternative is the time consuming Targeting Local Lesions IN Genomes (TILLING) methodology for identifying ENU-induced mutations in specific genes of interest, that necessitates a huge sequencing capacity (85, 86). A recent alternative is the heritable disruption of target genes by the use of zinc finger nucleases (87, 88). The application of this technology is predicted to have a huge impact on the use of the zebrafish model, including its use as an eye disease model.

A widely used efficient alternative is the transient knock-down of a protein of choice by morpholino antisense technology. Here, modified (morpholino) antisense nucleotides are injected into the fertilized egg. These nucleotides are designed in such a way that they either base pair with the translational start site blocking translation initiation, or with splice sites, leading to aberrant splicing and hence nonfunctional mRNA (89). Morpholino anti-sense nucleotides are stable and nontoxic. Off target toxicity may be ameliorated by concomitant p53 downregulation (90). The knockdown by morpholino technology gets progressively less effective at later stages, likely due to dilution in the growing embryos. However, we and others had good success in knocking down proteins as late as 5 dpf, sufficient to quantitatively analyze visual behavior (91, 92).

In the following, we illustrate these different approaches and highlight the advantages of the zebrafish model by briefly discussing ciliopathies and pathologies of the visual cycle.

A number of mutants isolated in the large-scale chemical mutagenesis screens showed a combination of pronephric cysts and photoreceptor degeneration, an association observed in human ciliopathies such as Bardet–Biedl (BBS) and Senior–Loken syndrome. A detailed morphological analysis of the retina of these three mutants (*oval*, *elipsa*, *fleer*) revealed a block in outer segment formation (93, 94). Subsequent positional cloning of the *oval* locus identified a mutation in the ortholog of the intraflagellar transport protein 88 (IFT88) of *Chlamydomonas* (95). The IFT complex is involved in the generation and maintenance of ciliated structures, of which the outer segments of photoreceptors and the sensory cilia of auditory hair cells and olfactory sensory neurons are among many other homologous structures in the body. A careful analysis of the IFT88 mutant in combination with morpholino-mediated downregulation of other IFT proteins, revealed that members of the IFT complex B, such as IFT88, IFT52, and IFT57, are essential for the maintenance but not the initial assembly of sensory cell cilia. The loss of a member of the IFT complex A (IFT140) produced only a mild phenotype (95). This conclusion was corroborated by the isolation of two retroviral insertion mutants in genes coding for IFT complex B genes, IFT57 and IFT172, which also displayed disrupted photoreceptor outer segments (83). Subsequent elegant biochemical studies showed that IFT88 is essential for outer segment formation, while IFT57 is rather required for efficient intraflagellar transport. It mediates association of IFT20, another IFT complex A component, to enable joining the IFT particle (96). These sets of experiments delineated the importance of IFT complex B in the maintenance of photoreceptor outer segments and give important functional evidence for the causation of human vision impairments such as Bardet–Biedl syndrome.

In the first step of seeing, the visual chromophore 11-*cis*-retinal gets photoisomerized to all-*trans*-retinal. To ensure continuous vision, the visual pigment needs to be regenerated in the visual cycle. In vertebrates the canonical visual cycle is situated in two compartments, the photoreceptor cell and cells of the retinal pigment epithelium (RPE). Mutations in genes coding for any of the components of the visual cycle can lead to retinal degeneration (97, 98). In the zebrafish mutant *fading vision*, which harbors a mutation in the *silver* gene, the visual cycle is restricted due to a defect in melanosome biogenesis in the RPE. Photoreceptor outer segments in the mutant are much shorter and misaligned. Ultrastructural analysis of the RPE reveals large vacuoles, comparable to inclusion bodies observed in the RPE of retina dystrophy patients (99). Recently evidence has accumulated that the canonical visual cycle is not the only biochemical pathway to regenerate visual pigment. Data from studies of the diurnal chicken and zebrafish retina support a parallel cone-specific visual cycle (100–102). One evidence was the surprising finding that cone vision in zebrafish is only marginally affected by the lack of RPE65 activity. RPE65-depleted zebrafish retinas display altered morphology of rod photoreceptors, similar to defects reported in the mouse knock-out mutant (103), while cone photoreceptors proved to be both morphologically and functionally intact (102). Cone photoreceptors are likely served by an alternative visual cycle, which involves Müller glia cells. So far, the only defined molecular component is the retinoid-binding protein CRALBP (cellular retinaldehyde-binding protein) (100). These studies provided evidence for a cone private visual pigment recycling pathway, while the canonical visual cycle serves both rod and cone photoreceptors. Since the human CRALBP gene is expressed in both Müller glia and RPE cells, the relative importance of a cone-specific visual cycle for human vision is still debated. The requirement of diurnal animals to keep cone vision functional at rod saturating light intensities may argue that at least diurnal animals must have evolved two visual cycles acting in parallel. The cone dominant zebrafish model is an ideal model to decipher the relative contribution of this alternative visual cycle for cone vision and its possible contribution to human visual impairments.

4. Conclusion

The study of retinal physiology and pathophysiology in a large number of different animal models has led to a good understanding of retinal development, maintenance, and function on a molecular, cellular, and tissue-specific level. This knowledge is used to identify causative mechanisms in degenerative diseases of the retina and to establish various therapeutical approaches to treat patients suffering from retinal degeneration, several of which are currently in clinical trials with promising results.

References

1. Hartong DT, Berson EL, Dryja TP (2006) Retinitis pigmentosa. Lancet 368:1795–1809
2. Berson EL, Rosner B, Sandberg MA et al (1993) A randomized trial of vitamin A and vitamin E supplementation for retinitis pigmentosa. Arch Ophthalmol 111:761–772
3. D'Cruz PM, Yasumura D, Weir J et al (2000) Mutation of the receptor tyrosine kinase gene Mertk in the retinal dystrophic RCS rat. Hum Mol Genet 9:645–651
4. Yuge K, Nambu H, Senzaki H et al (1996) *N*-methyl-*N*-nitrosourea-induced photoreceptor apoptosis in the mouse retina. In Vivo 10: 483–488
5. Nambu H, Yuge K, Nakajima M et al (1997) Morphologic characteristics of *N*-methyl-*N*-nitrosourea-induced retinal degeneration in C57BL mice. Pathol Int 47:377–383
6. Grimm C, Wenzel A, Hafezi F, Yu S, Redmond TM, Reme CE (2000) Protection of Rpe65-deficient

mice identifies rhodopsin as a mediator of light-induced retinal degeneration. Nat Genet 25:63–66
7. He L, Poblenz AT, Medrano CJ, Fox DA (2000) Lead and calcium produce rod photoreceptor cell apoptosis by opening the mitochondrial permeability transition pore. J Biol Chem 275:12175–12184
8. Shiraga S, Adamus G (2002) Mechanism of CAR syndrome: anti-recoverin antibodies are the inducers of retinal cell apoptotic death via the caspase 9- and caspase 3-dependent pathway. J Neuroimmunol 132:72–82
9. Cascio C, Guarneri R, Russo D et al (2002) A caspase-3-dependent pathway is predominantly activated by the excitotoxin pregnenolone sulfate and requires early and late cytochrome c release and cell-specific caspase-2 activation in the retinal cell death. J Neurochem 83:1358–1371
10. Zhang Y, Fortune B, Atchaneeyasakul LO et al (2008) Natural history and histology in a rat model of laser-induced photothrombotic retinal vein occlusion. Curr Eye Res 33:365–376
11. Noell WK, Walker VS, Kang BS, Berman S (1966) Retinal damage by light in rats. Invest Ophthalmol 5:450–473
12. Portera-Cailliau C, Sung CH, Nathans J, Adler R (1994) Apoptotic photoreceptor cell death in mouse models of retinitis pigmentosa. Proc Natl Acad Sci U S A 91:974–978
13. Reme CE, Grimm C, Hafezi F, Iseli HP, Wenzel A (2003) Why study rod cell death in retinal degenerations and how? Doc Ophthalmol 106:25–29
14. Reme CE, Grimm C, Hafezi F, Marti A, Wenzel A (1998) Apoptotic cell death in retinal degenerations. Prog Retin Eye Res 17:443–464
15. Reme CE, Grimm C, Hafezi F, Wenzel A, Williams TP (2000) Apoptosis in the Retina: the silent death of vision. News Physiol Sci 15:120–124
16. Paskowitz DM, LaVail MM, Duncan JL (2006) Light and inherited retinal degeneration. Br J Ophthalmol 90:1060–1066
17. Wu J, Seregard S, Algvere PV (2006) Photochemical damage of the retina. Surv Ophthalmol 51:461–481
18. La Vail MM (1976) Survival of some photoreceptor cells in albino rats following long-term exposure to continuous light. Invest Ophthalmol 15:64–70
19. Organisciak DT, Darrow RM, Barsalou L et al (1998) Light history and age-related changes in retinal light damage. Invest Ophthalmol Vis Sci 39:1107–1116
20. Hao W, Wenzel A, Obin MS et al (2002) Evidence for two apoptotic pathways in light-induced retinal degeneration. Nat Genet 32:254–260
21. Hafezi F, Steinbach JP, Marti A et al (1997) The absence of c-fos prevents light-induced apoptotic cell death of photoreceptors in retinal degeneration in vivo. Nat Med 3:346–349
22. Wenzel A, Grimm C, Marti A et al (2000) c-fos controls the "private pathway" of light-induced apoptosis of retinal photoreceptors. J Neurosci 20:81–88
23. Wenzel A, Grimm C, Seeliger MW et al (2001) Prevention of photoreceptor apoptosis by activation of the glucocorticoid receptor. Invest Ophthalmol Vis Sci 42:1653–1659
24. Joly S, Lange C, Thiersch M, Samardzija M, Grimm C. (2008) Leukemia inhibitory factor extends the lifespan of injured photoreceptors in vivo. J Neurosci 28:13965–13774
25. Samardzija M, Wenzel A, Aufenberg S, Thiersch M, Reme C, Grimm C (2006) Differential role of Jak-STAT signaling in retinal degenerations. FASEB J 20:2411–2413
26. Keller C, Grimm C, Wenzel A, Hafezi F, Reme C (2001) Protective effect of halothane anesthesia on retinal light damage: inhibition of metabolic rhodopsin regeneration. Invest Ophthalmol Vis Sci 42:476–480
27. Wenzel A, Reme CE, Williams TP, Hafezi F, Grimm C (2001) The Rpe65 Leu450Met variation increases retinal resistance against light-induced degeneration by slowing rhodopsin regeneration. J Neurosci 21:53–58
28. Grimm C, Wenzel A, Hafezi F, Reme CE (2000) Gene expression in the mouse retina: the effect of damaging light. Mol Vis 6:252–260
29. Bowes C, Li T, Danciger M, Baxter LC, Applebury ML, Farber DB (1990) Retinal degeneration in the rd mouse is caused by a defect in the beta subunit of rod cGMP-phosphodiesterase. Nature 347:677–680
30. Naash MI, Ripps H, Li S, Goto Y, Peachey NS (1996) Polygenic disease and retinitis pigmentosa: albinism exacerbates photoreceptor degeneration induced by the expression of a mutant opsin in transgenic mice. J Neurosci 16:7853–7858
31. Samardzija M, Wenzel A, Naash M, Reme CE, Grimm C (2006) Rpe65 as a modifier gene for inherited retinal degeneration. Eur J NeuroSci 23:1028–1034
32. Farber DB, Park S, Yamashita C (1988) Cyclic GMP-phosphodiesterase of rd retina: biosynthesis and content. Exp Eye Res 46:363–374
33. Fox DA, Poblenz AT, He L (1999) Calcium overload triggers rod photoreceptor apoptotic

cell death in chemical-induced and inherited retinal degenerations. Ann N Y Acad Sci 893: 282–285

34. Paquet-Durand F, Azadi S, Hauck SM, Ueffing M, van Veen T, Ekstrom P (2006) Calpain is activated in degenerating photoreceptors in the rd1 mouse. J Neurochem 96:802–814
35. Wu TH, Ting TD, Okajima TI et al (1998) Opsin localization and rhodopsin photochemistry in a transgenic mouse model of retinitis pigmentosa. Neuroscience 87:709–717
36. Surgucheva I, Ninkina N, Buchman VL, Grasing K, Surguchov A (2005) Protein aggregation in retinal cells and approaches to cell protection. Cell Mol Neurobiol 25:1051–1066
37. Frederick JM, Krasnoperova NV, Hoffmann K et al (2001) Mutant rhodopsin transgene expression on a null background. Invest Ophthalmol Vis Sci 42:826–833
38. Gurne DH, Tso MO, Edward DP, Ripps H (1991) Antiretinal antibodies in serum of patients with age-related macular degeneration. Ophthalmology 98:602–607
39. Heckenlively JR, Fawzi AA, Oversier J, Jordan BL, Aptsiauri N (2000) Autoimmune retinopathy: patients with antirecoverin immunoreactivity and panretinal degeneration. Arch Ophthalmol 118:1525–1533
40. Edwards AO, Ritter Rr, Abel KJ, Manning A, Panhuysen C, Farrer LA (2005) Complement factor H polymorphism and age-related macular degeneration. Science 308:421–424
41. Hageman GS, Anderson DH, Johnson LV et al (2005) A common haplotype in the complement regulatory gene factor H (HF1/CFH) predisposes individuals to age-related macular degeneration. Proc Natl Acad Sci U S A 102:7227–7232
42. Haines JL, Hauser MA, Schmidt S et al (2005) Complement factor H variant increases the risk of age-related macular degeneration. Science 308:419–421
43. Rohrer B, Demos C, Frigg R, Grimm C (2007) Classical complement activation and acquired immune response pathways are not essential for retinal degeneration in the rd1 mouse. Exp Eye Res 84:82–91
44. Samardzija M, Wenzel A, Thiersch M, Frigg R, Reme C, Grimm C (2006) Caspase-1 ablation protects photoreceptors in a model of autosomal dominant retinitis pigmentosa. Invest Ophthalmol Vis Sci 47:5181–5190
45. Kisseleva T, Bhattacharya S, Braunstein J, Schindler CW (2002) Signaling through the JAK/STAT pathway, recent advances and future challenges. Gene 285:1–24
46. Rattner A, Nathans J (2005) The genomic response to retinal disease and injury: evidence for endothelin signaling from photoreceptors to glia. J Neurosci 25:4540–4549
47. Ueki Y, Wang J, Chollangi S, Ash JD (2008) STAT3 activation in photoreceptors by leukemia inhibitory factor is associated with protection from light damage. J Neurochem 105:784–796
48. Jin M, Li S, Moghrabi WN, Sun H, Travis GH (2005) Rpe65 is the retinoid isomerase in bovine retinal pigment epithelium. Cell 122:449–459
49. Moiseyev G, Chen Y, Takahashi Y, Wu BX, Ma JX (2005) RPE65 is the isomerohydrolase in the retinoid visual cycle. Proc Natl Acad Sci U S A 102:12413–12418
50. Redmond TM, Poliakov E, Yu S, Tsai JY, Lu Z, Gentleman S (2005) Mutation of key residues of RPE65 abolishes its enzymatic role as isomerohydrolase in the visual cycle. Proc Natl Acad Sci U S A 102:13658–13663
51. Wenzel A, Grimm C, Samardzija M, Reme CE (2003) The genetic modifier Rpe65Leu(450): effect on light damage susceptibility in c-Fos-deficient mice. Invest Ophthalmol Vis Sci 44:2798–2802
52. Bainbridge JW, Smith AJ, Barker SS et al (2008) Effect of gene therapy on visual function in Leber's congenital amaurosis. N Engl J Med 358:2231–2239
53. Cideciyan AV, Aleman TS, Boye SL et al (2008) Human gene therapy for RPE65 isomerase deficiency activates the retinoid cycle of vision but with slow rod kinetics. Proc Natl Acad Sci U S A 105:15112–15117
54. Hauswirth W, Aleman TS, Kaushal S et al (2008) Treatment of leber congenital amaurosis due to RPE65 mutations by ocular subretinal injection of adeno-associated virus gene vector: short-term results of a phase I trial. Hum Gene Ther 19:979–990
55. Maguire AM, Simonelli F, Pierce EA et al (2008) Safety and efficacy of gene transfer for Leber's congenital amaurosis. N Engl J Med 358:2240–2248
56. Samardzija M, von Lintig J, Tanimoto N et al (2008) R91W mutation in Rpe65 leads to milder early-onset retinal dystrophy due to the generation of low levels of 11-cis-retinal. Hum Mol Genet 17:281–292
57. Samardzija M, Kostic C, Tanimoto N et al. (2009) In conditions of limited chromophore supply, rods trap 11-cis retinal leading to loss of cone function and cell death 18:1266–1275
58. MacDonald IM, Sauve Y, Sieving PA (2007) Preventing blindness in retinal disease: ciliary neurotrophic factor intraocular implants. Can J Ophthalmol 42:399–402

59. Ahuja P, Caffe AR, Holmqvist I et al (2001) Lens epithelium-derived growth factor (LEDGF) delays photoreceptor degeneration in explants of rd/rd mouse retina. Neuroreport 12:2951–2955
60. Cayouette M, Smith SB, Becerra SP, Gravel C (1999) Pigment epithelium-derived factor delays the death of photoreceptors in mouse models of inherited retinal degenerations. Neurobiol Dis 6:523–532
61. Frasson M, Picaud S, Leveillard T et al (1999) Glial cell line-derived neurotrophic factor induces histologic and functional protection of rod photoreceptors in the rd/rd mouse. Invest Ophthalmol Vis Sci 40:2724–2734
62. Leveillard T, Mohand-Said S, Lorentz O et al (2004) Identification and characterization of rod-derived cone viability factor. Nat Genet 36:755–759
63. O'Driscoll C, O'Connor J, O'Brien CJ, Cotter TG (2008) Basic fibroblast growth factor-induced protection from light damage in the mouse retina in vivo. J Neurochem 105:524–536
64. LaVail MM, Yasumura D, Matthes MT et al (1998) Protection of mouse photoreceptors by survival factors in retinal degenerations. Invest Ophthalmol Vis Sci 39:592–602
65. Grimm C, Wenzel A, Groszer M et al (2002) HIF-1-induced erythropoietin in the hypoxic retina protects against light-induced retinal degeneration. Nat Med 8:718–724
66. Grimm C, Wenzel A, Stanescu D et al (2004) Constitutive overexpression of human erythropoietin protects the mouse retina against induced but not inherited retinal degeneration. J Neurosci 24:5651–5658
67. Thiersch M, Raffelsberger W, Frigg R et al (2008) Analysis of the retinal gene expression profile after hypoxic preconditioning identifies candidate genes for neuroprotection. BMC Genomics 9:73
68. Nawrocki L, BreMiller R, Streisinger G, Kaplan M (1985) Larval and adult visual pigments of the zebrafish, *Brachydanio rerio.* Vision Res 25:1569–1576
69. Robinson J, Schmitt EA, Harosi FI, Reece RJ, Dowling JE (1993) Zebrafish ultraviolet visual pigment: absorption spectrum, sequence, and localization. Proc Natl Acad Sci U S A 90:6009–6012
70. Bilotta J, Saszik S (2001) The zebrafish as a model visual system. Int J Dev Neurosci 19:621–629
71. Driever W, Solnica-Krezel L, Schier AF et al (1996) A genetic screen for mutations affecting embryogenesis in zebrafish. Development 123:37–46
72. Haffter P, Granato M, Brand M et al (1996) The identification of genes with unique and essential functions in the development of the zebrafish, Danio rerio. Development 123:1–36
73. Malicki J, Neuhauss SC, Schier AF et al (1996) Mutations affecting development of the zebrafish retina. Development 123:263–273
74. Fadool JM, Brockerhoff SE, Hyatt GA, Dowling JE (1997) Mutations affecting eye morphology in the developing zebrafish (*Danio rerio*). Dev Genet 20:288–295
75. Huang YY, Neuhauss SC (2008) The optokinetic response in zebrafish and its applications. Front Biosci 13:1899–1916
76. Branchek T (1984) The development of photoreceptors in the zebrafish, *Brachydanio rerio.* II. Function. J Comp Neurol 224:116–122
77. Brockerhoff SE, Hurley JB, Janssen-Bienhold U, Neuhauss SC, Driever W, Dowling JE (1995) A behavioral screen for isolating zebrafish mutants with visual system defects. Proc Natl Acad Sci U S A 92:10545–10549
78. Makhankov YV, Rinner O, Neuhauss SC (2004) An inexpensive device for non-invasive electroretinography in small aquatic vertebrates. J Neurosci Methods 135:205–210
79. Saszik S, Bilotta J (1999) The effects of temperature on the dark-adapted spectral sensitivity function of the adult zebrafish. Vision Res 39:1051–1058
80. Seeliger MW, Rilk A, Neuhauss SC (2002) Ganzfeld ERG in zebrafish larvae. Doc Ophthalmol 104:57–68
81. Amsterdam A, Hopkins N (2006) Mutagenesis strategies in zebrafish for identifying genes involved in development and disease. Trends Genet 22:473–478
82. Gaiano N, Amsterdam A, Kawakami K, Allende M, Becker T, Hopkins N (1996) Insertional mutagenesis and rapid cloning of essential genes in zebrafish. Nature 383: 829–832
83. Gross JM, Perkins BD, Amsterdam A et al (2005) Identification of zebrafish insertional mutants with defects in visual system development and function. Genetics 170:245–261
84. Nagayoshi S, Hayashi E, Abe G et al (2008) Insertional mutagenesis by the Tol2 transposon-mediated enhancer trap approach generated mutations in two developmental genes: tcf7 and synembryn-like. Development 135: 159–169
85. Sood R, English MA, Jones M et al (2006) Methods for reverse genetic screening in zebrafish by resequencing and TILLING. Methods 39:220–227

86. Wienholds E, van Eeden F, Kosters M, Mudde J, Plasterk RH, Cuppen E (2003) Efficient target-selected mutagenesis in zebrafish. Genome Res 13:2700–2707
87. Doyon Y, McCammon JM, Miller JC et al (2008) Heritable targeted gene disruption in zebrafish using designed zinc-finger nucleases. Nat Biotechnol 26:702–708
88. Meng X, Noyes MB, Zhu LJ, Lawson ND, Wolfe SA (2008) Targeted gene inactivation in zebrafish using engineered zinc-finger nucleases. Nat Biotechnol 26:695–701
89. Nasevicius A, Ekker SC (2000) Effective targeted gene 'knockdown' in zebrafish. Nat Genet 26:216–220
90. Robu ME, Larson JD, Nasevicius A et al (2007) p53 activation by knockdown technologies. PLoS Genet 3:e78
91. Bahadori R, Rinner O, Schonthaler HB et al (2006) The Zebrafish fade out mutant: a novel genetic model for Hermansky-Pudlak syndrome. Invest Ophthalmol Vis Sci 47: 4523–4531
92. Rinner O, Makhankov YV, Biehlmaier O, Neuhauss SC (2005) Knockdown of cone-specific kinase GRK7 in larval zebrafish leads to impaired cone response recovery and delayed dark adaptation. Neuron 47:231–242
93. Bahadori R, Huber M, Rinner O et al (2003) Retinal function and morphology in two zebrafish models of oculo-renal syndromes. Eur J Neurosci 18:1377–1386
94. Doerre G, Malicki J (2002) Genetic analysis of photoreceptor cell development in the zebrafish retina. Mech Dev 110:125–138
95. Tsujikawa M, Malicki J (2004) Intraflagellar transport genes are essential for differentiation and survival of vertebrate sensory neurons. Neuron 42:703–716
96. Krock BL, Perkins BD (2008) The intraflagellar transport protein IFT57 is required for cilia maintenance and regulates IFT-particle-kinesin-II dissociation in vertebrate photoreceptors. J Cell Sci 121:1907–1915
97. Rando RR (2001) The biochemistry of the visual cycle. Chem Rev 101:1881–1896
98. Travis GH, Golczak M, Moise AR, Palczewski K (2007) Diseases caused by defects in the visual cycle: retinoids as potential therapeutic agents. Annu Rev Pharmacol Toxicol 47: 469–512
99. Schonthaler HB, Lampert JM, von Lintig J, Schwarz H, Geisler R, Neuhauss SC (2005) A mutation in the silver gene leads to defects in melanosome biogenesis and alterations in the visual system in the zebrafish mutant fading vision. Dev Biol 284:421–436
100. Fleisch VC, Schonthaler HB, von Lintig J, Neuhauss SC (2008) Subfunctionalization of a retinoid-binding protein provides evidence for two parallel visual cycles in the cone-dominant zebrafish retina. J Neurosci 28:8208–8216
101. Mata NL, Radu RA, Clemmons RC, Travis GH (2002) Isomerization and oxidation of vitamin a in cone-dominant retinas: a novel pathway for visual-pigment regeneration in daylight. Neuron 36:69–80
102. Schonthaler HB, Lampert JM, Isken A et al (2007) Evidence for RPE65-independent vision in the cone-dominated zebrafish retina. Eur J Neurosci 26:1940–1949
103. Redmond TM, Yu S, Lee E et al (1998) Rpe65 is necessary for production of 11-cis-vitamin A in the retinal visual cycle. Nat Genet 20:344–351
104. Weng J, Mata NL, Azarian SM, Tzekov RT, Birch DG, Travis GH (1999) Insights into the function of Rim protein in photoreceptors and etiology of Stargardt's disease from the phenotype in abcr knockout mice. Cell 98:13–23
105. Klement JF, Matsuzaki Y, Jiang QJ et al (2005) Targeted ablation of the abcc6 gene results in ectopic mineralization of connective tissues. Mol Cell Biol 25:8299–8310
106. Gorgels TG, Hu X, Scheffer GL et al (2005) Disruption of Abcc6 in the mouse: novel insight in the pathogenesis of pseudoxanthoma elasticum. Hum Mol Genet 14: 1763–1773
107. Ramamurthy V, Niemi GA, Reh TA, Hurley JB (2004) Leber congenital amaurosis linked to AIPL1: a mouse model reveals destabilization of cGMP phosphodiesterase. Proc Natl Acad Sci U S A 101:13897–13902
108. Liu X, Bulgakov OV, Wen XH et al (2004) AIPL1, the protein that is defective in Leber congenital amaurosis, is essential for the biosynthesis of retinal rod cGMP phosphodiesterase. Proc Natl Acad Sci U S A 101: 13903–13908
109. Dyer MA, Donovan SL, Zhang J et al (2004) Retinal degeneration in Aipl1-deficient mice: a new genetic model of Leber congenital amaurosis. Brain Res Mol Brain Res 132: 208–220
110. Arsov T, Larter CZ, Nolan CJ et al (2006) Adaptive failure to high-fat diet characterizes steatohepatitis in Alms1 mutant mice. Biochem Biophys Res Commun 342:1152–1159
111. Collin GB, Cyr E, Bronson R et al (2005) Alms1-disrupted mice recapitulate human Alstrom syndrome. Hum Mol Genet 14: 2323–2333

112. Li G, Vega R, Nelms K et al (2007) A role for Alstrom syndrome protein, alms1, in kidney ciliogenesis and cellular quiescence. PLoS Genet 3:e8
113. Yoo SY, Pennesi ME, Weeber EJ et al (2003) SCA7 knockin mice model human SCA7 and reveal gradual accumulation of mutant ataxin-7 in neurons and abnormalities in short-term plasticity. Neuron 37:383–401
114. Davis RE, Swiderski RE, Rahmouni K et al (2007) A knockin mouse model of the Bardet-Biedl syndrome 1 M390R mutation has cilia defects, ventriculomegaly, retinopathy, and obesity. Proc Natl Acad Sci U S A 104:19422–19427
115. Nishimura DY, Fath M, Mullins RF et al (2004) Bbs2-null mice have neurosensory deficits, a defect in social dominance, and retinopathy associated with mislocalization of rhodopsin. Proc Natl Acad Sci U S A 101:16588–16593
116. Mykytyn K, Mullins RF, Andrews M et al (2004) Bardet-Biedl syndrome type 4 (BBS4)-null mice implicate Bbs4 in flagella formation but not global cilia assembly. Proc Natl Acad Sci U S A 101:8664–8669
117. Kulaga HM, Leitch CC, Eichers ER et al (2004) Loss of BBS proteins causes anosmia in humans and defects in olfactory cilia structure and function in the mouse. Nat Genet 36:994–998
118. Marmorstein LY, Wu J, McLaughlin P et al (2006) The light peak of the electroretinogram is dependent on voltage-gated calcium channels and antagonized by bestrophin (best-1). J Gen Physiol 127:577–589
119. Guziewicz KE, Zangerl B, Lindauer SJ et al (2007) Bestrophin gene mutations cause canine multifocal retinopathy: a novel animal model for best disease. Invest Ophthalmol Vis Sci 48:1959–1967
120. Shah GN, Ulmasov B, Waheed A et al (2005) Carbonic anhydrase IV and XIV knockout mice: roles of the respective carbonic anhydrases in buffering the extracellular space in brain. Proc Natl Acad Sci U S A 102: 16771–16776
121. Haeseleer F, Imanishi Y, Maeda T et al (2004) Essential role of Ca2+-binding protein 4, a Cav1.4 channel regulator, in photoreceptor synaptic function. Nat Neurosci 7:1079–1087
122. Chang B, Heckenlively JR, Bayley PR et al (2006) The nob2 mouse, a null mutation in Cacna1f: anatomical and functional abnormalities in the outer retina and their consequences on ganglion cell visual responses. Vis Neurosci 23:11–24
123. Mansergh F, Orton NC, Vessey JP et al (2005) Mutation of the calcium channel gene Cacna1f disrupts calcium signaling, synaptic transmission and cellular organization in mouse retina. Hum Mol Genet 14:3035–3046
124. Wycisk KA, Budde B, Feil S et al (2006) Structural and functional abnormalities of retinal ribbon synapses due to Cacna2d4 mutation. Invest Ophthalmol Vis Sci 47:3523–3530
125. Wilson SM, Householder DB, Coppola V et al (2001) Mutations in Cdh23 cause nonsyndromic hearing loss in waltzer mice. Genomics 74:228–233
126. Di Palma F, Holme RH, Bryda EC et al (2001) Mutations in Cdh23, encoding a new type of cadherin, cause stereocilia disorganization in waltzer, the mouse model for Usher syndrome type 1D. Nat Genet 27:103–107
127. Radice GL, Ferreira-Cornwell MC, Robinson SD et al (1997) Precocious mammary gland development in P-cadherin-deficient mice. J Cell Biol 139:1025–1032
128. Chang B, Khanna H, Hawes N et al (2006) In-frame deletion in a novel centrosomal/ciliary protein CEP290/NPHP6 perturbs its interaction with RPGR and results in early-onset retinal degeneration in the rd16 mouse. Hum Mol Genet 15:1847–1857
129. Pickering MC, Cook HT, Warren J et al (2002) Uncontrolled C3 activation causes membranoproliferative glomerulonephritis in mice deficient in complement factor H. Nat Genet 31:424–428
130. Coffey PJ, Gias C, McDermott CJ et al (2007) Complement factor H deficiency in aged mice causes retinal abnormalities and visual dysfunction. Proc Natl Acad Sci U S A 104:16651–16656
131. van den Hurk JA, Hendriks W, van de Pol DJ et al (1997) Mouse choroideremia gene mutation causes photoreceptor cell degeneration and is not transmitted through the female germline. Hum Mol Genet 6:851–858
132. Tolmachova T, Anders R, Abrink M et al (2006) Independent degeneration of photoreceptors and retinal pigment epithelium in conditional knockout mouse models of choroideremia. J Clin Invest 116:386–394
133. Seigel GM, Lotery A, Kummer A et al (2002) Retinal pathology and function in a Cln3 knockout mouse model of juvenile Neuronal Ceroid Lipofuscinosis (batten disease). Mol Cell Neurosci 19:515–527
134. Cotman SL, Vrbanac V, Lebel LA et al (2002) Cln3(Deltaex7/8) knock-in mice with the common JNCL mutation exhibit

progressive neurologic disease that begins before birth. Hum Mol Genet 11:2709–2721

135. Biel M, Seeliger M, Pfeifer A et al (1999) Selective loss of cone function in mice lacking the cyclic nucleotide-gated channel CNG3. Proc Natl Acad Sci U S A 96: 7553–7557
136. Huttl S, Michalakis S, Seeliger M et al (2005) Impaired channel targeting and retinal degeneration in mice lacking the cyclic nucleotide-gated channel subunit CNGB1. J Neurosci 25:130–138
137. Sidjanin DJ, Lowe JK, McElwee JL et al (2002) Canine CNGB3 mutations establish cone degeneration as orthologous to the human achromatopsia locus ACHM3. Hum Mol Genet 11:1823–1833
138. Seegmiller R, Fraser FC, Sheldon H (1971) A new chondrodystrophic mutant in mice. Electron microscopy of normal and abnormal chondrogenesis. J Cell Biol 48: 580–593
139. Garofalo S, Vuorio E, Metsaranta M et al (1991) Reduced amounts of cartilage collagen fibrils and growth plate anomalies in transgenic mice harboring a glycine-to-cysteine mutation in the mouse type II procollagen alpha 1-chain gene. Proc Natl Acad Sci U S A 88:9648–9652
140. Li SW, Prockop DJ, Helminen H et al (1995) Transgenic mice with targeted inactivation of the Col2 alpha 1 gene for collagen II develop a skeleton with membranous and periosteal bone but no endochondral bone. Genes Dev 9:2821–2830
141. Donahue LR, Chang B, Mohan S et al (2003) A missense mutation in the mouse Col2a1 gene causes spondyloepiphyseal dysplasia congenita, hearing loss, and retinoschisis. J Bone Miner Res 18:1612–1621
142. Fassler R, Schnegelsberg PN, Dausman J et al (1994) Mice lacking alpha 1 (IX) collagen develop noninflammatory degenerative joint disease. Proc Natl Acad Sci U S A 91:5070–5074
143. Mehalow AK, Kameya S, Smith RS et al (2003) CRB1 is essential for external limiting membrane integrity and photoreceptor morphogenesis in the mammalian retina. Hum Mol Genet 12:2179–2189
144. van de Pavert SA, Kantardzhieva A, Malysheva A et al (2004) Crumbs homologue 1 is required for maintenance of photoreceptor cell polarization and adhesion during light exposure. J Cell Sci 117:4169–4177
145. van de Pavert SA, Meuleman J, Malysheva A et al (2007) A single amino acid substitution (Cys249Trp) in Crb1 causes retinal degeneration and deregulates expression of pituitary tumor transforming gene Pttg1. J Neurosci 27:564–573
146. Furukawa T, Morrow EM, Li T, Davis FC, Cepko CL (1999) Retinopathy and attenuated circadian entrainment in Crx-deficient mice. Nat Genet 23:466–470
147. LANE PW (1963) Whirler mice: a recessive behavior mutation in linkage group VIII. J Hered 54:263–266
148. Pillers DA, Weleber RG, Green DG et al (1999) Effects of dystrophin isoforms on signal transduction through neural retina: genotype-phenotype analysis of duchenne muscular dystrophy mouse mutants. Mol Genet Metab 66:100–110
149. Marmorstein LY, McLaughlin PJ, Peachey NS, Sasaki T, Marmorstein AD (2007) Formation and progression of sub-retinal pigment epithelium deposits in Efemp1 mutation knock-in mice: a model for the early pathogenic course of macular degeneration. Hum Mol Genet 16:2423–2432
150. McLaughlin PJ, Bakall B, Choi J et al (2007) Lack of fibulin-3 causes early aging and herniation, but not macular degeneration in mice. Hum Mol Genet 16:3059–3070
151. Fu L, Garland D, Yang Z et al (2007) The R345W mutation in EFEMP1 is pathogenic and causes AMD-like deposits in mice. Hum Mol Genet 16:2411–2422
152. McMahon A, Butovich IA, Mata NL et al (2007) Retinal pathology and skin barrier defect in mice carrying a Stargardt disease-3 mutation in elongase of very long chain fatty acids-4. Mol Vis 13:258–272
153. Vasireddy V, Jablonski MM, Mandal MN et al (2006) Elovl4 5-bp-deletion knock-in mice develop progressive photoreceptor degeneration. Invest Ophthalmol Vis Sci 47:4558–4568
154. Li W, Chen Y, Cameron DJ et al (2007) Elovl4 haploinsufficiency does not induce early onset retinal degeneration in mice. Vision Res 47:714–722
155. Yokokura S, Wada Y, Nakai S et al (2005) Targeted disruption of FSCN2 gene induces retinopathy in mice. Invest Ophthalmol Vis Sci 46:2905–2915
156. Wang Y, Huso D, Cahill H, Ryugo D, Nathans J (2001) Progressive cerebellar, auditory, and esophageal dysfunction caused by targeted disruption of the frizzled-4 gene. J Neurosci 21:4761–4771
157. Calvert PD, Krasnoperova NV, Lyubarsky AL et al (2000) Phototransduction in transgenic mice after targeted deletion of the rod transducin alpha-subunit. Proc Natl Acad Sci U S A 97:13913–13918

158. Chang B, Dacey MS, Hawes NL et al (2006) Cone photoreceptor function loss-3, a novel mouse model of achromatopsia due to a mutation in Gnat2. Invest Ophthalmol Vis Sci 47:5017–5021
159. McMillan DR, White PC (2004) Loss of the transmembrane and cytoplasmic domains of the very large G-protein-coupled receptor-1 (VLGR1 or Mass1) causes audiogenic seizures in mice. Mol Cell Neurosci 26:322–329
160. Chen CK, Burns ME, Spencer M et al (1999) Abnormal photoresponses and light-induced apoptosis in rods lacking rhodopsin kinase. Proc Natl Acad Sci U S A 96:3718–3722
161. Masu M, Iwakabe H, Tagawa Y et al (1995) Specific deficit of the ON response in visual transmission by targeted disruption of the mGluR6 gene. Cell 80:757–765
162. Mendez A, Burns ME, Sokal I et al (2001) Role of guanylate cyclase-activating proteins (GCAPs) in setting the flash sensitivity of rod photoreceptors. Proc Natl Acad Sci U S A 98:9948–9953
163. Howes KA, Pennesi ME, Sokal I et al (2002) GCAP1 rescues rod photoreceptor response in GCAP1/GCAP2 knockout mice. EMBO J 21:1545–1554
164. Olshevskaya EV, Calvert PD, Woodruff ML et al (2004) The Y99C mutation in guanylyl cyclase-activating protein 1 increases intracellular Ca2+ and causes photoreceptor degeneration in transgenic mice. J Neurosci 24:6078–6085
165. Yang RB, Robinson SW, Xiong WH, Yau KW, Birch DG, Garbers DL (1999) Disruption of a retinal guanylyl cyclase gene leads to cone-specific dystrophy and paradoxical rod behavior. J Neurosci 19:5889–5897
166. Walz A, Feinstein P, Khan M, Mombaerts P (2007) Axonal wiring of guanylate cyclase-D-expressing olfactory neurons is dependent on neuropilin 2 and semaphorin 3F. Development 134:4063–4072
167. Hu J, Zhong C, Ding C et al (2007) Detection of near-atmospheric concentrations of CO2 by an olfactory subsystem in the mouse. Science 317:953–957
168. Semple-Rowland SL, Lee NR, Van Hooser JP, Palczewski K, Baehr W (1998) A null mutation in the photoreceptor guanylate cyclase gene causes the retinal degeneration chicken phenotype. Proc Natl Acad Sci U S A 95:1271–1276
169. Gu JJ, Tolin AK, Jain J, Huang H, Santiago L, Mitchell BS (2003) Targeted disruption of the inosine 5′-monophosphate dehydrogenase type I gene in mice. Mol Cell Biol 23:6702–6712
170. Yokoyama T, Copeland NG, Jenkins NA, Montgomery CA, Elder FF, Overbeek PA (1993) Reversal of left-right asymmetry: a situs inversus mutation. Science 260:679–682
171. Xue Y, Gao X, Lindsell CE et al (1999) Embryonic lethality and vascular defects in mice lacking the Notch ligand Jagged1. Hum Mol Genet 8:723–730
172. Batten ML, Imanishi Y, Maeda T et al (2004) Lecithin-retinol acyltransferase is essential for accumulation of all-trans-retinyl esters in the eye and in the liver. J Biol Chem 279: 10422–10432
173. Liu L, Gudas LJ (2005) Disruption of the lecithin: retinol acyltransferase gene makes mice more susceptible to vitamin A deficiency. J Biol Chem 280:40226–40234
174. Ruiz A, Ghyselinck NB, Mata N et al (2007) Somatic ablation of the Lrat gene in the mouse retinal pigment epithelium drastically reduces its retinoid storage. Invest Ophthalmol Vis Sci 48:5377–5387
175. Holmen SL, Giambernardi TA, Zylstra CR et al (2004) Decreased BMD and limb deformities in mice carrying mutations in both Lrp5 and Lrp6. J Bone Miner Res 19:2033–2040
176. Xia CH, Liu H, Cheung D et al (2008) A model for familial exudative vitreoretinopathy caused by LPR5 mutations. Hum Mol Genet 17:1605–1612
177. Camenisch TD, Koller BH, Earp HS, Matsushima GK (1999) A novel receptor tyrosine kinase, Mer, inhibits TNF-alpha production and lipopolysaccharide-induced endotoxic shock. J Immunol 162: 3498–3503
178. Chen C, Li Q, Darrow AL et al (2004) Mer receptor tyrosine kinase signaling participates in platelet function. Arterioscler Thromb Vasc Biol 24:1118–1123
179. Lu Q, Gore M, Zhang Q et al (1999) Tyro-3 family receptors are essential regulators of mammalian spermatogenesis. Nature 398:723–728
180. Hawes NL, Chang B, Hageman GS et al (2000) Retinal degeneration 6 (rd6): a new mouse model for human retinitis punctata albescens. Invest Ophthalmol Vis Sci 41:3149–3157
181. Fath MA, Mullins RF, Searby C et al (2005) Mkks-null mice have a phenotype resembling Bardet-Biedl syndrome. Hum Mol Genet 14:1109–1118
182. Ross AJ, May-Simera H, Eichers ER et al (2005) Disruption of Bardet-Biedl syndrome ciliary proteins perturbs planar cell polarity in vertebrates. Nat Genet 37:1135–1140

183. Raabe M, Flynn LM, Zlot CH et al (1998) Knockout of the abetalipoproteinemia gene in mice: reduced lipoprotein secretion in heterozygotes and embryonic lethality in homozygotes. Proc Natl Acad Sci U S A 95: 8686–8691
184. Gibson F, Walsh J, Mburu P et al (1995) A type VII myosin encoded by the mouse deafness gene shaker-1. Nature 374:62–64
185. Berger W, van de Pol D, Bachner D et al (1996) An animal model for Norrie disease (ND): gene targeting of the mouse ND gene. Hum Mol Genet 5:51–59
186. Takahashi H, Ueyama Y, Hibino T et al (1986) A new mouse model of genetically transmitted polycystic kidney disease. J Urol 135:1280–1283
187. Akhmedov NB, Piriev NI, Chang B et al (2000) A deletion in a photoreceptor-specific nuclear receptor mRNA causes retinal degeneration in the rd7 mouse. Proc Natl Acad Sci U S A 97:5551–5556
188. Webber AL, Hodor P, Thut CJ et al (2008) Dual role of Nr2e3 in photoreceptor development and maintenance. Exp Eye Res 87: 35–48
189. Mears AJ, Kondo M, Swain PK et al (2001) Nrl is required for rod photoreceptor development. Nat Genet 29:447–452
190. Pardue MT, McCall MA, LaVail MM, Gregg RG, Peachey NS (1998) A naturally occurring mouse model of X-linked congenital stationary night blindness. Invest Ophthalmol Vis Sci 39:2443–2449
191. Wang T, Lawler AM, Steel G, Sipila I, Milam AH, Valle D (1995) Mice lacking ornithine-delta-aminotransferase have paradoxical neonatal hypoornithinaemia and retinal degeneration. Nat Genet 11:185–190
192. Davies VJ, Hollins AJ, Piechota MJ et al (2007) Opa1 deficiency in a mouse model of autosomal dominant optic atrophy impairs mitochondrial morphology, optic nerve structure and visual function. Hum Mol Genet 16:1307–1318
193. Smallwood PM, Olveczky BP, Williams GL et al (2003) Genetically engineered mice with an additional class of cone photoreceptors: implications for the evolution of color vision. Proc Natl Acad Sci U S A 100:11706–11711
194. Kuo YM, Duncan JL, Westaway SK et al (2005) Deficiency of pantothenate kinase 2 (Pank2) in mice leads to retinal degeneration and azoospermia. Hum Mol Genet 14:49–57
195. Torres M, Gomez-Pardo E, Dressler GR, Gruss P (1995) Pax-2 controls multiple steps of urogenital development. Development 121:4057–4065
196. Alur RP, Cox TA, Crawford MA, Gong X, Brooks BP (2008) Optic nerve axon number in mouse is regulated by PAX2. J AAPOS 12:117–121
197. RJn H-W, Ahmed ZM, Kjellstrom S et al (2006) Ames Waltzer deaf mice have reduced electroretinogram amplitudes and complex alternative splicing of Pcdh15 transcripts. Invest Ophthalmol Vis Sci 47:3074–3084
198. Chang B, Hawes NL, Pardue MT et al (2007) Two mouse retinal degenerations caused by missense mutations in the beta-subunit of rod cGMP phosphodiesterase gene. Vision Res 47:624–633
199. Brites P, Motley AM, Gressens P et al (2003) Impaired neuronal migration and endochondral ossification in Pex7 knockout mice: a model for rhizomelic chondrodysplasia punctata. Hum Mol Genet 12:2255–2267
200. Acland GM, Ray K, Mellersh CS et al (1998) Linkage analysis and comparative mapping of canine progressive rod-cone degeneration (prcd) establishes potential locus homology with retinitis pigmentosa (RP17) in humans. Proc Natl Acad Sci U S A 95:3048–3053
201. Shmelkov SV, Butler JM, Hooper AT et al (2008) CD133 expression is not restricted to stem cells, and both CD133+ and CD133– metastatic colon cancer cells initiate tumors. J Clin Invest 118:2111–2120
202. Graziotto JJ, Inglehearn CF, Pack MA, Pierce EA (2008) Decreased levels of the RNA splicing factor Prpf3 in mice and zebrafish do not cause photoreceptor degeneration. Invest Ophthalmol Vis Sci 49:3830–3838
203. Ma J, Norton JC, Allen AC et al (1995) Retinal degeneration slow (rds) in mouse results from simple insertion of a t haplotype-specific element into protein-coding exon II. Genomics 28:212–219
204. Kedzierski W, Nusinowitz S, Birch D et al (2001) Deficiency of rds/peripherin causes photoreceptor death in mouse models of digenic and dominant retinitis pigmentosa. Proc Natl Acad Sci U S A 98:7718–7723
205. McNally N, Kenna PF, Rancourt D et al (2002) Murine model of autosomal dominant retinitis pigmentosa generated by targeted deletion at codon 307 of the rds-peripherin gene. Hum Mol Genet 11:1005–1016
206. Ding XQ, Nour M, Ritter LM, Goldberg AF, Fliesler SJ, Naash MI (2004) The R172W mutation in peripherin/rds causes a cone-rod dystrophy in transgenic mice. Hum Mol Genet 13:2075–2087
207. Nour M, Ding XQ, Stricker H, Fliesler SJ, Naash MI (2004) Modulating expression of

peripherin/rds in transgenic mice: critical levels and the effect of overexpression. Invest Ophthalmol Vis Sci 45:2514–2521

208. Faust PL, Hatten ME (1997) Targeted deletion of the PEX2 peroxisome assembly gene in mice provides a model for Zellweger syndrome, a human neuronal migration disorder. J Cell Biol 139:1293–1305
209. Jacks T, Fazeli A, Schmitt EM, Bronson RT, Goodell MA, Weinberg RA (1992) Effects of an Rb mutation in the mouse. Nature 359:295–300
210. Sun H, Chang Y, Schweers B et al (2006) An E2F binding-deficient Rb1 protein partially rescues developmental defects associated with Rb1 nullizygosity. Mol Cell Biol 26:1527–1537
211. Quadro L, Blaner WS, Salchow DJ et al (1999) Impaired retinal function and vitamin A availability in mice lacking retinol-binding protein. EMBO J 18:4633–4644
212. Chang B, Heckenlively JR, Hawes NL, Roderick TH (1993) New mouse primary retinal degeneration (rd-3). Genomics 16:45–49
213. Friedman JS, Chang B, Kannabiran C et al (2006) Premature truncation of a novel protein, RD3, exhibiting subnuclear localization is associated with retinal degeneration. Am J Hum Genet 79:1059–1070
214. Maeda A, Maeda T, Imanishi Y et al (2006) Retinol dehydrogenase (RDH12) protects photoreceptors from light-induced degeneration in mice. J Biol Chem 281: 37697–37704
215. Kurth I, Thompson DA, Ruther K et al (2007) Targeted disruption of the murine retinal dehydrogenase gene Rdh12 does not limit visual cycle function. Mol Cell Biol 27:1370–1379
216. Driessen CA, Winkens HJ, Hoffmann K et al (2000) Disruption of the 11-cis-retinol dehydrogenase gene leads to accumulation of cis-retinols and cis-retinyl esters. Mol Cell Biol 20:4275–4287
217. Shang E, Lai K, Packer AI et al (2002) Targeted disruption of the mouse cis-retinol dehydrogenase gene: visual and nonvisual functions. J Lipid Res 43:590–597
218. Chen P, Hao W, Rife L et al (2001) A photic visual cycle of rhodopsin regeneration is dependent on Rgr. Nat Genet 28:256–260
219. Chen CK, Burns ME, He W, Wensel TG, Baylor DA, Simon MI (2000) Slowed recovery of rod photoresponse in mice lacking the GTPase accelerating protein RGS9-1. Nature 403:557–560
220. Keresztes G, Martemyanov KA, Krispel CM et al (2004) Absence of the RGS9.Gbeta5 GTPase-activating complex in photoreceptors of the R9AP knockout mouse. J Biol Chem 279:1581–1584
221. Olsson JE, Gordon JW, Pawlyk BS et al (1992) Transgenic mice with a rhodopsin mutation (Pro23His): a mouse model of autosomal dominant retinitis pigmentosa. Neuron 9:815–830
222. Naash MI, Hollyfield JG, al-Ubaidi MR, Baehr W (1993) Simulation of human autosomal dominant retinitis pigmentosa in transgenic mice expressing a mutated murine opsin gene. Proc Natl Acad Sci U S A 90:5499–5503
223. Humphries MM, Rancourt D, Farrar GJ et al (1997) Retinopathy induced in mice by targeted disruption of the rhodopsin gene. Nat Genet 15:216–219
224. Lem J, Krasnoperova NV, Calvert PD et al (1999) Morphological, physiological, and biochemical changes in rhodopsin knockout mice. Proc Natl Acad Sci U S A 96:736–741
225. Wang Z, Wen XH, Ablonczy Z, Crouch RK, Makino CL, Lem J (2005) Enhanced shutoff of phototransduction in transgenic mice expressing palmitoylation-deficient rhodopsin. J Biol Chem 280:24293–24300
226. Imai H, Kefalov V, Sakurai K et al (2007) Molecular properties of rhodopsin and rod function. J Biol Chem 282:6677–6684
227. Li T, Snyder WK, Olsson JE, Dryja TP (1996) Transgenic mice carrying the dominant rhodopsin mutation P347S: evidence for defective vectorial transport of rhodopsin to the outer segments. Proc Natl Acad Sci U S A 93:14176–14181
228. Petters RM, Alexander CA, Wells KD et al (1997) Genetically engineered large animal model for studying cone photoreceptor survival and degeneration in retinitis pigmentosa. Nat Biotechnol 15:965–970
229. Lewin AS, Drenser KA, Hauswirth WW et al (1998) Ribozyme rescue of photoreceptor cells in a transgenic rat model of autosomal dominant retinitis pigmentosa. Nat Med 4:967–971
230. Liu C, Li Y, Peng M, Laties AM, Wen R (1999) Activation of caspase-3 in the retina of transgenic rats with the rhodopsin mutation s334ter during photoreceptor degeneration. J Neurosci 19:4778–4785
231. Tam BM, Moritz OL (2007) Dark rearing rescues P23H rhodopsin-induced retinal degeneration in a transgenic Xenopus laevis model of retinitis pigmentosa: a chromophore-dependent mechanism characterized by

production of N-terminally truncated mutant rhodopsin. J Neurosci 27:9043–9053

232. Schoch S, Castillo PE, Jo T et al (2002) RIM1alpha forms a protein scaffold for regulating neurotransmitter release at the active zone. Nature 415:321–326
233. Saari JC, Nawrot M, Kennedy BN et al (2001) Visual cycle impairment in cellular retinaldehyde binding protein (CRALBP) knockout mice results in delayed dark adaptation. Neuron 29:739–748
234. Clarke G, Goldberg AF, Vidgen D et al (2000) Rom-1 is required for rod photoreceptor viability and the regulation of disk morphogenesis. Nat Genet 25:67–73
235. Gao J, Cheon K, Nusinowitz S et al (2002) Progressive photoreceptor degeneration, outer segment dysplasia, and rhodopsin mislocalization in mice with targeted disruption of the retinitis pigmentosa-1 (Rp1) gene. Proc Natl Acad Sci U S A 99:5698–5703
236. Liu Q, Lyubarsky A, Skalet JH, Pugh ENJ, Pierce EA (2003) RP1 is required for the correct stacking of outer segment discs. Invest Ophthalmol Vis Sci 44:4171–4183
237. Acland GM, Blanton SH, Hershfield B, Aguiree GD (1994) XLPRA: a canine retinal degeneration inherited as an X-linked trait. Am J Med Genet 52:27–33
238. Pang JJ, Chang B, Hawes NL et al (2005) Retinal degeneration 12 (rd12): a new, spontaneously arising mouse model for human Leber congenital amaurosis (LCA). Mol Vis 11:152–162
239. Hong DH, Pawlyk BS, Shang J, Sandberg MA, Berson EL, Li T (2000) A retinitis pigmentosa GTPase regulator (RPGR)-deficient mouse model for X-linked retinitis pigmentosa (RP3). Proc Natl Acad Sci U S A 97:3649–3654
240. Hong DH, Pawlyk BS, Adamian M, Li T (2004) Dominant, gain-of-function mutant produced by truncation of RPGR. Invest Ophthalmol Vis Sci 45:36–41
241. Zhao Y, Hong DH, Pawlyk B et al (2003) The retinitis pigmentosa GTPase regulator (RPGR)- interacting protein: subserving RPGR function and participating in disk morphogenesis. Proc Natl Acad Sci U S A 100:3965–3970
242. Mellersh CS, Boursnell ME, Pettitt L et al (2006) Canine RPGRIP1 mutation establishes cone-rod dystrophy in miniature longhaired dachshunds as a homologue of human Leber congenital amaurosis. Genomics 88:293–301
243. Vierkotten J, Dildrop R, Peters T, Wang B, Ruther U (2007) Ftm is a novel basal body protein of cilia involved in Shh signalling. Development 134:2569–2577
244. Weber BH, Schrewe H, Molday LL et al (2002) Inactivation of the murine X-linked juvenile retinoschisis gene, Rs1h, suggests a role of retinoschisin in retinal cell layer organization and synaptic structure. Proc Natl Acad Sci U S A 99:6222–6227
245. Zeng Y, Takada Y, Kjellstrom S et al (2004) RS-1 gene delivery to an adult Rs1h knockout mouse model restores ERG b-wave with reversal of the electronegative waveform of X-linked retinoschisis. Invest Ophthalmol Vis Sci 45:3279–3285
246. Jablonski MM, Dalke C, Wang X et al (2005) An ENU-induced mutation in Rs1h causes disruption of retinal structure and function. Mol Vis 11:569–581
247. Xu J, Dodd RL, Makino CL, Simon MI, Baylor DA, Chen J (1997) Prolonged photoresponses in transgenic mouse rods lacking arrestin. Nature 389:505–509
248. Rice DS, Huang W, Jones HA et al (2004) Severe retinal degeneration associated with disruption of semaphorin 4A. Invest Ophthalmol Vis Sci 45:2767–2777
249. Chen Z, Friedrich GA, Soriano P (1994) Transcriptional enhancer factor 1 disruption by a retroviral gene trap leads to heart defects and embryonic lethality in mice. Genes Dev 8:2293–2301
250. Sawada A, Kiyonari H, Ukita K, Nishioka N, Imuta Y, Sasaki H (2008) Redundant roles of Tead1 and Tead2 in notochord development and the regulation of cell proliferation and survival. Mol Cell Biol 28:3177–3189
251. Weber BH, Lin B, White K et al (2002) A mouse model for Sorsby fundus dystrophy. Invest Ophthalmol Vis Sci 43:2732–2740
252. Janssen A, Hoellenriegel J, Fogarasi M et al (2008) Abnormal vessel formation in the choroid of mice lacking tissue inhibitor of metalloprotease-3. Invest Ophthalmol Vis Sci 49:2812–2822
253. Jishage K, Arita M, Igarashi K et al (2001) Alpha-tocopherol transfer protein is important for the normal development of placental labyrinthine trophoblasts in mice. J Biol Chem 276:1669–1672
254. Ikeda S, Shiva N, Ikeda A et al (2000) Retinal degeneration but not obesity is observed in null mutants of the tubby-like protein 1 gene. Hum Mol Genet 9:155–163
255. Kobayashi A, Higashide T, Hamasaki D et al (2000) HRG4 (UNC119) mutation found in cone-rod dystrophy causes retinal degeneration in a transgenic model. Invest Ophthalmol Vis Sci 41:3268–3277
256. Ishiba Y, Higashide T, Mori N et al (2007) Targeted inactivation of synaptic HRG4

(UNC119) causes dysfunction in the distal photoreceptor and slow retinal degeneration, revealing a new function. Exp Eye Res 84:473–485

257. Johnson KR, Gagnon LH, Webb LS et al (2003) Mouse models of USH1C and DFNB18: phenotypic and molecular analyses of two new spontaneous mutations of the Ush1c gene. Hum Mol Genet 12: 3075–3086
258. Lefevre G, Michel V, Weil D et al (2008) A core cochlear phenotype in USH1 mouse mutants implicates fibrous links of the hair bundle in its cohesion, orientation and differential growth. Development 135:1427–1437
259. Lentz J, Pan F, Ng SS, Deininger P, Keats B (2007) Ush1c216A knock-in mouse survives Katrina. Mutat Res 616:139–144
260. Kitamura K, Kakoi H, Yoshikawa Y, Ochikubo F (1992) Ultrastructural findings in the inner ear of Jackson shaker mice. Acta Otolaryngol 112:622–627
261. Kikkawa Y, Shitara H, Wakana S et al (2003) Mutations in a new scaffold protein Sans cause deafness in Jackson shaker mice. Hum Mol Genet 12:453–461
262. Liu X, Bulgakov OV, Darrow KN et al (2007) Usherin is required for maintenance of retinal photoreceptors and normal development of cochlear hair cells. Proc Natl Acad Sci U S A 104:4413–4418
263. Yamamura H, Zhang M, Markwald RR, Mjaatvedt CH (1997) A heart segmental defect in the anterior-posterior axis of a transgenic mutant mouse. Dev Biol 186:58–72
264. Ishihara H, Takeda S, Tamura A et al (2004) Disruption of the WFS1 gene in mice causes progressive beta-cell loss and impaired stimulus-secretion coupling in insulin secretion. Hum Mol Genet 13:1159–1170
265. Riggs AC, Bernal-Mizrachi E, Ohsugi M et al (2005) Mice conditionally lacking the Wolfram gene in pancreatic islet beta cells exhibit diabetes as a result of enhanced endoplasmic reticulum stress and apoptosis. Diabetologia 48:2313–2321
266. Neuhauss SC, Seeliger MW, Schepp CP, Biehlmaier O (2003) Retinal defects in the zebrafish bleached mutant. Doc Ophthalmol 107:71–78
267. Drummond IA, Majumdar A, Hentschel H et al (1998) Early development of the zebrafish pronephros and analysis of mutations affecting pronephric function. Development 125:4655–4667
268. Biehlmaier O, Neuhauss SC, Kohler K (2003) Synaptic plasticity and functionality at the cone terminal of the developing zebrafish retina. J Neurobiol 56:222–236
269. Biehlmaier O, Makhankov Y, Neuhauss SC (2007) Impaired retinal differentiation and maintenance in zebrafish laminin mutants. Invest Ophthalmol Vis Sci 48:2887–2894
270. Lee J, Gross JM (2007) Laminin beta1 and gamma1 containing laminins are essential for basement membrane integrity in the zebrafish eye. Invest Ophthalmol Vis Sci 48: 2483–2490
271. Tsujikawa M, Omori Y, Biyanwila J, Malicki J (2007) Mechanism of positioning the cell nucleus in vertebrate photoreceptors. Proc Natl Acad Sci U S A 104:14819–14824
272. Del Bene F, Wehman AM, Link BA, Baier H (2008) Regulation of neurogenesis by interkinetic nuclear migration through an apical-basal notch gradient. Cell 134: 1055–1065
273. Li L, Dowling JE (1997) A dominant form of inherited retinal degeneration caused by a non-photoreceptor cell-specific mutation. Proc Natl Acad Sci U S A 94:11645–11650
274. Li L, Dowling JE (2000) Disruption of the olfactoretinal centrifugal pathway may relate to the visual system defect in night blindness b mutant zebrafish. J Neurosci 20: 1883–1892
275. Maaswinkel H, Mason B, Li L (2003) ENU-induced late-onset night blindness associated with rod photoreceptor cell degeneration in zebrafish. Mech Ageing Dev 124: 1065–1071
276. Maaswinkel H, Riesbeck LE, Riley ME et al (2005) Behavioral screening for nightblindness mutants in zebrafish reveals three new loci that cause dominant photoreceptor cell degeneration. Mech Ageing Dev 126:1079–1089
277. Brockerhoff SE, Rieke F, Matthews HR et al (2003) Light stimulates a transducin-independent increase of cytoplasmic Ca2+ and suppression of current in cones from the zebrafish mutant nof. J Neurosci 23: 470–480
278. Taylor MR, Kikkawa S, Diez-Juan A et al (2005) The zebrafish pob gene encodes a novel protein required for survival of red cone photoreceptor cells. Genetics 170:263–273

Chapter 5

Animal Models for Age-Related Macular Degeneration

Joe G. Hollyfield and Lisa Kuttner-Kondo

Abstract

Age-related macular degeneration (AMD) has a number of characteristic features including late onset and accumulation of deposits (drusen) below the retinal pigment epithelium on Bruch's membrane in the macula. A progressive increase in these deposits (in some individuals) leads to macular blindness, following either the local loss of the retinal pigment epithelium (geographic atrophy) or the hemorrhage of new blood vessels that originate in the choroid and invade the compartment between the photoreceptors and retinal pigment epithelium (choroidal neovascularization). Over the last few years a number of mouse models for AMD have been described that replicate some of the changes manifest in the human disease. This chapter begins with a description of the hallmarks of AMD, discusses some of the ideas about the underlying mechanisms and then summarizes the features of AMD found in experimental animals that are purported to model this disorder.

Key words: Age-related macular degeneration (AMD), Drusen, Bruch's membrane, Geographic atrophy, Choroidal neovascularization, AMD animal model

1. Age-related macular degeneration: The Human Condition

Age-related macular degeneration (AMD) constitutes a major cause of legal blindness in the elderly population of developed countries. Millions of individuals over 55 years of age are blind from AMD in Europe and North America with over 300,000 new AMD cases presenting annually (1, 2). In some individuals, in addition to the normal age-related thickening of Bruch's membrane, focal accumulation of debris on Bruch's membrane causes nodular elevations of the RPE that are recognized in an eye exam as drusen (3–5). Clinicians have long recognized that the presence of drusen in the macula, their size, density, and the area encompassed by these deposits represent early stages in the AMD disease process. While the initial drusen deposits are not associated with

Iok-Hou Pang and Abbot F. Clark (eds.), *Animal Models for Retinal Diseases*, Neuromethods, vol. 46
DOI 10.1007/978-1-60761-541-5_5, © Springer Science+Business Media, LLC 2010

macular blindness, individuals with drusen are considered at risk for developing the advanced forms of AMD that are accompanied by loss of macular vision (6–8). Although progression to end-stage AMD does not occur in most individuals, this disease is generally considered to be a progressive disorder with small numbers of drusen giving way to larger numbers and ultimately leading to end-stage AMD and loss of central vision.

End-stage AMD occurs in one of two forms: (a) geographic atrophy, or (b) choroidal neovascularization. Geographic atrophy (also known as the end-stage of the "dry form" of AMD) develops slowly and is characterized by the loss (atrophy) of a localized area of RPE below the fovea (9). Without the RPE the outer blood–retinal barrier is compromised, focal swelling follows, foveal photoreceptors degenerate and central vision is lost. Choroidal neovascularization (also called the "wet form" of AMD) is characterized by the development of new blood vessels originating in the choroid that break through Bruch's membrane and the RPE invading the interphotoreceptor matrix (subretinal space). These new blood vessels are leaky and can hemorrhage. The leaky vessels are thought to be responsible for macular edema causing reduced vision. Blindness is immediate when a hemorrhage occurs in the compartment between the foveal photoreceptors and RPE. Over time, the ensuing clot that forms will be replaced by a fibrovascular scar, but photoreceptors degenerate causing loss of sensory input from the region of the retina adjacent to the clot (6, 7).

AMD is distinctly different from the early onset forms of macular degeneration that have simple Mendelian inheritance patterns. These so-called juvenile forms of macular degeneration include: Stargardt's disease, caused by mutations in the gene for the ABCR transporter (10), or the ELOVL4 gene (11); Malattia Leventinese, caused by mutations in the EFEMP1 gene (12); Sorsby's fundus dystrophy, caused by mutations in the TIMP-3 gene (13), and Best's macular dystrophy, caused by mutations in an RPE-specific chloride transporter gene (14, 15). While each of these diseases results in a distinct form of macular degeneration, and gene-specific targeted defects in mice result in many of the features found in humans with these mutations, major attempts to link these early onset diseases to AMD have been unsuccessful (16). Animal models for these early onset forms of macular degeneration are not included in this review, although some of the purported models of AMD mentioned below show early onset.

Before the report by Hageman and Mullins (17) that proteins involved in the complement defense pathway were present in drusen, there was little focus on a common pathway that might be responsible for AMD. Since their initial report a large number of studies have expanded our understanding of the drusen and focused interest on the immune system's involvement in this disease (17–22).

Some of the proteins now recognized as components of drusen include C1q, C3, C4, C5, C7, C8, C9, C5b-9 complex (the neoepitope that appears with the assembly of five proteins into the membrane attack complex), complement factor H (CFH), complement receptor 1 (CD35), membrane cofactor protein (CD46), apolipoprotein E, vitronectin, clusterin, Aß-protein, and C-reactive protein. The identifications of these complement proteins in AMD eye tissue brought a much needed focus for additional work in this area.

The possibility that genetic factors are involved in AMD has been recognized for some time, primarily from studies showing the similarity in time of disease onset in identical twins (23–26). However, because the disease usually occurs so late in life and is not preceded by symptoms of the disease that can be recognized in young individuals it was not possible to perform classical genetic linkage analysis to study AMD. Recent genetic studies using single nucleotide polymorphism (SNPs) analysis and candidate gene approach have identified a number of mutations/polymorphisms in AMD patients to genes coding for complement pathway proteins (complement component C3, C2, and factor B) or its regulators (factor H, factor H-related proteins and SERPING1). It is estimated that characteristic polymorphisms/mutations in these genes may be present in approximately one-half of all AMD patients (27–32). Collectively these studies strongly indicate that AMD has a genetic component and that the complement pathway is involved in the pathology of AMD.

The buildup of fluorescent deposits in the RPE, referred to as lipofuscin, is also suggested to be causally involved in the AMD disease process. Lipofuscin is a heterogeneous storage/waste material containing bis-retinoid (A2E) fluorescent pigments that accumulate with age. Lipofuscin can be monitored in vivo by imaging fundus autofluorescence. The abundance of lipofuscin in and around lesions developing in the RPE are noted in two early forms of macular degeneration (Best's macular dystrophy and Stargardt's disease) and at the expanding margin of RPE lesions in geographic atrophy. This association has led to the notion that lipofuscin may be toxic and is actively involved in RPE pathology (33–35). Experimental studies demonstrate that RPE cells fed lipofuscin granules render these cells susceptible to phototoxicity (34, 36–39). Lipofuscin may also be involved in activation of complement (40).

The complement system plays an essential role in inflammation and immune responses. Soluble complement proteins are present in the blood in precursor forms and require activation to fulfill their specific physiological roles. Activated complement has diverse functions, including the initiation of inflammation, recruitment of leukocytes, clearance of immune complexes, neutralization of pathogens, regulation of antibody responses, and disruption of cell membranes.

The complement cascade can be activated by one of three initiating pathways. Activation through the classical pathway depends on assembly of complement factors at sites of antigen–antibody interaction involving complement component-1 and -4 (C1 and C4). Activation through the lectin pathway is initiated by a mannan-binding lectin interacting with pathogen surfaces. Activation through the alternative pathway is triggered by a variety of pathogen surfaces and requires the interaction of complement component-3 (C3), factor B, and factor D. Regardless of the pathway, activation leads to the cleavage of C3. This generates the smaller, proinflammatory C3a fragment and the larger C3b fragment. C3b together with other activated proteins form the important convertases required for the terminal part of the complement cascade, culminating in the assembly of the membrane attack complex (MAC) that mediates cell membrane disruption. Because many of these complement pathway proteins are present in drusen and sub-RPE deposits (17, 18, 20–22, 41), along with the findings that genes coding for several complement pathway proteins or regulators of this pathway are linked to AMD (27–32), the evidence is compelling that complement is involved in the AMD disease process.

But what is the inflammatory signal from the outer retina that directs a complement mediated attack specifically to this tissue interface? Mutations in factor H, factor B, and C3 genes cause defects in the proteins coded by these genes throughout all organ systems in the body. If AMD is caused by an altered functional response in the proteins coded by these defective genes one would expect altered complement function throughout the body, with disease manifest in many organs. But AMD is targeted to the outer retina, appearing first on the basal side of the RPE along Bruch's membrane. For this interface to be targeted for a complement mediated attack, there must be an inflammatory signal coming from the cells/tissue along this interface that directs an immune response to this site.

One potential source of an inflammatory signal from the outer retina is docosahexaenoic acid (DHA) (42). DHA is the most oxidizable of all long chain fatty acids with 22 carbons and six double bonds (22:6ω3). DHA is highly concentrated in photoreceptors and the RPE, accounting for approximately 40% of the fatty acids in photoreceptor outer segment membranes (43, 44). These DHA-rich membranes reside in a compartment rich in oxygen and light, conditions that are highly permissive for oxidative damage (43–45). Proteins isolated from drusen and Bruch's membrane are adducted with carboxyethylpyrrole (CEP), a novel adduct that can only be produced by the oxidative fragmentation of DHA (21). CEP adducts are found on albumin, pyruvate kinase, and glutathione S-transferase, with a number of other proteins showing CEP-immunoreactivity (46).

In addition, CEP-adducted proteins are more prevalent in RPE/Bruch's membrane/choroid tissues from AMD donor eyes than from these tissues in normal age-matched donor eyes (21). CEP-adducted proteins are also present in plasma, with AMD plasma containing 40% higher levels of CEP-adducted protein and autoantibodies to CEP than are present in plasma from healthy age-matched donors without AMD (42). The presence of adducts generated by the specific oxidation of DHA are particularly intriguing because of the long-recognized association of AMD with oxidative damage (47–51). As will be described below, this CEP-adduct is sufficient to cause an AMD-like disease when mice are sensitized to this hapten by immunization (52).

2. AMD: In Animal Models

Based on our current knowledge of the course and patterns of AMD in the human condition, what hallmarks of AMD should be expected in an animal model of this disease? From the above description several features should be found. The pathology should be progressive, occur late in the life of the animal, be restricted to the outer retina, show features of drusen formation, with complement system involvement, RPE changes and/or choroidal neovascularization. How many of these features of AMD are found in the animal models for AMD that have been described to date? In this section, a short description of the features of several of the published models will be provided with emphasis on those that show some of the hallmarks present in the human disease. In fairness to the authors of some of these studies, it should be mentioned that several of the models were published before it was known that inflammation and the complement pathway are involved in the pathology of AMD.

2.1. A Mouse Model for AMD Based on Dietary Fat and Blue Light

This model relies on a feeding regimen of a diet high in saturated and unsaturated fats over a period of 4–5 months to normal C57BL/6 mice. Several variables were evaluated. The severity of lesions below the RPE were dependent on the age of the mice when placed on the diet, sex of the animals, and exposure to low levels of blue-green light. If mice were 9 months or older at the time of initiation of diet they frequently developed basal laminar deposits below the RPE of moderate severity, but only the 16-month-old mice developed more severe deposits after exposure to blue-green light. Some mice in the older group developed choroidal neovascularization through Bruch's membrane. Aged mice fed with normal diets developed only mild subretinal deposits. These studies show that the age of the animal increases the capacity of dietary fat, especially in the presence of environmental light, to result in basal laminar deposits below the RPE (53, 54). A hormonal component was also demonstrated in this model. When female mice were made estrogen deficient in middle-age, and then challenged with the high-fat diet, the mice showed increased severity of sub-RPE deposit formation (54). It should be noted that sub-RPE deposits have been noted in senescence-accelerated mice but these animals have many systemic changes and are not included in this listing of mouse models (55).

2.2. A Mouse Model for AMD Based on Cigarette Smoke

Cigarette smoking is a major risk factor for developing AMD (47, 56). This mouse model builds on the dietary fat model discussed above (53), but, in addition, maintains mice in a smoking chamber for several hours each day or evaluates mice fed hydroquinone (HQ), a potent oxidant present in cigarette smoke. C57BL/6 female mice, 16 months old, were fed a high-fat diet (HFD) for 4.5 months. Mice were then divided into two major groups, one to examine the effects of cigarette smoke and one to study the effects of a defined cigarette smoke component such as HQ. In the first group, mice eyes were exposed to blue-green light (positive controls) or to whole cigarette smoke. Mice at the same age, but with no treatment served as a negative control. In the second group, animals received a diet that included HQ (0.8%) with low or high fat content for 4.5 months. Most of the animals fed a HFD without other oxidant exposure demonstrated normal morphology or, in a few cases, small nodular basal laminar deposits below the RPE, as described previously (54). Eyes of mice exposed to cigarette smoke or to HQ in their diet demonstrated variable degrees of basal laminar deposits and diffusely thickened Bruch's membrane. The choriocapillaris was also variably hypertrophic. From these results the conclusion was made that exposure to cigarette smoke or the smoke-related oxidant molecule, HQ, results in the formation of sub-RPE deposits, thickening of Bruch's membrane, and accumulation of debris within Bruch's membrane in aged mice (57).

2.3. A Mouse Model for AMD in an Apolipoprotein E Transgenic Line

Apolipoprotein E is a known genetic risk factor for AMD (58). This model is based on a combination of three known AMD risk factors: advanced age, high fat cholesterol-rich (HF-C) diet, and the presence of the human apolipoprotein E (apoE) transgene. Eyes of aged mice from lines with targeted replacement of human apoE2, apoE3, or apoE4 (C57BL/6 apoE TR) that were maintained on a HF-C diet show apoE isoform-dependent pathologies of differential severity. ApoE4 mice were the most severely affected. They developed multiple changes including: diffuse deposits, drusenoid deposits below the RPE, thickened Bruch's membrane, and hypopigmentation, hyperpigmentation, or atrophy of the RPE. Some of the apoE4 mice also developed choroidal neovascularization, but this was an inconsistent finding (59). These studies demonstrate that mice expressing the human apolipoprotein E transgene are more susceptible to AMD-like lesions as they age and are subjected to a high-fat diet.

2.4. A Mouse Model for AMD in a Cathepsin D Transgenic Line

This model is based on targeting the disruption of function of the lysosomal system of the RPE by using transgenic technology to express an inactive form of cathepsin D in the RPE (60). AMD-like lesions/changes were observed in 11- to 12-month-old mcd/mcd mice, which included RPE cell proliferation and atrophy as well as pigmentary changes, shortening of photoreceptor outer segments, buildup of photoreceptor debris in RPE cells, basal laminar and basal linear deposits below the RPE, and decreased electroretinogram amplitudes. While there is no indication that the gene for cathepsin D or other lysosomal enzyme genes are targets for mutations in AMD, this transgenic line shows that alteration in the function of the lysosomal degradation pathway is accompanied by the development of AMD-like lesions.

2.5. A Mouse Model for AMD in Macrophage Chemokine/Chemokine Receptor Deficient Lines

This model is based on disruption of function of the macrophage chemoattractant pathway. Mice deficient in either monocyte chemoattractant protein-1 (*Ccl-2*; also known as MCP-1) or its cognate chemokine receptor-2 (*Ccr-2*) that function in recruitment of monocytes develop a number of features of AMD. These include accumulation of lipofuscin in the RPE, drusen deposition on Bruch's membrane, photoreceptor atrophy, and choroidal neovascularization. Complement and IgG deposition are also observed below the RPE and associated choroid during aging in these mice. Impaired monocyte/macrophage recruitment could allow accumulation of extracellular material present in these mice, which, in turn, may be causally involved in development of choroidal neovascularization. These two knockout lines clearly demonstrate the importance of macrophage recruitment for the ongoing removal of extracellular material below the RPE and suggest a role for macrophage dysfunction in AMD (61).

2.6. Mouse Model for AMD in a Double Knockout with Ccl2(−/−)/Cx3cr1(−/−) Deficiencies

Loss-of-function SNPs within the CX3CR1 gene, which encodes the specific receptor for the CX3CL1/fractalkine chemokine, is also reported to be associated with AMD (62). When double knockout mice with *Ccl2(−/−)/Cx3cr1(−/−)* deficiencies were generated, AMD lesions develop that are reported to be more characteristic and reproducible than in animals missing only one of these genes (63). By 6 weeks of age, all double knockout mice developed AMD-like lesions, including drusen, RPE changes, and photoreceptor loss. In addition, choroidal neovascularization occurred in 15% of the mice. These lesions progressed with age. A2E, a major lipofuscin fluorophore that accumulates in the RPE during AMD progression, was significantly higher in the *Ccl2(−/−)/Cx3cr1(−/−)* retina than in the wild-type retina. Membrane cofactor protein (CD46), a regulator of complement activation, had a greater area of expression in the *Ccl2(−/−)/Cx3cr1(−/−)* than in WT mice RPE. In the double KO, CD46 was observed on the apical side of the RPE, leading the authors to speculate that it was upregulated due to increased complement activity (63). Additionally, four proteins including calcium-binding 140k protein, ERp29 precursor, and RIKEN cDNA 2210010C04 were found to be differentially expressed in *Ccl2(−/−)/Cx3cr1(−/−)* retina compared with controls. The authors concluded from these observations that *Ccl2(−/−)/Cx3cr1(−/−)* mice develop a broad spectrum of AMD-like abnormalities but the onset of these lesions are severe and occur very early.

2.7. A Mouse Model for AMD with an Mdm1 Mutation

This model is in a mouse line with a naturally occurring mutation that shows AMD-like lesions with aging (64). At 6 months of age these mice have normal fundus, ERG amplitudes and retinal histology, but by 14 months the retinal blood vessels are attenuated, the RPE shows atrophy and pigmentary abnormalities that progress to complete loss of photoreceptors and extinguished ERG by 22 months of age. Genetic analysis revealed that the retinal degeneration in this line segregates in an autosomal recessive manner and the disease gene is located on mouse chromosome 10. Positional cloning identified a nonsense mutation in the mouse double minute-1 gene (*Mdm1*), which caused the truncation of the putative protein from 718 amino acids to 398. The novel transcript of the *Mdm1* gene was identified in the neural retina where it is localized to the nuclear lamina. These findings suggest that truncation of this *Mdm1* transcript may underlie the mechanism leading to late-onset retinal degeneration with AMD-like features in this mutant mouse. Analysis of a cohort of patients with age-related macular degeneration (AMD) where the susceptibility locus maps to chromosome 12q, a region bearing the human ortholog to Mdm1, did not reveal association between human MDM1 and AMD.

2.8. A Mouse Model for AMD in a Superoxide Dismutase Transgenic Line

Oxidative stress has long been linked to the pathogenesis of neurodegenerative diseases (47–50); however, whether it is a cause or merely accelerates the degenerative process is not well-understood. AMD-like features are reported in mice deficient in the Cu, Zn-superoxide dismutase (Sod1). Analysis of *Sod1(−/−)* mice of different ages demonstrate that older animals have drusen, thickened Bruch's membrane, and choroidal neovascularization. The number of drusen increased with age, and exposure of young *Sod1(−/−)* mice to excess light showed increased drusen formation. The RPE cells of *Sod1(−/−)* mice showed disruption of beta-catenin-mediated cell–cell junctional complexes, indicating that oxidative stress may alter the barrier integrity of the RPE. These observations suggest that oxidative stress may play a causal role in age-related retinal degeneration (65). Moreover, in one study, in contrast to WT C57BL/6 retinas, *Sod1*(−/−) retinas stained positive for CD46, C5, and Ig in drusen and/or surrounding tissues, leading to a potential connection between oxidative stress and complement activation (65). It should be mentioned that disorders of the outer retina are not the only changes present in these SOD1-deficient mice as changes are also found in many other organs as well.

2.9. A Mouse Model for AMD Based on Ceruloplasmin/ Hephaestin Deficiency

Iron is an essential element in metabolism but also is a potent generator of oxidative damage. This model is based on a deficiency in both the ferroxidase ceruloplasmin (*Cp*) and its homologue hephaestin (*Heph*) in double knockout mice. These mice exhibit reproducible age-dependent iron overload, which plateaued at 6 months of age, with subsequent progressive retinal degeneration continuing to at least 12 months, the oldest mice examined. The degeneration developing in these mice shared some features of AMD, including RPE hypertrophy and hyperplasia, lipofuscin accumulation, photoreceptor degeneration, sub-RPE deposits of activated C3, and subretinal neovascularization (66). The changes described in this model are quite severe, but mimic some of the changes found in AMD. Because these genes have not been found as targets in AMD, it is likely that this model best demonstrates the vulnerability of the outer retina to iron overload. Whether this pathway is causally involved in the etiology of AMD remains to be demonstrated.

2.10. Neovascularization in Transgenic Mice Expressing Prokineticin 1

Prokineticin 1 (hPK1) is a mitogen for fenestrated endothelium. When transgenic mice were developed with the *hPK1* gene driven by a rhodopsin promoter, a line of mice was established that developed a substantially enlarged choroidal vascular bed that, according to the authors, resembled CNV (67). The choroidal changes occurred without any morphological changes in the retinal vasculature. In addition A2E, the major fluorophore of lipofuscin, was present at approximately twofold higher amounts in the transgenic mouse eyes compared to controls. While the *hPK1* transgenic mice did not show any invasion of choroidal blood vessels through Bruch's membrane, the use of this line, in combination with some of the other transgenic or knockout lines showing small percentages of choroidal neovascularization cited above, might be expected to increase the frequency of choroidal neovascularization.

2.11. Mouse Model for AMD Based on a Deficiency in Factor H

Polymorphisms in the gene encoding complement factor H (CFH) are associated with high risk for development of AMD (27–31). In humans, these polymorphisms do not lead to the absence of CFH, but rather to potentially altered function (68–70). However, the relationship between the absence of CFH and retinal disease was evaluated in 2-year-old CFH-deficient (*cfh*–/–) mice (71). A strong CFH signal was observed at the level of the RPE/Bruch's membrane in control mice, indicating the likely importance of CFH in complement regulatory function at this site. Perhaps as a result of CFH deficiency, the *cfh*–/– animals had reduced rod response amplitudes as compared with age-matched controls. Retinal imaging showed an increase in autofluorescent and subretinal deposits, whereas the fundus and vasculature appeared normal. Microscopic analysis of tissue sections showed an accumulation of complement C3 in the neural retina of the *cfh*–/– mice, together with a decrease in electron-dense material, thinning of Bruch's membrane, changes in the cellular distribution of organelles in the RPE, and disorganization of rod outer segments. Collectively, these data suggest that *cfh* is critically required for the long-term functional health of the mouse retina.

2.12. Mouse Model for AMD Based on an Oxidative Damage-Induced Inflammatory Signal

Oxidative damage and inflammation are postulated to be involved in AMD. However, the molecular signal(s) linking oxidation to inflammation in AMD are unknown. AMD-like lesions were generated in normal C57BL/6 mice following immunization with mouse serum albumin adducted with carboxyethylpyrrole, an oxidation fragment of docosahexaenoic acid previously found on proteins in eye tissue and plasma from AMD patients (21, 72). Mice developed antibodies to this hapten; deposited C3 in Bruch's membrane; accumulated basal laminar deposits below the RPE during aging; and developed lesions in the RPE mimicking geographic atrophy. Immunization followed by multiple boosts caused rapid development of the pathology within 3 months. A single immunization was accompanied by the appearance of basal laminar deposits within 1 year. Immunization of these mice with the CEP-hapten sensitizes the animal to CEP-adduct generation in the outer retina. The specificity of this tissue as the disease target comes from the fact that the source of this hapten is docosahexaenoic acid, which is most abundant in photoreceptors and the RPE where conditions for oxidative damage are highly permissive (52).

2.13. A Monkey Model for AMD

A number of papers have been published that investigate the etiology of AMD-like features in strains of cynomolgus monkey (*Macaca fascicularis*) that show either early or late stage features of AMD including the development of drusen (73–76). The drusen contained apolipoprotein E, amyloid P component, complement C5 and C5b-9, and vitronectin (73, 74). A variety of biochemical, physiological, and genetic studies have been performed on these animals, but the underlying disease mechanism has not yet been identified. However, in rhesus macaques (*Macaca mulatta*), a recent study reported an association of a HTRA1 polymorphism with formation of drusen (77), while a second study reported a potential AMD-associated polymorphism in LOC387715 and another in HTRA1 (78).

3. Comparison of the AMD Animal Models

Table 1 summarizes the features of AMD that are found in the purported animal models of this disease covered in the above section. With regards to the time of onset when AMD-like changes are observed, most of the mouse models show disease features only late in the life of the animal. The exception is the *Ccl2*(–/–)/*Cx3cr1*(–/–) double knockout mice, where sub-RPE deposits are observed as early as 6 weeks postpartum. It should be mentioned that *Ccl2* and *Cx3cr1* are genes that code for lymphocyte/macrophage chemokines or their receptors. Neither of these genes has thus far been implicated as a target for AMD. While the focus of the studies on these single and double knockout mice was on the AMD-like disease features present in these lines, it should be pointed out that these papers are the first to clearly demonstrate the importance of the macrophage in maintenance of the outer retina. The involvement of macrophages and the role of chemokines in their recruitment is in itself a major advance in our understanding the importance of these cells in the normal health and maintenance of this outer retina interface.

Each of the mouse models cited has some form of sub-RPE deposit, with the exception of the Prokineticin1 transgenic line and the *Mdm1* mutation. In AMD sub-RPE deposits can be present as punctate drusen, basal laminar deposits or basal linear deposits (79). Basal laminar deposits are composed of granular material with long-spaced collagen located between the plasma membrane and the basal lamina of the RPE. Basal linear deposits are composed of material with coated and noncoated vesicles and some membranous material located external to the RPE basal lamina in the inner collagenous zone of Bruch's membrane. Most of the mouse models described have basal laminar deposits but some also show discrete punctate drusen.

RPE changes were the most consistent AMD feature described in each of the mouse animal models. These include increased autofluorescence and A2E accumulation, pyknosis, hypertrophy and vacuolation of RPE cells, and in some instances, loss of expanses of the RPE mimicking geographic atrophy.

With the exception of the expanded choroidal vascular supply evident in all the prokineticin1 transgenic mice studied, choroidal

Table 1
Features of AMD found in the mammalian models for this disease cited in this chapter

Model	Early/late onset	Sub-RPE deposits	RPE changes	CNV	Complement involvement	Risk factor/ gene target involved in AMD
Smoking (57)	Late	BLD	Yes	No	Not studied	Yes
Apolipoprotein E-TG (59)	Late	BLD	Yes	Yes	Not studied	Yes
Cathepsin D-TG (60)	Late	BLD	Yes	No	Not studied	No
Prokineticin1-TG (67)	Early	No	Yes	Yes	Not studied	No
Ccl2-Ccr2(−/−) (61)	Late	Drusen	Yes	Yes	Yes	No
Ccl2(−/−)/*Cx3cr1*(−/−) (63)	Early	Drusen	Yes	Yes	Yes	No
Mdm1 mutation (64)	Late	No	Yes	No	Not studied	No
*Sod*1(−/−) (65)	Late	BLD, drusen	Yes	Yes	Yes	No
Cp(−/−)/*Heph*(−/−) (66)	Early	BLD	Yes	Yes	Yes	No
Factor H(−/−) (71)	Late	BLD	Yes	No	Yes	Yes
CEP-hapten (52)	Early & late	BLD	Yes	No	Yes	Yes
Mouse (Mus)						
Fat diet/light (53, 54)	Late	BLD	Yes	Some	Not studied	Yes
Primate (Macaca)						
Spontaneous (73–78, 80–82)	Early & late	Drusen	Yes	Yes	Yes	Yes

TG transgenic, *BLD* basal laminar deposits

neovascularization was absent in most of the mouse models and when observed was an inconsistent feature. The prokineticin1 transgenic perhaps should not be considered an AMD model because the most prominent feature is an increased expanse of choroidal blood vessels, which were not reported to undergo subretinal neovascularization.

Complement pathway involvement in the disease process was found in every mouse AMD model in which this pathway was specifically evaluated. The findings usually involved deposition of C3 and/or C5 in Bruch's membrane. It is likely that some of the earlier reported models will also show complement involvement, but this must wait further study.

Only two of the target genes in the genetically based models have been linked to AMD. These are the transgenic mice expressing the human apoE gene (59), and the factor H knockout mouse (71). Neither of these models is perfect, although both replicate some features of AMD. The apoE transgenic line shows only mild changes in the outer retina and requires substantial dietary fat and light, in addition to advanced age, to cause prominent pathology (59). It should be mentioned that high fat intake and light was used in one of the first AMD-like models described and these challenges generate AMD-like features in normal mice (53). The factor H knockout mouse shows only minor changes, which occur very late in life (at 2 years of age). The absence of factor H does not replicate any known condition in AMD families where factor H changes have been reported. The mutation present in factor H in humans (Y402H) will be difficult to create in the mouse since there is poor homology in this region of the factor H gene between these two species. Human and mouse CFH share only about 61% amino acid identity. This compares to an approximately 84% amino acid identity between human and mouse factor B, and a 77% identity between human and mouse C3. These homologies were established using Blast 2 alignments between gi31965 and gi387181 for CFH, gi168985957 and gi218156289 for factor B, and gi119370332 and gi1352102 for C3 (83). Creating a mouse with complement pathway features representative of the human condition may require knockout–knockin technology to humanize the mouse complement pathway. In addition to humanizing factor H, humanization of C3 and factor B may also be required.

If complement is involved in the AMD disease process, an inflammatory signal coming from the outer retina should be a requirement for the immune system to target this tissue. Presently only one of the models exploits an inflammatory antigen (CEP-hapten) that is produced in the outer retina, found in drusen in AMD donor eyes (21) and the antigen and autoantibodies are present at progressively higher levels in the plasma of AMD patients as the disease progresses to more advanced stages (72). When normal mice are immunized with this CEP-hapten adducted to mouse serum albumin they make antibodies to the hapten, deposit C3 in Bruch's membrane, and show complement-mediated attack causing lesions in the RPE and sub-RPE deposits (52). These studies demonstrate that the interaction of this CEP-hapten with the mouse immune system is sufficient to target the outer retina and cause AMD-like changes as the immune system responds to more of the CEP-hapten as it is generated over time in these cells. This model will be useful in studies of the early events in immune-mediated attack on the outer retina as well as a preclinical platform for testing therapeutics that are being developed to prevent or retard the AMD disease process.

A superior mouse model for AMD might be created from a combination of the models outlined above. Such a model might benefit from a primary stimulus that would generate an inflammatory or autoimmune response at the RPE/choroid interface, (for example, the CEP-hapten model) coupled with the *cfh–/–* or alternatively deficient animal that has a reduced capacity to regulate complement activation and protect this sensitive area. Such a model would have additional benefits since it would provide multiple intervention targets, i.e., reducing or preventing the primary stimulus that initiates the inflammatory response as well as intervention later during the progression of the disease.

While photoreceptors and RPE in all vertebrate eyes share a large number of common molecular and cell biological features, because the mouse retina does not have a fovea at the center of a well-defined macula, there will always be critics who feel that modeling AMD in a mouse cannot be truly representative of the human disease. While a highly developed fovea is a common retinal feature in many reptilian and avian species, additional criticisms would be expected if a lizard or bird were chosen as a model for AMD. Among mammals, only some primates have a fovea. Several reports describe drusen formation and AMD-like features in nonhuman primates (80–82). The species most intensely studied is *Macaca fascicularis* (73–76). A variety of biochemical, physiological, and genetic studies have been performed on these animals. The underlying disease mechanism has not been identified. However, recent work in *Macaca mulatta* suggests HTRA1 and/or LOC387715 may be involved (77, 78). More studies will be required to see if this holds true for *Macaca fascicularis*.

In conclusion, this chapter has summarized a number of features of AMD in humans and has discussed several of the features of this disease in experimental and naturally occurring animal models. Animal models for this disorder will certainly be important for screening the efficacy of therapeutics being developed to treat AMD.

Acknowledgment

The authors research is supported by grants from the National Institutes of Health, National Eye Institute, Bethesda, MD; the State of Ohio, BRTT Program, Columbus, OH; the Foundation Fighting Blindness, Owings Mills, MD; and Research to Prevent Blindness, New York, NY. We thank Mary E. Rayborn for help with editing the manuscript.

References

1. Javitt J, Zhou Z, Maguire M, Fine S, Willke R (2003) Incidence of exudative age-related macular degeneration among elderly Americans. Ophthalmology 110:1534–1539
2. Augood C, Vingerling J, de Jong P et al (2006) Prevalence of age-related maculopathy in older Europeans: the European Eye Study (EUREYE). Arch Ophthalmol 124: 529–535
3. Curcio C, Millican C, Bailey T, Kruth H (2001) Accumulation of cholesterol with age in human Bruch's membrane. Invest Ophthalmol Vis Sci 42:265–274
4. Hussain A, Rowe L, Marshall J (2002) Age-related alterations in the diffusional transport of amino acids across the human Bruch's-choroid complex. J Opt Soc Am A Opt Image Sci Vis 19:166–172
5. Moore D, Hussain A, Marshall J (1995) Age-related variation in the hydraulic conductivity of Bruch's membrane. Invest Ophthalmol Vis Sci 36:1290–1297
6. Bressler S, Maguire M, Bressler N, Fine S (1990) Relationship of drusen and abnormalities of the retinal pigment epithelium to the prognosis of neovascular macular degeneration. The Macular Photocoagulation Study Group. Arch Ophthalmol 108:1442–1447
7. Sarks S, van Driel D, Maxwell L, Killingsworth M (1980) Softening of drusen and subretinal neovascularization. Trans Ophthalmol Soc UK 100:414–422
8. Vinding T (1990) Occurrence of drusen, pigmentary changes, and exudative changes in the macula with reference to age-related macular degeneration. An epidemiological study of 1000 aged individuals. Acta Ophthalmol 68:410–414
9. Holz F, Bellman C, Staudt S, Schutt F, Volcker H (2001) Fundus autofluorescence and development of geographic atrophy in age-related macular degeneration. Invest Ophthalmol Vis Sci 42:1051–1056
10. Martinez-Mir A, Paloma E, Allikmets R et al (1998) Retinitis pigmentosa caused by a homozygous mutation in the Stargardt disease gene ABCR. Nature Genet 18:11–12
11. Zhang K, Kniazeva M, Han M et al (2001) A 5-bp deletion in ELOVL4 is associated with two related forms of autosomal dominant macular dystrophy. Nat Genet 27:89–93
12. Stone E, Lotery A, Munier F et al (1999) A single EFEMP1 mutation associated with both Malattia Leventinese and Doyne honeycomb retinal dystrophy. Nat Genet 22: 199–202
13. Weber BH, Vogt G, Pruett RC, Stohr H, Felbor U (1994) Mutations in the tissue inhibitor of metalloproteinase-3 (TIMP-3) in patients with Sorsby's fundus dystrophy. Nat Genet 8:352–356
14. Petrukhin K, Koisti MJ, Bakall B et al (1998) Identification of the gene responsible for Best macular dystrophy. Nature Genet 19:241–247
15. Sun H, Tsunenari T, Yau K-W, Nathans J (2002) The vitelliform macular dystrophy protein defines a new family of chloride channels. Proc Natl Acad Sci USA 99(6): 4008–4013
16. Stone E, Sheffield V, Hageman G (2001) Molecular genetics of age-related macular degeneration. Hum Mol Genet 10: 2285–2292
17. Hageman G, Mullins R (1999) Molecular composition of drusen as related to substructural phenotype. Mol Vis 5:28–37
18. Johnson L, Leitner W, Staples M, Anderson D (2001) Complement activation and inflammatory processes in drusen formation and age related macular degeneration. Exp Eye Res 73:887–896
19. Anderson D, Mullins R, Hageman G, Johnson L (2002) A role for local inflammation in the formation of drusen in the aging eye. Am J Ophthalmol 134:411–431
20. Hageman G, Luthert PJ, Chong N, Johnson L, Anderson D, Mullins R (2001) An integrated hypothesis that considers drusen as biomarkers of immune-mediated processes at the RPE-Bruch's membrane interface in aging and age-related macular degeneration. Prog Retin Eye Res 20:705–732
21. Crabb J, Miyagi M, Gu X et al (2002) Drusen proteome analysis: an approach to the etiology of age-related macular degeneration. Proc Natl Acad Sci USA 99:14682–14687
22. Johnson L, Ozaki S, Staples M, Erickson P, Anderson D (2000) A potential role for immune complex pathogenesis in drusen formation. Exp Eye Res 70:441–449
23. Doyne RW (1989) Peculiar condition of choroiditis occurring in several members of the same family. Trans Ophthalmol Soc UK 19:71–77
24. Gass JDM (1987) Stereoscopic atlas of macular diseases. C.V. Mosby, St. Louis
25. Meyers SM, Zachary AA (1988) Monozygotic twins with age-related macular degeneration. Arch Ophthalmol 106:651–653
26. Seddon JM, Ajani U, Mitchell B (1997) Familial aggregation of age-related maculopathy. Am J Ophthalmol 123:199–206

27. Klein R, Zeiss C, Chew E et al (2005) Complement factor H polymorphism in age-related macular degeneration. Science 308: 385–389
28. Edwards A, Ritter R, Abel K, Manning A, Panhuysen C, Farrer L (2005) Complement factor H polymorphism and age-related macular degeneration. Science 308:421–424
29. Hageman G, Anderson D, Johnson L et al (2005) A common haplotype in the complement regulatory gene factor H (HF1/CFH) predisposes individuals to age-related macular degeneration. Proc Natl Acad Sci USA 102: 7227–7232
30. Haines J, Hauser M, Schmidt S et al (2005) Complement factor H variant increases the risk of age-related macular degeneration. Science 308:419–421
31. Gold B, Merriam J, Zernant J et al (2006) Variation in factor B (BF) and complement component 2 (C2) genes is associated with age-related macular degeneration. Nature Genet 38:458–462
32. Ennis S, Jomary C, Mullins R et al (2008) Association between the SERPING1 gene and age-related macular degeneration: a two-stage case-control study. Lancet 372: 1788–1789
33. Eldred G (1998) Lipofuscin and other lysosomal storage deposits in the retinal pigment epithelium. In: Marmor M, Wolfensberger T (eds) The retinal pigment epithelium. Oxford University Press, Oxford, pp 651–668
34. Sparrow J, Boulton M (2005) RPE lipofuscin and its role in retinal pathobiology. Exp Eye Res 80:595–606
35. Sparrow J (2007) RPE lipofuscin: formation, properties, and felevance to retinal degeneration. In: Tombran-Tink J, Barnstable CJ (eds) Retinal degenerations: biology, diagnostics and therapeutics. Humana Press, Totowa, NJ, pp 213–236
36. Boulton M, Dontsov A, Jarvis-Evans J, Ostrovsky M, Svistunenko D (1993) Lipofuscin is a photoinducible free radical generator. J Photochem Photobiol B 19:201–204
37. Wassell J, Davies S, Bardsley W, Boulton M (1999) The photoreactivity of the retinal age pigment lipofuscin. J Biol Chem 274: 23828–23832
38. Sparrow J, Nakanishi K, Parish C (2000) The lipofuscin fluorophore A2E mediates blue light-induced damage to retinal pigmented epithelial cells. Invest Ophthalmol Vis Sci 41:1981–1989
39. Sparrow J, Zhou J, Ben-Shabat S, Vollmer H, Itagaki Y, Nakanishi K (2002) Involvement of oxidative mechanisms in blue light induced damage to A2E-laden RPE. Invest Opthalmol Vis Sci 43:1222–1227
40. Zhou J, Jang Y, Kim S, Sparrow J (2006) Complement activation by photooxidation products of A2E, a lipofuscin constituent of the retinal pigment epithelium. Proc Natl Acad Sci USA 103:16182–16187
41. Johnson LV, Leitner WP, Rivest AJ, Staples MK, Radeke MJ, Anderson DH (2002) The Alzheimer's Abeta -peptide is deposited at sites of complement activation in pathologic deposits associated with aging and age-related macular degeneration. Proc Natl Acad Sci USA 99(18):11830–11835
42. Gu X, Meer S, Miyagi M et al (2003) Carboxyethylpyrrole protein adducts and autoantibodies, biomarkers for age-related macular degeneration. J Biol Chem 278:42027–42035
43. Anderson R, Lissandrello P, Maude M, Matthes M (1976) Lipids of bovine retinal pigment epithelium. Exp Eye Res 23: 149–157
44. Anderson RE (1970) Lipids of the ocular tissues. IV. A comparison of the phospholipids from the retina of six mammalian species. Exp Eye Res 10:339–344
45. Fliesler SJ, Anderson RE (1983) Chemistry and metabolism of lipids in the vertebrate retina. Prog Lipid Res 22:79–131
46. Hollyfield J, Crabb J, Salomon R (2003) Proteomic approaches to understanding age-related macular degeneration. Adv Exp Med Biol 533:83–89
47. Seddon JM, Willett WC, Speizer FE, Hankinson SE (1996) A prospective study of cigarette smoking and age-related macular degeneration in women. JAMA 276: 1141–1146
48. Klein R, Klein B, Cruickshanks K (1999) The prevalence of age-related maculopathy by geographic region and ethnicity. Prog Retin Eye Res 18:371–389
49. Snow K, Seddon J (1999) Do age-related macular degeneration and cardiovascular disease share common antecedents? Ophthalmic Epidemiol 6:125–143
50. Christen W, Glynn R, Manson J, Ajani U, Buring J (1996) A prospective study of cigarette smoking and risk of age-related macular degeneration in men. JAMA 276: 1147–1151
51. Solberg Y, Rosner M, Belkin M (1998) The association between cigarette smoking and ocular diseases. Surv Ophthalmol 42: 535–547

52. Hollyfield J, Bonilha V, Lu L et al (2008) Oxidative damage induced inflammation initiates age-related macular degeneration. Nat Med 14:194–198
53. Cousins S, Espinosa-Heidmann D, Alexandridou A, Sall J, Dubovy S, Csaky K (2002) The role of aging, high fat diet and blue light exposure in an experimental mouse model for basal laminar deposit formation. Exp Eye Res 75:543–553
54. Cousins S, Marin-Castaño M, Espinosa-Heidmann D, Alexandridou A, Striker L, Elliot S (2003) Female gender, estrogen loss, and Sub-RPE deposit formation in aged mice. Invest Ophthalmol Vis Sci 44:1221–1229
55. Majji A, Cao J, Chang K et al (2000) Age-related retinal pigment epithelium and Bruch's membrane degeneration in senescence-accelerated mouse. Invest Ophthalmol Vis Sci 41:3936–3942
56. Smith W, Mitchell P, Leeder S (1996) Smoking and age-related maculopathy. The Blue Mountains Eye Study. Arch Ophthalmol 114:1518–1523
57. Espinosa-Heidmann D, Suner I, Catanuto P, Hernandez E, Marin-Castano M, Cousins S (2006) Cigarette smoke-related oxidants and the development of sub-RPE deposits in an experimental animal model of dry AMD. Invest Ophthalmol Vis Sci 47(2):729–737
58. Baird P, Guida E, Chu D, Vu H, Guymer R (2004) The 2 and 4 alleles of the apolipoprotein gene are associated with age-related macular degeneration. Invest Ophthalmol Vis Sci 45:1311–1315
59. Malek G, Johnson L, Mace B et al (2005) Apolipoprotein E allele-dependent pathogenesis: a model for age-related retinal degeneration. Proc Natl Acad Sci USA 102: 11900–11905
60. Rakoczy P, Zhang D, Robertson T et al (2002) Progressive age-related changes similar to age-related macular degeneration in a transgenic mouse model. Am J Pathol 181: 1515–1524
61. Ambati J, Anand A, Fernandez S et al (2003) An animal model of age-related macular degeneration in senescent Ccl-2- or Ccr-2-deficient mice. Nature Med 9:1390–1397
62. Tuo J, Smith B, Bojanowski C et al (2004) The involvement of sequence variation and expression of CX3CRl in the pathogenesis of age-related macular degeneration. FASEB J 18:1297–1299
63. Tuo J, Bojanowski C, Zhou M et al (2007) Murine ccl2/cx3crl deficiency results in retinal lesions mimicking human age-related macular degeneration. Invest Ophthalmol Vis Sci 48:3827–3836
64. Chang B, Mandal M, Chavali V et al (2008) Age-related retinal degeneration (arrd2) in a novel mouse model due to a nonsense mutation in the Mdml gene. Hum Mol Genet 17:3929–3941
65. Imamura Y, Noda S, Hashizume K et al (2006) Drusen, choroidal neovascularization, and retinal pigment epithelium dysfunction in SOD1-deficient mice: a model of age-related macular degeneration. Proc Natl Acad Sci USA 103:11282–11287
66. Hadziahmetovic M, Dentchev T, Song Y et al (2008) Ceruloplasmin/hephaestin knockout mice model morphologic and molecular features of AMD. Invest Ophthalmol Vis Sci 49:2728–2736
67. Tanaka N, Ikawa M, Mata N, Verma I (2006) Choroidal neovascularization in transgenic mice expressing prokineticin 1: an animal model for age-related macular degeneration. Mol Ther 13:609–616
68. Clark S, Higman V, Mulloy B et al (2006) His-384 allotypic variant of factor H associated with age-related macular degeneration has different heparin binding properties from the non-disease-associated form. J Biol Chem 281:24713–24720
69. Prosser B, Johnson S, Roversi P et al (2007) Structural basis for complement factor H linked age-related macular degeneration. J Exp Med 204:2277–2283
70. Haapasalo K, Jarva H, Siljander T, Tewodros W, Vuopio-Varkila J, Jokiranta T (2008) Complement factor H allotype 402H is associated with increased C3b opsonization and phagocytosis of *Streptococcus pyogenes*. Mol Microbiol 70:583–594
71. Coffey P, Gias C, McDermott C et al (2007) Complement factor H deficiency in aged mice causes retinal abnormalities and visual dysfunction. Proc Natl Acad Sci USA 104:16651–16656
72. Gu X, Sun M, Hazen S, Crabb J, Salomon R (2003) Oxidatively truncated docosahexaenoate phosphoplipids: total synthesis, generation and peptide adduction chemistry. J Org Chem 68:3749–3761
73. Umeda S, Suzuki M, Okamoto H et al (2005) Molecular composition of drusen and possible involvement of anti-retinal autoimmunity in two different forms of macular degeneration in cynomolgus monkey (*Macaca fascicularis*). FASEB J 19:1683–1685
74. Umeda S, Ayyagari R, Allikmets R, Suzuki MT, Karoukis AJ, Ambasudhan R, Zernant J,

Okamoto H, Ono F, Terao K, Mizota A, Yoshikawa Y, Tanaka Y, Iwata T (2005) Early-onset macular degeneration with drusen in a cynomolgus monkey (*Macaca fascicularis*) pedigree: exclusion of 13 candidate genes and loci. Invest Ophthalmol Vis Sci 46:683–691
75. Iwata T (2007) Complement activation of drusen in primate model (*Macaca fascicularis*) for age-related macular degeneration. Adv Exp Med Biol 598:251–259
76. Kaidzu S, Tanito M, Ohira A et al (2008) Immunohistochemical analysis of aldehyde-modified proteins in drusen in cynomolgus monkeys (*Macaca fascicularis*). Exp Eye Res 86:856–859
77. Singh K, Krawczak M, Dawson W, Schmidtke J (2009) Association of HTRA1 and ARMS2 gene variation with drusen formation in rhesus macaques. Exp Eye Res 88(3):479–482
78. Francis P, Appukuttan B, Simmons E et al (2008) Rhesus monkeys and humans share common susceptibility genes for age-related macular disease. Hum Mol Genet 17: 2673–2680
79. Green WR, Enger C (1993) Age-related macular degeneration histopathologic studies: the 1992 Lorenz E. Zimmerman lecture. Ophthalmology 100:1519–1535
80. El-Mofty A, Gouras P, Eisner G, Balazs E (1978) Macular degeneration in rhesus monkey (*Macaca mulatta*). Exp Eye Res 27:499–502
81. Hope G, Dawson W, Engel H, Ulshafer R, Kessler M, Sherwood M (1992) A primate model for age related macular drusen. Br J Ophthalmol 76:11–16
82. Gouras P, Ivert L, Mattison J, Ingram D, Neuringer M (2008) Drusenoid maculopathy in rhesus monkeys: autofluorescence, lipofuscin and drusen pathogenesis. Graefes Arch Clin Exp Ophthalmol 246:1403–1411
83. Tatusova TA, Madden TL (1999) Blast 2 sequences, a new tool for comparing protein and nucleotide sequences. FEMS Microbiol Lett 174:247–250

Chapter 6

Animal Models of Retinopathy of Prematurity

Susan E. Yanni and John S. Penn

Abstract

Retinopathy of prematurity (ROP), a condition affecting premature infants, is characterized by pathological angiogenesis, or neovascularization (NV), of the retina. Much of what is known about the development of the retinal vasculature and the progression of ROP has been learned through the use of animal models of oxygen-induced retinopathy (OIR), which approximate the human condition. Animal models of OIR have provided a wealth of information regarding the cellular and molecular pathogenesis of ROP. Moreover, this information has contributed to a better understanding of other, nonocular, neovascular conditions. This chapter describes the various animal models of OIR, and explores their contributions to the understanding and treatment of ROP.

Key words: Angiogenesis, Animal models, Oxygen-induced retinopathy (OIR), Retinopathy of prematurity (ROP)

1. Human Retinopathy of Prematurity

Theodore Terry first described Retinopathy of Prematurity (ROP) in 1942 as a disease of prematurity characterized by retinal neovascularization (NV) (1). An epidemic of ROP occurred during the 1950s, exposing the need for research focused on the identification and characterization of its pathogenesis. In 1951, Campbell proposed that the incidence of ROP was correlated with the delivery of supplemental oxygen to premature infants with underdeveloped pulmonary function (2). During the 1950s, several convincing studies correlated the use of supplemental oxygen with the incidence and progression of ROP (3–7). This led to more rigorous monitoring of the oxygen being given to premature infants. Consequently, the percentage of blindness attributed to ROP dropped from 50% in 1950 to 4% in 1965 (8). The prevalence of ROP rose again during the 1970s and 1980s (9),

Iok-Hou Pang and Abbot F. Clark (eds.), *Animal Models for Retinal Diseases*, Neuromethods, vol. 46
DOI 10.1007/978-1-60761-541-5_6, © Springer Science+Business Media, LLC 2010

presumably due to the increased survival of very low birth-weight premature infants requiring supplemental oxygen.

According to the most recent estimates of the National Eye Institute, each year approximately 15,000 premature infants (weighing ≤1,250 g) that are born before 31 weeks of gestation develop some stage of ROP. Of these infants, 400–600 will suffer from ROP-induced blindness (10). ROP is the leading cause of childhood blindness in the developed world (10). For this and other reasons, research focused on understanding physiological retinal vascular development and pathological retinal NV is highly significant.

Several animal models that approximate human ROP have been developed. To emphasize the differences between human ROP and experimentally induced retinopathy in animals, the term oxygen-induced retinopathy (OIR) is often used to describe these models. Animal models of OIR have been, and continue to be, widely used to study the cellular and molecular aspects of physiological and pathological retinal NV. Their usefulness, however, is dependent on several factors, most notably the degree of similarity between the animal model and the human condition.

The retina is one of the last organ systems of the developing fetus to undergo vascularization, beginning at approximately 16 weeks of gestation (11). At this time, vasculogenesis (the de novo formation of blood vessels from mesodermal precursor cells) occurs, starting in the most posterior region of the superficial retina (the optic disk) and proceeding in a wave-like manner to the periphery of the retina. At 25 weeks of gestation, angiogenesis (the development of new capillaries from pre-existing blood vessels) begins, proceeding from the optic disk in a peripheral wave, resulting in the development of a deeper (more sclerad) vessel network (11). It is believed that the hypoxic uterine environment (30 mm Hg) drives retinal vascularization during normal gestation. In utero, physiological retinal hypoxia leads to the production of proangiogenic growth factors that stimulate the growth of retinal blood vessels, which satisfy the increasing oxygen demands of the developing retina. Vascularization of the retina is complete at 36–40 weeks of gestation, and the relatively hyperoxic (55–80 mm Hg) postnatal environment effectively prevents growth factor production and, thus, persistent vasoproliferation (12, 13).

Retinal vascular development is altered in premature infants that develop ROP. The pathogenesis of ROP is biphasic. The first phase of ROP occurs while the premature infant is receiving supplemental oxygen and results in vasoattenuation, or cessation of retinal vascular development. During oxygen therapy, the oxygen tension within the retina inhibits the production and secretion of vascular growth factors that would normally occur in utero. Diminished growth factor production and attenuated vessel growth result in an incompletely vascularized retinal periphery, a hallmark of ROP. Peripheral retinal avascularity, in turn, leads to retinal tissue hypoxia, particularly when supplemental oxygen is discontinued. This tissue hypoxia is exacerbated by the development and maturation of the neural retina and its growing oxygen demands. Thus, retinal ischemia induces retinal hypoxia which, in turn, induces the onset of the second, vasoproliferative phase of ROP. The second phase of ROP is characterized by retinal NV (7, 14–16). This NV is best described as unregulated angiogenesis, resulting in the production of fragile, nonpatent vascular structures that grow through the inner limiting membrane of the retina into the vitreous cavity. These abnormal vascular structures are often referred to as preretinal neovascular tufts, and they predispose affected infants to intravitreal hemorrhages, retinal detachment, and subsequent vision loss.

The severity of ROP is inversely proportional to the gestational age of the infant (17). Normally, retinal vascularization is complete at, or near, full term. Again, premature infants exhibit an underdeveloped and, therefore, incompletely vascularized retina, with peripheral retinal avascularity. Infants who receive supplemental oxygen due to pulmonary immaturity suffer attenuation of the already sparse retinal vasculature. As described above, return to a relatively hypoxic room air environment after oxygen therapy is terminated causes subsequent development of NV and retinopathy. More prematurity correlates to a larger avascular retinal area at birth, which leads to more severe retinal hypoxia on return to room air and, thus, more severe ROP.

Notably, however, a large clinical trial demonstrated that approximately half of the infants developed NV before removal from oxygen therapy (18). Hypoxemia related to removal from therapy, therefore, is not necessary for NV, nor is it the sole determinant of ROP pathogenesis. Developmental timing may regulate the responses of the immature retina to oxygen (19). Human ROP involves a complex sequence of pathological events with the potential to be influenced by temporal patterns of gene expression as well as environmental factors related to clinical care.

2. The Cat Model

The kitten's retinal vasculature develops from spindle-shaped mesenchymal precursor cells derived from the adventitia of the hyaloid artery (20–22). As in human retinal vascular development, the kitten's retinal vasculature apparently develops via vasculogenesis, de novo formation of blood vessels from the vascular precursor cells. The retina of a newborn kitten resembles the retina of a premature infant at 28 weeks of gestation, an infant at risk for developing ROP (23).

Much of our early knowledge regarding the role of oxygen in the pathogenesis of ROP was obtained using the kitten model. In 1954, Norman Ashton first used a kitten model of OIR (7). Ashton evaluated the effects of age, oxygen concentration, and length of oxygen exposure on the severity of retinopathy. Immature retinal vessels are exquisitely sensitive to tissue oxygen concentration. Ashton demonstrated that, as they mature, retinal blood vessels lose their sensitivity to hyperoxia. In the kitten, retinal vascular maturity is acquired by 21 days of age. After this time, the kitten retina is relatively unresponsive to high oxygen. Ashton also demonstrated that the severity of the OIR response was directly proportional to oxygen concentration and length of exposure. Oxygen concentrations between 70 and 80% produce the most severe retinopathy, whereas oxygen concentrations below 35% have little or no effect on the most sensitive age group. Arnall Patz and Robert Flower both used the kitten to investigate the effects of hyperoxia on the retinal vasculature (24, 25). These and other studies have contributed fundamental knowledge about the role of oxygen-induced vasoconstriction in the pathogenesis of ROP.

More recently, Chan-Ling and colleagues have used the kitten to study the cellular and vascular changes that occur during, and as a result of, retinopathy (26–29). In this model, within hours after birth, newborn kittens are placed in 70–80% oxygen for 4 days. This leads to cessation of retinal vascular growth, leaving the retina completely avascular. After 4 days, the kittens are removed to room air, where they remain for several days. Eleven days after return to the relatively hypoxic room air environment, rampant NV is observed (27).

The idea that supplemental oxygen may be used to treat ROP was first tested in the kitten. Kittens that had been exposed to hyperoxia were allowed to recover in room air or in a variable hyperoxic-hypoxic environment. The kittens that were exposed to variable oxygen during the postexposure period demonstrated less severe retinopathy than kittens recovered in room air (30). The results of these and other studies led to the supplemental therapeutic oxygen protocol for retinopathy of prematurity (STOP-ROP) clinical trial (18).

Although the kitten model has yielded information of much importance to the scientific community, it suffers several drawbacks. The kitten model does produce intraretinal and intravitreal NV, but the location and pattern of vasoproliferation seen in kittens is different from that seen in human infants with ROP (31). In the kitten, blood vessels extend from the optic disk into the vitreous, whereas in the human, preretinal vessels arise only from the advancing edge of retinal capillary formation. On the other hand, however, the kitten does demonstrate iris vascular engorgement and pupillary rigidity, anterior segment abnormalities similar to those that occur in human ROP.

Additionally, the kitten model does not develop retinal detachment, and the NV that does develop in the kitten eventually regresses over time. Furthermore, the costs associated with purchasing, breeding, and maintaining cats for extended periods of time are substantially higher than the costs associated with using rodents (discussed in detail later).

3. The Dog Model

As is the case for the human and kitten retinas, research suggests that the retinal vasculature of the beagle puppy derives from immature vascular precursor cells, or angioblasts (32). At full term, as much as 66% of the beagle puppy's retina is vascularized. Blood vessels reach the periphery of the retina by day 14, and vascular remodeling is complete by day 28 (32). The retinal vasculature of a newborn beagle puppy is comparable to the immature retinal vasculature of a premature infant, in that vascular development is incomplete.

In 1954, Arnall Patz first used a canine model of OIR (33). McLeod and colleagues have since used the model to detail the response of the canine retina to high oxygen (34–36). The model produces a vascular response similar to the response seen in kittens exposed to high oxygen (34). Newborn beagle puppies are exposed to extreme hyperoxia (95–100% oxygen). Four days later, the animals are returned to room air, where they remain until 22–45 days of age. The retina's vasoproliferative response peaks between 3 and 10 days after return to room air (37). Fifteen days after oxygen exposure, ADPase-stained retinas demonstrate dilated and tortuous vessels, an incompletely vascularized periphery, intravitreal hemorrhages, and NV. Retinal folds have also been observed in this model, although it is not entirely clear whether these are the consequence of tractional elements or are the result of dysplastic retinal development (36).

The investigators using this model suggest that it is an appropriate model of human ROP and that it is superior to others because it is the only model in which a high percentage of animals develop retinal NV with subsequent retinal detachment. However, retinal detachment only occurs in a small percentage of infants in whom ROP develops. Additionally, retinal folds and detachments have been demonstrated to occur spontaneously in dogs (38–40). Furthermore, the cost of purchasing, breeding, and maintaining dogs is a disadvantage of the species.

4. The Mouse Model

The contributions of vasculogenesis and angiogenesis to the formation of the retinal vasculature of the mouse remain controversial. Some research suggests that the retinal vasculature of the mouse derives from a vasculogenic process, while numerous other studies suggest that angiogenesis is the means by which the retina becomes vascularized (41). Thus, the contributions of vasculogenesis and angiogenesis to retinal vascularization may be species-specific (42). Regardless, the retinal vasculature of a newborn mouse is comparable to that of an infant at 25 weeks of gestation, an infant at risk for developing ROP (43). Accordingly, the retinal vasculature of a newborn mouse pup presents an appropriate venue for studies of the retinal vasculature of a premature infant.

In 1954, Gyllensten and Hellstrom exposed newborn mouse pups to 100% oxygen for 1–3 weeks. The eyes were examined after oxygen treatment. Approximately one-third of the animals experienced hemorrhages in both the vitreous and anterior chambers. Subsequent studies demonstrated that exposing the pups to 100% oxygen followed by 5 days in room air induced vasoproliferation of the retinal vessels, the presumed source of vitreous hemorrhages (43). Removal to room air was required for induction of the ROP-like vasoproliferative changes (44).

While the mouse provided the research community with a means by which to explore ROP in greater detail, early studies were inconclusive, yielding inconsistent results. These attempts to model ROP were confounded by the fact that hyperoxic exposure of newborn mice, followed by removal to room air, resulted in proliferation and engorgement of the hyaloid (5, 45). Reasoning that the "hyaloidopathy" might explain the variability in the early attempts to model ROP, Smith and colleagues proposed a novel method for inducing retinopathy in the mouse, a model that sought to minimize this confounding feature (46).

The Smith model produces a consistent and reliable model of OIR. Mice at postnatal day 7 (P7) are exposed to 75% oxygen for 5 days to induce vasoattenuation and atrophy of the central retinal vascular bed (Fig. 1a). Removal of the mice to room air for variable lengths of time induces retinal vasoproliferation and revascularization of the central retina. Typically, eyes are analyzed for the presence of NV at P17–P21.

Since the advent and widespread use of this mouse model, extensive research has been conducted on the susceptibility of various inbred strains of mice to pathological retinal NV. To evaluate genetic heterogeneity in angiogenic susceptibility, D'Amato and colleagues implanted a pellet containing an angiogenic protein, basic fibroblast growth factor (bFGF), into the corneas of 25

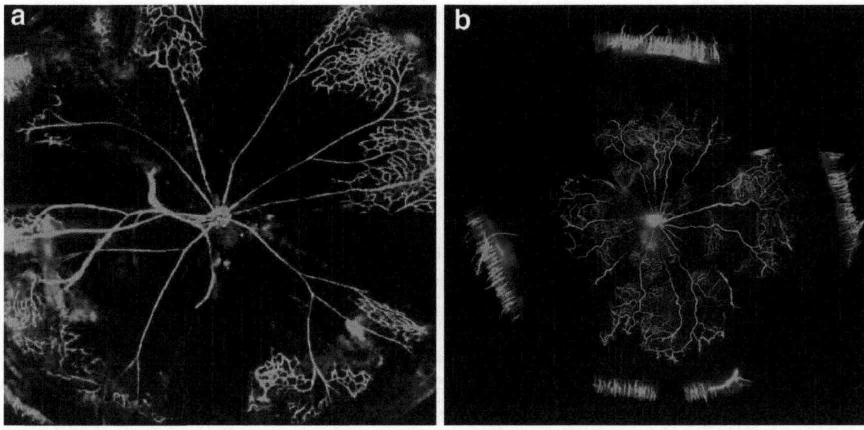

Fig. 1. A P12 FITC-dextran-infused mouse retina (**a**). Five days at 75% oxygen induces vasoattenuation and atrophy of the central retinal beds, a pattern in direct contrast to the peripheral avascularity characterizing the human condition. A P14 FITC-dextran infused rat retina (**b**). Fourteen days of variable oxygen induces vasoattenuation and peripheral retinal avascularity. This pattern mimics the pathology of human ROP

strains of mice (47). Normally, the blood vessels of the limbus do not grow into the avascular cornea. Strain differences in angiogenic response were assessed by analyzing the growth of blood vessels into the cornea on bFGF stimulation. A tenfold range of responsiveness was observed, with 129/SvImJ mice eliciting the most potent angiogenic response, while the commonly used C57BL/6J mice fell in the middle of the response profile. Subsequent studies revealed that vascular endothelial growth factor (VEGF) elicited a response profile that closely correlated with that of bFGF (48).

Following D'Amato's report, Hinton and colleagues analyzed strain-related differences in retinal angiogenesis using the mouse model of OIR (49). The strain-related differences in retinal angiogenesis mirrored the results obtained by D'Amato. Differential analysis of the angiogenic responsiveness of mice with different genetic backgrounds has allowed for the identification of various pro- and antiangiogenic factors involved in the pathogenesis of ROP.

When considering relevance to ROP, the mouse model of OIR suffers several drawbacks. The spatial pattern of vascular pathology observed in mice is the opposite of that observed in humans with ROP. In the human condition, the central retina is vascularized, while the peripheral retina is avascular. In contrast, the mouse exhibits a pattern of central avascularity, and the peripheral retina is vascularized. Claxton and Fruttiger studied the vascular patterning of the retinas of mice that had been exposed to hyperoxia (50). They hypothesized that because the retinal arteries and hyaloidal blood supply pass through the optic nerve head, the proximal retina has a relatively high oxygen tension. VEGF is a survival factor for endothelial cells and is downregulated in response to high oxygen tension, explaining the pruning of peri-arterial capillaries around the optic nerve head as well as around the central retinal arteries in the mouse. Hyperoxic exposure further increases the retinal oxygen tension, expanding regions of decreased retinal VEGF, inducing endothelial cell apoptosis and vascular atrophy, and resulting in the expansion of capillary-free zones within the central retina.

In addition, in contrast to human ROP, retinal detachment is not observed in the mouse model. This may be due to the fact that the lens occupies 40% of the mouse eye, effectively reducing the vitreous volume that is available to contract in response to NV and fibrovascular scar formation.

The mouse model is the most commonly used model in studies of retinal angiogenesis. Mice reliably produce large litters, are relatively inexpensive to purchase and maintain, and consistently produce a neovascular response. The mouse model has provided much of what is currently known about the pathogenesis of ROP, its progression, and potential means by which to prevent and/or ameliorate the disease. Importantly, the ability to manipulate the mouse genome has facilitated our understanding of potential genetic influences to ROP, as well as our examination of the wide variety of molecular regulators of angiogenesis.

5. The Rat Model

As in the human, the retinal vasculature of the rat appears to derive from spindle-shaped mesenchymal precursor cells deriving from the adventitia of the hyaloid artery. Vascularization of a superficial network of arteries and veins occurs first, followed by angiogenic growth of a deeper capillary network. Whereas in the human, retinal vascularization is usually complete at the time of birth, in the rat, the process is completed at around P15. The retinal vasculature of a newborn rat pup resembles that of a preterm infant and a newborn mouse: virtually avascular and susceptible to OIR.

Early on, Patz, Ashton, and Gole attempted to model ROP by exposing newborn rat pups to a constant level of extreme hyperoxia (24, 51, 52). This resulted in substantial vasoattenuation, but an inconsistent vasoproliferative response. These studies were informative, though it was not until 1993 that Penn and colleagues developed a protocol that consistently produced proliferative retinopathy in the rat (53). Penn noted that variable oxygenation is more likely to produce retinal angiogenesis than is constant hyperoxia. Notably, variable oxygenation more closely mimics the fluctuating lung function and subsequent change in arterial blood oxygen partial pressure, PaO_2, of a neonatal infant in the NICU. In Penn's 1993 study, exposing rat pups to 80% oxygen, followed by a period of postexposure in room air, rarely induced preretinal NV. However, a variable oxygen exposure (cycling between 80 and 40% oxygen every 12 h), in combination with a postexposure period of return to room air, induced preretinal NV in 66% of the rats. Subsequent experiments by Penn led to the rat model of OIR that is used today (54). In this model, newborn rats are cycled between 50 and 10% oxygen every 24 h for 14 days. This oxygen profile, which more accurately reflects the fluctuating lung function and PaO_2 of a preterm infant in the NICU, resulted in a high incidence (97%) of retinopathy. Additionally, the angiogenic pattern seen in the rat retina precisely mimics the pattern of human ROP (Fig. 1b). Both species exhibit an avascular peripheral retina and both develop NV at the boundary of vascular and avascular retina (Fig. 2). Thus, the rat provides a highly relevant model with which to address ROP-related questions.

Like the mouse model, the rat model is subject to strain-related differences in susceptibility to retinal NV and, within strains, to vendor-related differences. Ma and colleagues compared the differential susceptibilities of two rat strains, Brown Norway and Sprague Dawley, to ischemia-induced retinopathy (55). Using a constant oxygen exposure paradigm (like that developed by Smith and colleagues for the mouse), Ma found

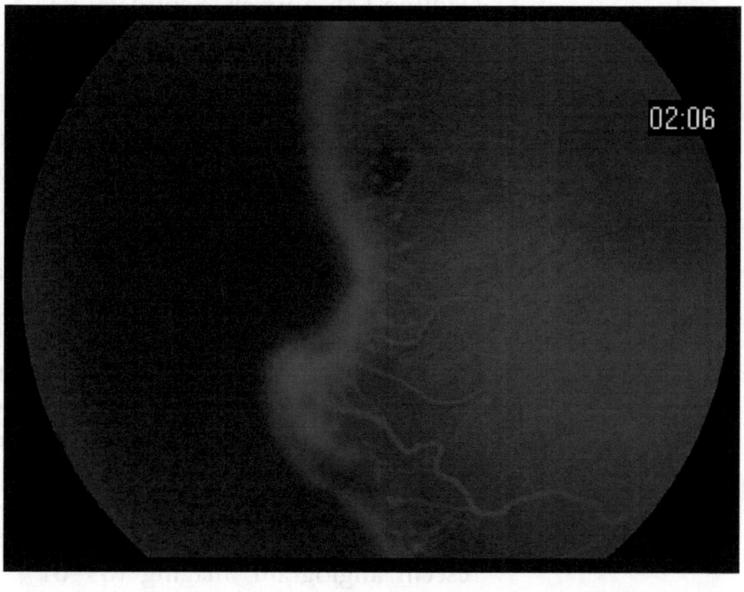

Fig. 2. A fluorescein angiogram depicting retinal pathology in a human infant with ROP. Neovascularization is seen at the boundary between the vascular central retina and avascular peripheral retina. (Image courtesy of Franco Recchia, MD)

that, at the time of removal to room air, the Brown Norway rats exhibited an avascular area approximately four times greater than that of the Sprague Dawley rats. The Brown Norway rats subsequently developed three times the amount of preretinal NV. Later studies confirmed the above findings by demonstrating that Brown Norway rats exhibited an increased amount and duration of retinal vascular permeability relative to Sprague Dawley rats exposed to the same paradigm (56). The difference between the two strains is likely due to differences in retinal expression of pro- and antiangiogenic factors, as demonstrated in the mouse (49, 55). These two studies, though informative, were conducted under conditions of constant, extreme hyperoxia, instead of the more clinically relevant variable oxygen protocol. To address this issue, Holmes and colleagues used a modified protocol of cyclic hyperoxia and hypoxia. Brown Norway rats again demonstrated a higher incidence and severity of NV than did the Sprague Dawley rats; only a few Sprague Dawley rats, as opposed to all of the Brown Norway rats, developed NV (57).

There are also vendor-related differences in susceptibility to retinal NV within the same strain of rat. Penn (unpublished observations) identified differences in the pathological response of a single rat strain obtained from several different vendors. When exposed to the OIR protocol, Sprague Dawley rats from Charles River (Charles River Laboratories, Wilmington, MA) produced a twofold greater area of NV than those from Zivic-Miller (Zivic Laboratories, Pittsburg, PA). Sprague Dawley rats obtained from Harlan (Harlan, Indianapolis, IN) and Hilltop (Hilltop Lab Animals, Scottdale, PA) demonstrated intermediate levels of pathology compared to Charles River and Zivic-Miller rats. Holmes and colleagues similarly tested the OIR response of Sprague Dawley rats obtained from Harlan and Charles River (58). Notably, the Charles River rats demonstrated a 62% greater susceptibility to, and more severe, oxygen-induced NV. Thus, susceptibility to NV depends on genetic variation, environment, and oxygen treatment paradigm.

The rat also has the advantage of large litter sizes (typically twice the size of mouse litters) and relatively inexpensive maintenance costs. Most importantly, unlike the mouse, the rat model consistently produces human-like patterns of vasoattenuation and vasoproliferation. For these reasons, the rat constitutes an ideal model of retinal NV that is often used to test the efficacy of antiangiogenic compounds for application in both ocular and non-ocular pathologies. Notably, the rat model has also been used to demonstrate the progression of the disease in real time with fluorescein angiogram imaging (59–61), and to demonstrate the speed and objectivity of pathology assessment by computer-assisted image analysis (62).

6. Conclusions

The development and use of large and small animal models of OIR have contributed greatly to our understanding of both physiological and pathological retinal vascular development. Because of the ease with which retinopathy outcome can be altered by manipulation of environmental or genetic influences, animal models of OIR have also provided a commonly used means to study fundamental aspects of the angiogenic process and its molecular regulation. In addition, OIR models have advantages over tumor angiogenesis models, in that eye tissue is readily accessible, and its vasculature is easy to visualize and assess. OIR models have shed light on the cellular and molecular basis of angiogenesis. They have also provided an important venue for testing novel treatment strategies and therapeutic targets. Continued refinement of the models and knowledge gained through their use will facilitate the development of therapies directed against neovascular diseases of the human eye.

References

1. Terry TL (1942) Extreme prematurity and fibroblastic overgrowth of persistent vascular sheath behind each crystalline lens: I, preliminary report. Am J Ophthalmol 25:203–204
2. Campbell K (1951) Intensive oxygen therapy as a possible cause of retrolental fibroplasia; a clinical approach. Med J Aust 2:48–50
3. Patz A, Hoeck LE, De La Cruz E (1952) Studies on the effect of high oxygen administration in retrolental fibroplasia: I, nursery observations. Am J Ophthalmol 35:1248–1253
4. Gyllensten LJ, Hellstrom BE (1952) Retrolental fibroplasias: animal experiments. Acta Paediatr 41:577–582
5. Patz A, Eastham A, Higginbotham D et al (1953) Oxygen studies in retrolental fibroplasia: II, the production of the microscopic changes of retrolental fibroplasia in experimental animals. Am J Ophthalmol 36:1511–1522
6. Ashton N, Ward B, Serpell G (1953) Role of oxygen in the genesis of retrolental fibroplasia: a preliminary report. Br J Ophthalmol 37:513–520
7. Ashton N, Ward B, Serpell G (1954) Effect of oxygen on developing retinal vessels with particular reference to the problem of retrolental fibroplasias. Br J Ophthalmol 38:397–432
8. Hatfield EM (1972) Blindness in infants and young children. Sight Sav Rev 42:69–89
9. Gibson DL, Sheps SB, Schechter MT et al (1989) Retinopathy of prematurity: a new epidemic. Pediatrics 83:486–492
10. http://www.nei.nih.gov/health/rop/
11. Hughes S, Yang H, Chan-Ling T (2000) Vascularization of the human fetal retina: roles of vasculogenesis and angiogenesis. Invest Ophthalmol Vis Sci 41:1217–1228
12. Madan A (2003) Angiogenesis and antiangiogenesis in the neonate: relevance to retinopathy. NeoReviews 4:356–363
13. Madan A, Penn JS (2003) Animal models of oxygen-induced retinopathy. Front Biosci 8:1030–1043
14. Roth AM (1977) Retinal vascular development in premature infants. Am J Ophthalmol 84:636–640
15. Foos R, Kopelow S (1973) Development of retinal vasculature in paranatal infants. Surv Ophthalmol 18:117–127
16. Michaelson I (1948) The mode of development of the vascular system of the retina, with some observations in its significance for certain retinal diseases. Trans Ophthalmol Soc U K 68:137–180
17. Coats DK, Paysse EA, Steinkuller PG (2000) Threshold retinopathy of prematurity in neonates less than 25 weeks' estimated gestational age. J AAPOS 4:183–185

18. (2000) Supplemental therapeutic oxygen for prethreshold retinopathy of prematurity (STOP-ROP), a randomized, controlled trial. I: primary outcomes. Pediatrics 105: 295–310
19. Flynn JT, Chan-Ling T (2006) Retinopathy of prematurity: two distinct mechanisms that underlie zone 1 and zone 2 disease. Am J Ophthalmol 142:46–59
20. Ashton N, Gaymore C, Pedler C (1957) Studies in developing retinal vessels V. Mechanism of vaso-obliteration. Br J Ophthalmol 41:449–460
21. Ashton N (1970) Retinal angiogenesis in the human embryo. Br Med Bull 26:103–106
22. Halasz P, Stone J (1984) Development of retinal vasculature in the cat. J Anat 139:187
23. Patz A (1957) The role of oxygen in retrolental fibroplasia. Pediatrics 19:504–524
24. Patz A (1954) Oxygen studies in retrolental fibroplasia IV. Clinical and experimental observations. Am J Ophthalmol 38: 291–308
25. Flower RW, Patz A (1971) Oxygen studies in retrolental fibroplasia. IX. The effects of elevated arterial oxygen tension on retinal vascular dynamics in the kitten. Arch Ophthalmol 85:197–203
26. Chan-Ling T, Gock B, Stone J (1995) The effect of oxygen on vasoformative cell division. Evidence that 'physiological hypoxia' is the stimulus for normal retinal vasculogenesis. Invest Ophthalmol Vis Sci 36:1201–1214
27. Chan-Ling T, Tout S, Hollander H, Stone J (1992) Vascular changes and their mechanisms in the feline model of retinopathy of prematurity. Invest Ophthalmol Vis Sci 33:2128–2147
28. Chan-Ling T, Stone J (1992) Degeneration of astrocytes in feline retinopathy of prematurity causes failure of the blood-retinal barrier. Invest Ophthalmol Vis Sci 33:2148–2159
29. Tailoi CL, Gock B, Stone J (1995) Supplemental oxygen therapy. Basis for noninvasive treatment of retinopathy of prematurity. Invest Ophthalmol Vis Sci 36: 1215–1230
30. Phelps DL, Rosenbaum AL (1987) Effects of variable oxygenation and gradual withdrawal of oxygen during the recovery phase in oxygen-induced retinopathy: kitten model. Pediatr Res 22:297–301
31. Kremer I, Kissun R, Nissenkorn I et al (1987) Oxygen-induced retinopathy in newborn kittens. A model for ischemic vasoproliferative retinopathy. Invest Ophthalmol Vis Sci 28:126–130
32. Flower RW, McLeod DS, Lutty GA et al (1985) Post-natal retinal vascular development of the puppy. Invest Ophthalmol Vis Sci 26:957–968
33. Patz A (1954) Clinical and experimental studies on role of oxygen in retrolental fibroplasia. Trans Am Acad Ophthalmol Otolaryngol 58:45–50
34. McLeod DS, Brownstein R, Lutty GA (1996) Vaso-obliteration in the canine model of oxygen-induced retinopathy. Invest Ophthalmol Vis Sci 37:300–311
35. McLeod DS, Crone SN, Lutty GA (1996) Vasoproliferation in the neo-natal dog model of oxygen-induced retinopathy. Invest Ophthalmol Vis Sci 37:1322–1333
36. McLeod DS, D'Anna SA, Lutty GA (1998) Clinical and histopathological features of canine oxygen-induced proliferative retinopathy. Invest Ophthalmol Vis Sci 39:1918
37. Taomoto M, McLeod DS, Merges C, Lutty GA (2000) Localization of adenosine A2a receptor in retinal development and oxygen-induced retinopathy. Invest Ophthalmol Vis Sci 41:230–243
38. Roberts S (1959) Detachment of the retina in animals. J Am Vet Med Assoc 135:423–431
39. Flower R (1983) A mechanism for oxygen damage to the immature retinal vasculature. New York Press, New York, NY
40. Bossi E, Koerner F (1982) Retinopathy of prematurity (retrolental fibroplasia): old and new facts. Helv Paediatr Acta 37:413–420
41. Fruttiger M (2007) Development of the retinal vasculature. Angiogenesis 10:77–88
42. Chan-Ling T, McLeod DS, Hughes S et al (2004) Astrocyte-endothelial cell relationships during human retinal vascular development. Invest Ophthalmol Vis Sci 45: 2020–2032
43. Gyllensten L, Hellstrom B (1954) Experimental approach to the pathogenesis of retrolental fibroplasias: I. changes of the eyes induced by exposure of newborn mice to concentrated oxygen. Acta Pediat 43: 131–148
44. Gyllensten L, Hellstrom B (1955) Experimental approach to the pathogenesis of retrolental fibroplasia II. The influence of the developmental maturity in oxygen-induced changes in the mouse eye. Am J Ophthalmol 39:475–488
45. Bischoff PM, Wajer SD, Flower RW (1983) Scanning electron microscopic studies of the hyaloid vascular system in newborn mice exposed to O_2 and CO_2. Graefes Arch Clin Exp Ophthalmol 220:257–263

46. Smith LE, Wesolowski E, McLellan A et al (1994) Oxygen-induced retinopathy in the mouse. Invest Ophthalmol Vis Sci 35:101–111
47. Rohan RM, Fernandez A, Udagawa T et al (2000) Genetic heterogeneity of angiogenesis in mice. FASEB J 14:871–876
48. Rogers MS, Rohan RM, Birsner AE et al (2003) Genetic loci that control vascular endothelial growth factor-induced angiogenesis. FASEB J 17:2112–2114
49. Chan CK, Pham LN, Zhou J et al (2005) Differential expression of pro- and antiangiogenic factors in mouse strain-dependent hypoxia-induced retinal neovascularization. Lab Invest 85:721–733
50. Claxton S, Fruttiger M (2003) Role of arteries in oxygen induced vaso-obliteration. Exp Eye Res 77:305–311
51. Gole GA (1985) Animal models of retinopathy of prematurity. In: Silverman WA, Flynn JT (eds) Contemporary issues in fetal and neonatal medicine 2: retinopathy of prematurity. Blackwell Scientific Publishers, Boston, pp 53–95
52. Ashton N, Blach R (1961) Studies in developing retinal vessels VIII: effect of oxygen on the vessels of the ratling. Br J Ophthalmol 45:321–340
53. Penn JS, Tolman BL, Lowery LA (1993) Variable oxygen exposure causes preretinal neovascularization in the newborn rat. Invest Ophthalmol Vis Sci 34:576–585
54. Penn JS, Henry MM, Tolman BL (1994) Exposure to alternating hypoxia and hyperoxia causes severe proliferative retinopathy in the newborn rat. Pediatr Res 36:724–731
55. Gao G, Li Y, Fant J et al (2002) Difference in ischemic regulation of vascular endothelial growth factor and pigment epithelium-derived factor in brown norway and sprague dawley rats contributing to different susceptibilities to retinal neovascularization. Diabetes 51:1218–1225
56. Zhang SX, Ma JX, Sima J et al (2005) Genetic difference in susceptibility to the blood-retina barrier breakdown in diabetes and oxygen-induced retinopathy. Am J Pathol 166:313–321
57. Floyd BN, Leske DA, Wren SM et al (2005) Differences between rat strains in models of retinopathy of prematurity. Mol Vis 11:524–530
58. Kitzmann A, Leske DA, Chen Y et al (2002) Incidence and severity of neovascularization in oxygen-and metabolic acidosis-induced retinopathy depend on rat source. Curr Eye Res 25:215–220
59. Larrazabal LI, Penn JS (1989) Study of ocular vasculature in the newborn rat by fluorescein angiography. J Ophthal Phot 11:49–52
60. Larrazabal LI, Penn JS (1990) Fluorescein angiography in the newborn rat: implications in oxygen-induced retinopathy. Invest Ophthalmol Vis Sci 31:810–818
61. Penn JS, Johnson BD (1993) Fluorescein angiography as a means of assessing retinal vascular pathology in oxygen-exposed newborn rats. Curr Eye Res 12:561–570
62. Penn JS, Tolman BL, Lowery LA et al (1992) Oxygen-induced retinopathy in the rat: hemorrhages and dysplasias may lead to retinal detachment. Curr Eye Res 11:939–953

Chapter 7

Animal Models of Diabetic Retinopathy

Adrian M. Timmers, Casey M. Miller, and Li Zhu

Abstract

Diabetic retinopathy threatens vision in millions of patients in the USA. Prolonged hyperglycemia causes irreversible pathological changes in the retina, leading to proliferative diabetic retinopathy with preretinal neovascularization and diabetic macular edema. Much of the disease progression appears similar between man and animal. Thus, animal models are essential in understanding the pathology of this disease and development of effective treatments. This chapter describes and discusses the use of the rat, mouse, and dog in diabetic retinopathy studies.

Key words: Diabetic retinopathy, Preretinal neovascularization, Diabetic macular edema, Streptozotocin, Rat model of diabetic retinopathy, Mouse model of diabetic retinopathy, Dog model of diabetic retinopathy

1. Introduction

Diabetes mellitus (DM) is a group of metabolic diseases characterized by hyperglycemia resulting from defects in insulin secretion, insulin action, or both. The chronic hyperglycemia of diabetes is associated with long-term damage, dysfunction, and failure of various organs, especially the eyes, kidneys, nerves, heart, and blood vessels (1).

1.1. Impact of Disease

The World Health Organization estimated the number of DM patients to reach 300 million in 2025. In the USA alone, 48 million people could be diabetics in 2050 (2). Based on data from the National Health Interview and US Census Bureau, the Center for Disease Control (CDC) predicts that in 2050 16 million people will have diabetic retinopathy (DR), and for 3.4 million DR will be severe enough to threaten vision (3). This forecasts a tripling of cases from 2004 to 2050; in 2004 an estimated $500 M

Iok-Hou Pang and Abbot F. Clark (eds.), *Animal Models for Retinal Diseases*, Neuromethods, vol. 46
DOI 10.1007/978-1-60761-541-5_7,

was spent on the medical expenses associated with DR alone (3). The diabetes epidemic underscores the need for novel efficient treatments for complications like DR.

1.2. Etiology and Pathology

Although the etiology of the two types of diabetes, Insulin-Dependent Diabetes Mellitus (IDDM) or Type 1 DM (T1DM; autoimmune destruction of the β-cells of the pancreas with consequent insulin deficiency) and Noninsulin-Dependent Diabetes Mellitus (NDDM) or Type 2 DM (T2DM; abnormalities that result in resistance to insulin action) differs, the natural course of pathologic events in the retina are similar (4). After approximately half a decade of diabetes, signs of retinopathy (vascular abnormalities) start to become clinically noticeable in T1DM patients, after two decades nearly all have retinopathy, 50% show proliferative diabetic retinopathy (PDR) with preretinal neovascularization. The incidence of PDR in T2DM patients is approximately half of those with T1DM. In addition to PDR, diabetic macular edema (DME) is the other major threat to vision. The incidence of DME is similar in both types of diabetic patients (5). It is proven indisputably that hyperglycemia is the root cause of the diabetic complications including DR (6–12). In man and animal, the histological vascular manifestations of early stages of DR are characterized by loss of intramural capillary pericytes with microaneurysms, capillary closure, acellular capillaries and thickening of the basement membrane. Before these vascular DR symptoms become evident, substantial time of diabetes passes (13). Clinical evidence available to date strongly suggests that the late stage of retinopathy, such as PDR or DME, is a direct consequence of the earlier pathologies (14). To date the pathological progression into diabetic retinopathy cannot be reversed with later correction of hyperglycemia in man and animal (15).

Clearly, the similarities in diabetic retinopathy progression in man and animal strongly corroborate the importance and value of animal models in understanding the pathogenicity of diabetic retinopathy and hunting for effective treatments.

2. Animal Models of Diabetic Retinopathy

This chapter will discuss animal models (mainly rodents) and the role they have played and continue to play in gaining insights into early nonproliferative diabetic retinopathy. Other models for DR including dogs, cats, and nonhuman primates have recently been discussed in reviews by Kern and colleagues (13, 16). Although most early retinal pathologies can be reproduced in animal models, none of the currently available animal models represent the full spectrum of diabetic retinopathy in man.

2.1. Chemical-Induced Hyperglycemia

2.1.1. TIDM Rat Models

Hyperglycemia can be induced by streptozotocin (STZ) or alloxan; both chemicals initiate destruction of beta cells in the pancreas. This causes insulin deficiency and Type 1 diabetes. In rats, STZ-induced diabetes has provided the principal model for diabetic retinopathy. Additionally, several rat strains have been described that spontaneously develop symptoms resembling diabetes mellitus in man. Typically, the more advanced DR phenotypes such as preretinal neovascularization, microaneurysms and intraretinal vascular abnormalities are not detected in rat models.

Vascular symptoms of NPDR (leukostasis/capillary occlusion, acellular capillaries, pericyte loss, and basement membrane thickening) develop reproducibly in rats with chemically induced diabetes (discussed in detail below). Significant changes in acellular capillaries, pericyte ghosts are not observed until at least 6 months from the onset of diabetes. The spontaneous T1DM rat models display substantial variability in timing and severity of the phenomenology of diabetic retinopathy. The BBW rat (17) presents NPDR similarly to the chemically induced diabetic rat models (18, 19). Manifestations of DR in the spontaneous diabetic Torii (SDT) rat (20) appear inconsistently with a much later onset at around 60 weeks. In 2005, Yamada and coworkers published retinal neovascularization without ischemia (21), more detailed studies mention tractional retinal detachment (22). The phenotypic differences of DR progression in chemically diabetic rats and man versus the Torii rat are thought to be the result of high levels of pigment epithelium-derived factor (PEDF) in the Torii rat retina (23, 24). The ocular effects in the Torii diabetic rat could be prevented by glycemic control (25) or treatment with soluble flt-1 (26). Miyamura and coworkers (27, 28) reported abnormal basement membranes in retinal capillaries in the WBN/Kob rat (29); however, the degeneration of the retina resembles retinitis pigmentosa rather than diabetes-induced retinal changes (30, 31).

2.1.2. T2DM Rat Models

Diabetic retinopathy in models of T2DM has not been studied as extensively as in T1DM animal models. The Zucker diabetic fatty (ZDF) rats (32) show some signs of NPDR (capillary basement membrane thickening) but capillary degeneration (acellular capillaries, pericyte ghosts) develops only minimally (33, 34). In contrast, Behl and colleagues reported similar levels of capillary degeneration in STZ rats and ZDF rats after 6 months of diabetes (35). Using corrosion casts of retinal vasculature of Otsuka Long-Evans Tokushima fatty (OLETF) diabetic rats (36), Miyamura et al. earlier reported thicker basement membranes and pericyte loss (37, 38). However, applying retinal trypsin digests, Matsuura and colleagues concluded later that the "OLETF rat is not a suitable animal model for the study of angiopathic diabetic retinopathy" (39). The GK rat (40) reportedly does not develop DR (41) although a study in 2006 measured acellular capillaries but no pericyte ghosts (42), confirming an earlier qualitative report (43). Substantial ERG changes (a-wave amplitude reduction) occurred in the GK rat very early (44). Transfer of the *lepr*^{fa} (previously known as the *fa* gene) gene from Zucker onto Wistar Kyoto rat resulted in the obese-hyperglycemic "Wistar fatty rat" (45), among which only the males develop hyperglycemia (45). In the BBZDR/Wor rat (46) carrying the Zucker *leptin*^{fa} gene, thickening of the basement membrane (47, 48), pericyte ghosts and capillary irregularities were noted in trypsin digests of the diabetic retina (19, 49, 50). The obese SHR Koletsky (SHROB) rat (51) develops retinal capillary dropout, vascular tortuosity and vascular permeability (52). The Spontaneous Hypertensive/NIH corpulent rat (SHR/N-cp) (53), a strain homozygous for a mutation in leptin receptor (corpulent gene) (54), when fed a high sucrose diet developed loss of pericytes and capillary lesions (49). Introduction of the *leptin*^{fa} allele of the Zucker fatty (ZF) rat into the SDT rat (55) resulted in a new model of obese T2DM, in which DR develop earlier (55). To date, no further information on changes in retinal vasculature of these rats is available.

2.2. Mouse Models

The small size of mouse retina presents unique challenges for research on DR. However, genetic manipulation in mice is much more advanced and provides excellent avenues to further unravel the pathology of DR. As in rats, T1DM can be induced chemically in mice. NPDR characteristics (such as acellular capillaries and pericyte ghosts) develop in the commonly used C57Bl/6 J mice at approximately 6 months after induction of diabetes with STZ. With extended duration of diabetes, the symptoms intensify (56). A mutation in the insulin gene in the *Ins2*^{Akita} mouse results in spontaneous diabetes in the males. Vascular pathology in the *Ins2*^{Akita} mice follows a similar pattern with increasing number of acellular capillaries and pericyte ghosts in 6 months (56–58).

Several spontaneous T2DM mice models have also been reported, with only a few strains explored for diabetic retinopathy phenotypes. Vascular DR symptoms have not been analyzed in the New Zealand obese mouse (59, 60), sand rat (61–63), nonobese diabetic (NOD) mouse (64), or ob/ob mouse (65). Retinopathy in NOD mice has been reported (276). According to Jackson Laboratories Database, hyperglycemia in ob/ob mouse is mild, transient and varies greatly between individual animals. For some strains, increase of retinal capillary basement thickening was reported in the spiny mouse after approximate 1.5 years hyperglycemia (66). Pericyte loss, acellular capillaries and loss of neuronal cells in ganglion cell layer and inner nuclear layer have been described in KK (and KKAy) mice (67). Pericyte loss, acellular capillaries, thickening of capillary basement membrane, and breakdown of blood–retinal barrier has been observed in the genetically diabetic db/db mouse (68–70).

Genetic background and rearing environment affect the severity of diabetes and the rate of progression of diabetic retinopathy, and such variability exists in the human population as well. Despite these differences, the type of vascular distortions appears to be rather general across strains and species.

2.3. Alternative Diabetic Retinopathy Models

In 1984, Engerman and Kern presented diabetic retinopathy phenomena in galactosemic dogs (71). The elevation of blood levels of D-galactose, an isomer of glucose, due to diet containing 30% galactose reproduced many of the effects of diabetes on the lens, cornea, and peripheral nerve. In the galactose-fed dogs, blood galactose varied throughout the day from values of approximately zero after an overnight fast to values as high as 150–250 mg/dL after consumption of the galactose-rich diet. The blood glucose and serum insulin levels remained normal in the galactosemic animals in contrast to the values for the diabetic animals (72). Diabetes in dogs and humans is marked also by abnormalities such as supranormal levels of fibrinogen, branched-chain amino acids, and lipids. Galactosemic dogs lacked these features of diabetes as well (72). Glycated hemoglobin levels increase significantly in galactosemic animals (72–75). The galactose-rich diet also produced diabetic like vascular abnormalities in rats and mice (76, 77). This model supports the notion that hyperhexosemia is the main culprit for diabetic retinopathy, a concept strongly agreed by the findings of the Diabetes Control and Complication Trial. However, important differences were noted as well. Aminoguanidine blocks a variety of processes thought to be important in development of DR, including nonenzymatic glycation (78, 79), oxidative stress (80, 81), activation of protein kinase C (81), induction of vascular endothelial growth factor (VEGF), nitric oxide synthase (81, 82) and cycloxygenase-2 (82). Interestingly, aminoguanidine does prevent DR in experimental diabetes but not in galactosemia-induced retinopathy (73, 75, 83). Despite the similarities of vascular phenomena in diabetic and galactosemic retinopathy, clearly there are some important differences for which the causes have not been identified. Perhaps the strong fluctuation of hyperhexosemia in galactosemic animals (peaks postprandial then returns to normal), or the anomaly in insulin metabolism in diabetes holds additional factors crucial for development of diabetic retinopathy.

2.4. STZ Animal Model for Diabetic Retinopathy

STZ (2-deoly-2-(3-methyl-3-nitrosourea) 1-D-glucopyranose) is a broad-based antibiotic that causes diabetes with a single injection by inducing β-cell necrosis in the pancreas (84). Being a derivative of glucose (85), STZ is transported across the β-cell membrane by GLUT2 receptors (86, 87). Death of the β-cells is mediated by alkylation of DNA via reactive carbonium ions (88, 89), the production of hydrogen peroxide (90, 91), and/or production of nitric oxide (NO) (92–94).

STZ can be administrated at doses varying from 35 to 100 mg/kg and with single or multiple dosing schedules (84, 95–102). Due to concerns of its stability, STZ is prepared immediately before injection in citrate buffer pH 4.5 (84, 103). Animals are fasted 1 day before STZ administration, the diabetic condition is assessed by blood glucose levels the next morning. Depending on the experimental paradigm, insulin is typically only administered to reverse weight loss and prevent premature death of the diabetic animal.

2.4.1. Vascular Diabetic Retinopathy in STZ-Induced Diabetes

In diabetic patients, progressive leakage of albumin into retina with increasing severity of diabeties was detected (104). Increase retinal levels of intracellular adhesion molecule-1 (ICAM-1), VEGF, and leukocytes were also found in DR patients (105–110). Leukostasis, adherence of leukocytes to the endothelial lining mediated by ICAM-1, is thought to be one of the first steps toward vascular permeability and blood–retinal barrier (BRB) breakdown (111–115). VEGF upregulates expression of ICAM-1 (116–118).

In STZ rats, leukostasis occurs within weeks after onset of diabetes (23, 57, 99, 101, 114, 115, 117, 119–125). Leukostasis concurs with increased expression of ICAM-1 (23, 101, 126–129) and VEGF (101, 128–133), combined with the breakdown of the BRB demonstrated by increased permeability in the retina (117, 123, 132, 134–138) and dysfunction of endothelial tight junctions (100, 139–142). Perpetual leukostasis can lead to capillary occlusion, acellular capillaries and pericyte loss, which are processes thought to be precursors to ischemia and induction of neovascularization. Cytokines such as ICAM-1 and VEGF appear to contribute to the early vascular pathology in DR.

2.4.2. Neuronal Diabetic Retinopathy in STZ-Induced Diabetes

Classically diabetic retinopathy has been regarded as a disease of the microvasculature of the retina. The stages of DR have been based on clinically observable lesions that originated from the retinal vasculature. The focus on the diabetic vasculature pathology overshadowed the notion postulated in 1962 by Bloodworth that neuronal pathologies also are an integral part of the pathogenesis of early DR: "Diabetic retinopathy is a complex degenerative disease of all elements of the retina, probably due to a fundamental metabolic or enzymatic defect of the cells and is not related to vascular supply" (143). In human it takes 5–7 years after onset of diabetes to clinically image vascular irregularities, while in experimental animals the time ranges from 0.5 to 1 year.

2.4.3. Glial Components

Altered glial function with increased levels of glial fibrillary acidic protein (GFAP) (144, 145) becomes evident during first several months of STZ-induced diabetes (146–149). Glial cells support the functions of neurons and endothelial cells (150, 151). Muller glial cells regulate the neuronal extracellular environment by clearing glutamate, converting it to glutamine for resynthesis to glutamate in the neuronal cells. In the diabetic rat the glutamate-to-glutamine conversion is strongly reduced in both T1DM and T2DM experimental models of DR (146). This is accompanied by an overall increase in retinal glutamate, which is toxic to retinal neurons (152, 153). Glutamate levels in vitreous of patients with PDR are elevated (154). Cepko and coworkers linked reactive gliosis to vascular change (155). Muller cells are the major producers of VEGF, and its expression increases in Muller cells under hyperglycemic or hypoxic conditions (156). These observations present evidence for a correlation between gliosis and DR.

2.4.4. Neuronal Cell Death

Apoptotic pericyte and endothelial nuclei were detected in diabetic donor eyes using TUNEL (*t*erminal transferase d*U*TP *n*ick *e*nd *l*abeling) (157, 158). Bax, a regulator of apoptosis and a member of the Bcl-2 family (159), was significantly upregulated in retinas from diabetic donors (160). Bax expression (associated with neurons under hyperglycemic duress) (161–163), staining of neurons with Fluoro-Jade B (a marker for degenerative neurons), and caspase-9 and -3 increased significantly in retinas from diabetic donors (164), as were caspases-1, -3, -4 and -6 in eyes from T2DM donors (165). STZ rats (7.5 months) measured a reduced thickness of the inner plexiform (22%), inner nuclear layer (14%) and a reduction in surviving retinal ganglion cells (RGC) (10%) (158). In whole mount retinas of STZ rats after 1 month diabetes a tenfold increase in TUNEL-positive nuclei over control indicates the early onset of apoptosis (158). TUNEL-positive pericytes and endothelial cells in diabetic rats (31 weeks duration) preceded any histological evidence of retinopathy (157). Caspases-1, -2, -6, -8, and -9 were activated in diabetic mice as early as 2 months duration (165). The caspase activity profile changes with duration of diabetes, caspase-6 increases and caspase-3 can be detected when capillary cells start to show signs of apoptosis (165, 166). In mice diabetic for 14 weeks, caspase-3 activity and TUNEL-positive cells in RGC coincided with reduced (25%) RGC density (167). In diabetic rats (3 weeks) a significant increase of TUNEL-positive cells in the ganglion layer and increase of Bax expression in the retinal neuronal cells was noted (163). TUNEL-positive cells increased in the outer nuclear layer (ONL) after 1 week of diabetes peaking at 4–6 weeks with a significant reduction of nuclei in the ONL and retinal thickness (168). Diabetic rats analyzed from 1 to 24 weeks showed degeneration of the postsynaptic processes of horizontal cells loss of RGC at 4 weeks, and amacrine cells at 12 weeks (169). A large reduction in the ONL was noted at 24 weeks (169). Controversial results have also been observed in STZ mice with no obvious RGC loss after a considerable length of diabetes duration (56, 170).

2.4.5. Electroretinogram

ERG provides a useful tool to evaluate the impact of diabetes on visual function (171–178) and to predict the progression of DR (179, 180). Reduction in amplitudes or delay of implicit times of oscillatory potentials has been observed in diabetic patients (181–183). Fortune and coworkers showed delays in local multifocal ERG in eyes with DR (175). ERG measurements in STZ diabetic rats show a reduction of the a-wave (photoreceptor cells) and b-wave (amacrine cells) amplitudes from 2 weeks onward (184). Other studies also measured reduced b-wave amplitudes from 4 weeks forward (185–189). Substantial variability exists with regard to timing of reduction in oscillatory potential (OP). Li and coworkers noted significance after 2 weeks (184). Aizu et al., reported a significant reduction in OP at 4 weeks (189), Kohzaki and colleagues see significance at 8 and 11 weeks duration (190). Others present significance in OP reduction after 12 weeks (186–188, 191, 192). ERG functions such as scotopic threshold response (STR) were reduced at 4 weeks while this group did not see OP changes until 8 weeks (190). The leading edge amplitude (RmPIII) of the a-wave was significantly reduced after 12 weeks of diabetes (193).

2.4.6. Strain Differences

Spontaneous diabetic animals revealed variability in severity and onset of DR. Although STZ induction of diabetes allows perfect synchronization of onset, genetic makeup of the animals, however, could impact progression of DR symptoms. Comparison of Brown Norway (BN), Sprague Dawley (SD), Wistar, and Lewis rats have revealed some differences (135, 189, 194). Diabetic BN rats have significantly reduced a-wave and b-wave amplitudes and OP after 1 month (189). At their earliest timepoint of 4 months, Kern and colleagues noted that in Lewis, a-wave and b-wave amplitudes were significantly reduced while SD rats had no significant changes and in Wistar the b-wave was delayed. In BN rats, RGC loss was significant after 4 weeks of diabetes (185). Significant cell loss in the ganglion cell layer was not detected in Lewis rats until 2 months. At 4 months, VEGF, COX2 and the inducible isoform of NOS (iNOS) were increased in all strains, only significant for some. After 8 months, significant increase in acellular capillaries and pericyte ghosts was noted in Lewis and Wistar rats (194). SD and BN rats were compared by Zhang and coworkers utilizing a retinal vascular permeability (RVP) assay that assesses retinal permeability for albumin from the blood circulation. RVP in diabetic BN rats was significantly higher than in SD rats. Measurement of VEGF levels and sensitivity to injected VEGF revealed higher levels and sensitivity in BN rats (135). No information on the development of acellular capillaries or pericyte ghosts in BN rats is currently available.

3. Pathogenesis of DR and Treatment Approaches Learned from Animal Models of DR

3.1. Pathogenesis of DR

More and more evidence has been accumulated to support that chronic subclinical inflammation plays an essential role in the early development of many signature vascular abnormalities of DR, in addition to the previously extensively studied polyol pathway, nonenzymatic glycation and diacylglycerol-protein kinase C pathway (195–197). Inflammatory features that characterize DR include increased blood flow and vascular permeability, tissue (macular) edema, neovascularization (198), increased expression of inflammatory mediators (106, 110, 199–209), accelerated retinal neural (56, 158) and microvascular (157) cell death, macrophage infiltration (210, 211), microglial cell activation (56, 147, 212–214), increased leukocyte adhesion (117, 127), complement activation (215), Fas ligand upregulation (216–219), and acute phase response protein expression (215). Animal models, although imperfect, are essential in the efforts of resolving the nature and time course of molecular changes, especially crucial early events leading to destructive results in DR patients.

3.1.1. Proinflammatory Cytokines

Proinflammatory cytokines, including ICAM-1, tumor necrosis factor-α (TNF-α), interleukin-1β (IL-1β), interleukin-6 (IL-6), and VEGF (especially the isoform 164/165), have been found at elevated levels in the eyes of both human DR patients (107, 204, 220, 221) and diabetic animals (110, 208, 221–227). These molecules are produced locally in the eye by resident and infiltrating cells and exhibit pleiotropic effects on homeostasis of glia, neurons, and retinal vascular endothelial cells in the retina. Changes induced by chronic hyperglycemia lead to dysregulation of these cytokines and thus may contribute to chronic deteriorating effects that eventually lead to retinopathy.

An important component of the early DR development is leukostasis, which is directly linked with upregulation of ICAM-1 in retinas of diabetic animals (115, 128). Interaction of ICAM-1 with CD18 adhesion molecule on monocytes and neutrophils contributes to the diabetes-induced leukocyte adhesion to retinal vessels, which potentially causes capillary occlusion. STZ diabetic mice genetically deficient in either ICAM or CD18 have been shown to develop much less microvascular damage and capillary leakage than the wild-type controls (127).

3.1.2. TNF-α

In the diabetic retina, TNF-α potentially comes from astrocytes, Muller cells (228) and microglia (229). Levels of sTNFR1, receptor of TNF-α, have been found to be higher in T2DM patients with microvascular complications (230) and TNF-α is an independent serum marker for proliferative retinopathy in T1DM patients (226). A soluble TNF-α receptor/Fc construct, etanercept, has been shown to not only lower ICAM-1 level and DNA-binding capacity of nuclear factor-κB (NF-κB), but also reduce retinal leukostasis and BRB breakdown (228). Another recent study with pegsunercept, a TNF-α-specific inhibitor, has been shown to inhibit pericyte ghost formation and reduce acellular capillary formation in both STZ-induced diabetic rats (T1DM) and Zucker diabetic fatty rats (ZDF, T2DM model) (35). TNFR deficiency in diabetic mice protected against galactose-induced retinopathy (231).

3.1.3. Nitric Oxide Synthase

Excessive NO production was thought to be induced in the inflammatory response to hyperglycemia. Endothelial cells express constitutive NO synthase (eNOS) (232, 233) and in some studies retinal vascular eNOS level was reported to decrease in diabetes. It is suggested that iNOS (234) may be the main source of the harmful amount of NO production in DR development (114, 170, 235–237). Again, iNOS-defecient (*iNOS–/–*) mice made diabetic with STZ (114, 170) are protected, which shows that iNOS isoform plays a predominant role in leukostasis and BRB breakdown, via a mechanism that involves ICAM-1 upregulation and tight junction protein downregulation. Also it was found that diabetic retina microvascular lesions observed in STZ-induced diabetes in mice showed a much earlier onset in *eNOS–/–* mice as 3 months versus 6 months in wild-type STZ mice (Li QH et al., unpublished data). This NO abnormality can be corrected by treating STZ-induced diabetic rats or dogs with inhibitors of NO synthase (114, 236). Aminoguanidine (238) has been found to inhibit the development of retinopathy in diabetic dogs (239) and rats (73). The mechanism may be attributed to multiple pathways, including blockage of advanced glycation endproduct formation, iNOS, COX-2, activation of protein kinase C, induction of VEFG and oxidative stress.

3.1.4. Interleukin-1β

IL-1β induces the expression of the iNOS, which use L-arginine as substrate to overproduce NO. Mice deficient in the IL-1β receptor were protected from diabetes-induced caspase activation and retinal pathology at 7 months of diabetes, indicating that the caspases-1/IL-1β signaling pathway plays an important role in DR (240). It was also reported that IL-1 type 1 receptor-deficiency in genetically modified mice generally conferred neuroprotection, evidenced by reduced neuroinflammation and both cytotoxic and vasogenic edema in these mice (241).

3.1.5. Nuclear Factor-κB

NF-κB is also known to play an important role in the promotion of retinal vascular cell apoptosis (242) and downstream inflammation regulation in DR, being actively involved in multiple pathways. It has been shown that the NF-κB inhibitor can block the abnormal increases of mRNA and protein levels of angiotensinogen, angiotensin II type 1 and 2 receptors in the STZ-induced diabetic mice (243). Also, intermediate doses of salicylates that inhibit NF-κB activation in the retinas of diabetic rats have been shown to inhibit upregulation of inflammatory mediators like vascular cell adhesion molecule (VCAM), iNOS, ICAM-1, and COX2, as well as the signature microvascular abnormalities in the STZ rat models.

3.1.6. VEGF

VEGF is well-recognized as a causative factor in neovascularization and increased vascular permeability in retinopathies. Characterized in diabetic animals, it was found that VEGF accumulates in the retina early in diabetes (244–246). RGC, Muller cells and pericytes can all produce VEGF. Clinical trials of anti-VEGF therapies have shown promising results against late stage of DR, which was mimicked in VEGF-induced DR-like condition in monkeys with vascular tortuosity and microaneurysms (247, 248). Recently, it has been suggested that unbalanced VEGF and PEDF expression in diabetic retina may contribute to the pathogenesis of proliferative diabetic retinopathy (122, 249, 250). PEDF has been found in the vitreous, and its levels were decreased in angiogenic eye diseases although plasma level in advanced DR patients increases (251), suggesting that PEDF in the eye is functionally important in the pathogenesis of proliferative diabetic retinopathy (252–254). It has also been reported that low leukostasis and an absence of nonperfused area in SDT rats with long-term diabetes may be attributed to inhibition of the VEGF-induced leukostasis by PEDF in these rats, since they showed significantly higher levels of PEDF than STZ rats (23, 24).

3.1.7. Time-Dependent Changes

Characterization of the time course of early retinal damage in DR at the molecular level is also very important in understanding the pathogenesis. Gene expression changes found with whole genome microarray and qPCR revealed that at 3 months of STZ-induced diabetes in rats, vascular permeability and caspase-3 activity were increased, but not at 1 month. The importance of increasing duration of diabetes is further confirmed with significantly more and larger magnitude gene expression changes observed after 3 months than 1 month of diabetes. Recent long-term good blood glucose control studies in STZ-induced diabetic rat model revealed that diabetes duration before implementation of good blood glucose control (the diabetic “memory”) has an pivotal impact on the extent of DR disease state development (15), which is consistent with clinical studies with diabetic patients. Glyceraldehyde-3-phosphate dehydrogenase (GAPDH) was thought to contribute to diabetic retinopathy progression after cessation of hyperglycemia (15).

3.2. Treatment Approaches

Many potential therapies have been reported effective in experimental diabetic animals, mostly STZ-induced diabetic rats or mice. Aldose reductase inhibitors like sorbinil or ARI-809 were shown to protect retinas of STZ-rats from both neuronal and vascular damage (Sun et al. 2005 and ref. 126), while antiplatelet drug clopidogrel showed no effect (Sun et al. 2005). Benfotiamine was reported to inhibit both PKC activation by hyperglycemia and advanced glycation endproducts and thereon prevented diabetic retinopathy in STZ-rats (255). A tyrosine kinase inhibitor, genistein, was first shown to be protective in normalization of retinal vascular permeability in experimental diabetes (256, 257) and the PKC inhibitor, ruboxistaurin, was led into clinical trial from efficacy demonstrated in the diabetic animal models (258). Inhibition of COX, especially COX-2, with nepafenac or meloxicam both have been shown to inhibit diabetes-induced leukostasis, with nepafenac significantly decreased histological lesions in diabetic rat retinas with long-term topical dosing (101). Poly-ADP-ribose polymerase inhibitor PJ34 also showed successful inhibition of the diabetes-induced early retinal microvascular lesions while dosed 9 months to diabetic rats (259). Corticosteroids (260), such as dexamethasone delivered by intravitreal injection in STZ rats, have been demonstrated to improve diabetic retinal edema through inhibiting leukocyte recruitment in the diabetic retina (57). Three different nonsteroidal anti-inflammatory drugs, aspirin (239), sodium salicylate and sulfasalazine, were also found to inhibit the development of the diabetes-induced retinal lesion in STZ-induced diabetic rats (261). Minocycline was reported to reduce proinflammatory cytokine expression, microglial activation, and caspase-3 activation in the STZ-induced diabetic rat retina (213). Antioxidants dosing have been shown to partially inhibit these retinal lesion, with α-tocopherol (262) and α-lipoic acid (263, 264) being the most effective, possibly by partially inhibiting diabetes-induced inflammation. Micronutrients that were shown to reduce the risk of development of age-related macular degeneration in the Age-Related Eye Disease Study (AREDS) were shown to inhibit the development of diabetic retinopathy in STZ rats, possibly by inhibiting the oxidative and nitrative stress (265). Exogenous erythropoietin injected intravitreally to STZ-induced diabetic rat eyes were shown to protect both retinal vascular and neuronal cells in early diabetes (168). Most recently, Rho kinase inhibition by fasudil was shown to ameliorate diabetes-induced microvascular damage (266). Recent discovery of diabetic-retinopathy-like capillary and neural changes in dopamine beta hydroxylase knockout (*Dbh–/–*) mice demonstrated that loss of sympathetic neurotransmission also contributes to DR pathogenesis (267)

4. Future Outlook

Despite the progress made thus far, the molecular mechanism(s) of diabetic retinopathy and hence how effective therapies work remain largely unknown. It is possible that extended duration of hyperglycemia may trigger a series of signaling pathways that lead to vision loss which could complicate therapy development. The normal inflammatory response aims to limit tissue injury; however, in diabetes it may result in exacerbation. It becomes more obvious that many of these factors "cross-talk" and are part of interweaving signaling pathways. For instance, aldose reductase produces cytotoxic sorbitol in hyperglycemia that leads to NADPH depletion, which in turn reduces NO production. Inflammatory cytokines such as IL-1 and TNF-α have multiple effects on the endothelium, can lead to prothrombotic and proinflammatory states, and stimulate NO production through iNOS. It should be noted that in human patients, high amounts of TNF-α and IL-1β alone contribute to the deterioration of the hemodynamic condition, which results in hypotension, shock, and eventually tissue damage (268). The level of eNOS can be profoundly downregulated in the presence of TNF-α (269), as well as by pathophysiological stimuli such as hypoxia (232, 270) and endothelial proliferation (233). Apoptosis was induced in endothelial cell culture with TNF-α and was inhibited by simultaneous exposure of endothelial cells to VEGF (268). Simple inhibition of VEGF which is supposed to support cell survival may aggravate retinal neuronal or vascular cell death in diabetes, with seemingly improved clinical results (271). A new mechanism has been proposed in diabetic nephrology linking the uncoupling of VEGF and NO to the pathology (272). The protective effect of angiotensin-converting enzyme inhibitor on DR is associated with a decreased VEGF/PEDF ratio, which involves the mitochondria-ROS pathway through PPARγ-mediated changes of UCP-2 (273). The effects of aspirin and salicylates in preventing early retinal vascular lesions in diabetic rats were shown to inhibit NF-κB activation as well as retinal expression of ICAM, iNOS, COX-2 that are regulated by NF-κB-dependent transcription. Neither salicylates (274) nor aminoguanidine (275) treatment in STZ-induced diabetic rats reduced the formation of early nonenzymatic glycation products, although they both arrested other early retinal vascular abnormalities (261). Aminoguanidine treatment was able to inhibit retinal microvascular disease development in STZ-induced diabetic dogs and rats, but not in galactose-fed rats, probably due to the different biochemical pathways leading to the pathologies. The elucidation of the full picture of DR pathogenesis is dependent on combining in vitro and in vivo diabetic models developed and to be developed in the future.

5. Conclusions

Animal models have proven crucial in the progress made into the understanding of diabetic retinopathy, an increasingly important disease in light of the gathering diabetes epidemic. Clearly, diabetic retinopathy progresses along multiple pathways toward the loss of vision. Most of the current knowledge comes from work with animals (rodents) made diabetic with STZ and increasingly with combination of chemically made or spontaneous diabetic and genetically modified animals. Much more research using animals is required to identify the pivotal event(s) of the crucial gateway process and to identify reliable predictive early clinical indicators. Research with animal models will spawn insights into the design and testing of putative therapies to stop diabetic retinopathy early or reverse progress by "erasing" the "diabetes memory." Experimental diabetic animals remain seminal in the scientific battle against diabetic retinopathy.

References

1. (2007) Diagnosis and classification of diabetes mellitus. Diabetes Care 30(1):S42–S47
2. Narayan KM, Boyle JP, Geiss LS, Saaddine JB, Thompson TJ (2006) Impact of recent increase in incidence on future diabetes burden: U.S., 2005–2050. Diabetes Care 29(9):2114–2116
3. Saaddine JB, Honeycutt AA, Narayan KM, Zhang X, Klein R, Boyle JP (2008) Projection of diabetic retinopathy and other major eye diseases among people with diabetes mellitus: United States, 2005–2050. Arch Ophthalmol 126(12):1740–1747
4. Klein R, Davis MD, Moss SE, Klein BE, DeMets DL (1985) The Wisconsin epidemiologic study of diabetic retinopathy. A comparison of retinopathy in younger and older onset diabetic persons. Adv Exp Med Biol 189:321–335
5. Frank RN (2006) Etiologic mechanisms in diabetic retinopathy. In: Schachat AP, Ryan SJ (eds) Retina (Philadelphia, PA), 4th edn. Elsevier Inc, Philadelphia, pp 1241–1270
6. (1995) The effect of intensive diabetes treatment on the progression of diabetic retinopathy in insulin-dependent diabetes mellitus. The Diabetes Control and Complications Trial. Arch Ophthalmol 113(1):36–51
7. (1993) The effect of intensive treatment of diabetes on the development and progression of long-term complications in insulin-dependent diabetes mellitus. The Diabetes Control and Complications Trial Research Group. N Engl J Med 29(14):977–986
8. (1995) The relationship of glycemic exposure (HbA1c) to the risk of development and progression of retinopathy in the diabetes control and complications trial. Diabetes 44(8): 968–983
9. (1996) The absence of a glycemic threshold for the development of long-term complications: the perspective of the Diabetes Control and Complications Trial. Diabetes 45(10):1289–1298
10. Klein R, Klein BE, Moss SE (1996) Relation of glycemic control to diabetic microvascular complications in diabetes mellitus. Ann Intern Med 124(1 Pt 2):90–96
11. (1998) Effect of intensive blood-glucose control with metformin on complications in overweight patients with type 2 diabetes (UKPDS 34). UK Prospective Diabetes Study (UKPDS) Group. Lancet 352(9131): 854–865
12. (1998) Intensive blood-glucose control with sulphonylureas or insulin compared with conventional treatment and risk of complications in patients with type 2 diabetes (UKPDS 33). UK Prospective Diabetes Study (UKPDS) Group. Lancet 352(9131):837–853
13. Kern TS, Mohr S (2007) Nonproliferative stages of diabetic retinopathy: animal models of pathogenesis. In: Joussen AM, Gardner TW, Kirchhof B, Ryan SJ (eds) Retinal vascular disease. Springer-Verlag, Heidelberg, pp 303–316

14. Kern TS, Barber AJ (2008) Retinal ganglion cells in diabetes. J Physiol 586(Pt 18): 4401–4408
15. Kanwar M, Kowluru RA (2009) Role of glyceraldehyde 3-phosphate dehydrogenase in the development and progression of diabetic retinopathy. Diabetes 58(1): 227–234
16. Kern TS (2008) In vivo models of diabetic retinopathy. In: Duh EJ (ed) Contemporary diabetes: diabetic retinopathy. Humana Press, Totowa, NJ, pp 137–156
17. Nakhooda AF, Like AA, Chappel CI, Murray FT, Marliss EB (1977) The spontaneously diabetic Wistar rat. Metabolic and morphologic studies. Diabetes 26(2):100–112
18. Sima AA, Garcia-Salinas R, Basu PK (1983) The BB Wistar rat: an experimental model for the study of diabetic retinopathy. Metabolism 32(7 Suppl. 1):136–140
19. Sima AA, Chakrabarti S, Garcia-Salinas R, Basu PK (1985) The BB-rat – an authentic model of human diabetic retinopathy. Curr Eye Res 4(10):1087–1092
20. Shinohara M, Masuyama T, Shoda T et al (2000) A new spontaneously diabetic non-obese Torii rat strain with severe ocular complications. Int J Exp Diabetes Res 1(2): 89–100
21. Yamada H, Yamada E, Higuchi A, Matsumura M (2005) Retinal neovascularisation without ischaemia in the spontaneously diabetic Torii rat. Diabetologia 48(8):1663–1668
22. Kakehashi A, Saito Y, Mori K et al (2006) Characteristics of diabetic retinopathy in SDT rats. Diabetes Metab Res Rev 22(6):455–461
23. Matsuoka M, Ogata N, Minamino K, Matsumura M (2007) Leukostasis and pigment epithelium-derived factor in rat models of diabetic retinopathy. Mol Vis 13: 1058–1065
24. Matsuoka M, Ogata N, Minamino K, Higuchi A, Matsumura M (2006) High levels of pigment epithelium-derived factor in the retina of a rat model of type 2 diabetes. Exp Eye Res 82(1):172–178
25. Sasase T, Ohta T, Ogawa N et al (2006) Preventive effects of glycaemic control on ocular complications of Spontaneously Diabetic Torii rat. Diabetes Obes Metab 8(5):501–507
26. Ideno J, Mizukami H, Kakehashi A et al (2007) Prevention of diabetic retinopathy by intraocular soluble flt-1 gene transfer in a spontaneously diabetic rat model. Int J Mol Med 19(1):75–79
27. Miyamura N, Amemiya T (1998) Lens and retinal changes in the WBN/Kob rat (spontaneously diabetic strain). Electron-microscopic study. Ophthalmic Res 30(4): 221–232
28. Bhutto IA, Miyamura N, Amemiya T (1999) Vascular architecture of degenerated retina in WBN/Kob rats: corrosion cast and electron microscopic study. Ophthalmic Res 31(5): 367–377
29. Tsuchitani M, Saegusa T, Narama I, Nishikawa T, Gonda T (1985) A new diabetic strain of rat (WBN/Kob). Lab Anim 19(3):200–207
30. Saegusa T, Tsuchitani M, Narama I (1988) Retinal degeneration in the WBN/Kob rat. Jikken Dobutsu 37(3):333–337
31. Kiyosawa I, Aoki M, Imamura T et al (1995) Age-related changes in the retina of WBN/Kob rats – a pathological study. Exp Anim 43(5):651–661
32. Clark JB, Palmer CJ, Shaw WN (1983) The diabetic Zucker fatty rat. Proc Soc Exp Biol Med 173(1):68–75
33. Danis RP, Yang Y (1993) Microvascular retinopathy in the Zucker diabetic fatty rat. Invest Ophthalmol Vis Sci 34(7): 2367–2371
34. Yang YS, Danis RP, Peterson RG, Dolan PL, Wu YQ (2000) Acarbose partially inhibits microvascular retinopathy in the Zucker Diabetic Fatty rat (ZDF/Gmi-fa). J Ocul Pharmacol Ther 16(5):471–479
35. Behl Y, Krothapalli P, Desta T, Dipiazza A, Roy S, Graves DT (2008) Diabetes-enhanced tumor necrosis factor-{alpha} production promotes apoptosis and the loss of retinal microvascular cells in type 1 and type 2 models of diabetic retinopathy. Am J Pathol 172(5): 1411–1418
36. Kawano K, Hirashima T, Mori S, Saitoh Y, Kurosumi M, Natori T (1992) Spontaneous long-term hyperglycemic rat with diabetic complications. Otsuka Long-Evans Tokushima Fatty (OLETF) strain. Diabetes 41(11):1422–1428
37. Miyamura N, Bhutto IA, Amemiya T (1999) Retinal capillary changes in Otsuka Long-Evans Tokushima fatty rats (spontaneously diabetic strain). Electron-microscopic study. Ophthalmic Res 31(5):358–366
38. Lu ZY, Bhutto IA, Amemiya T (2003) Retinal changes in Otsuka long-evans Tokushima Fatty rats (spontaneously diabetic rat) – possibility of a new experimental model for diabetic retinopathy. Jpn J Ophthalmol 47(1):28–35

39. Matsuura T, Yamagishi S, Kodama Y, Shibata R, Ueda S, Narama I (2005) Otsuka Long-Evans Tokushima fatty (OLETF) rat is not a suitable animal model for the study of angiopathic diabetic retinopathy. Int J Tissue React 27(2):59–62
40. Goto Y, Suzuki K, Ono T, Sasaki M, Toyota T (1988) Development of diabetes in the non-obese NIDDM rat (GK rat). Adv Exp Med Biol 246:29–31
41. Takeo-Goto S, Doi M, Ma N, Goto R, Semba R, Uji Y (2002) Immunohistochemical localization of amino acids in the diabetic retina of Goto-Kakizaki rats. Ophthalmic Res 34(3):139–145
42. Yatoh S, Mizutani M, Yokoo T et al (2006) Antioxidants and an inhibitor of advanced glycation ameliorate death of retinal microvascular cells in diabetic retinopathy. Diabetes Metab Res Rev 22(1):38–45
43. Agardh CD, Agardh E, Zhang H, Ostenson CG (1997) Altered endothelial/pericyte ratio in Goto-Kakizaki rat retina. J Diabetes Complications 11(3):158–162
44. Matsubara H, Kuze M, Sasoh M, Ma N, Furuta M, Uji Y (2006) Time-dependent course of electroretinograms in the spontaneous diabetic Goto-Kakizaki rat. Jpn J Ophthalmol 50(3):211–216
45. Ikeda H, Shino A, Matsuo T, Iwatsuka H, Suzuoki Z (1981) A new genetically obese-hyperglycemic rat (Wistar fatty). Diabetes 30(12):1045–1050
46. Guberski DL, Butler L, Manzi SM, Stubbs M, Like AA (1993) The BBZ/Wor rat: clinical characteristics of the diabetic syndrome. Diabetologia 36(10):912–919
47. Chakrabarti S, Sima AA (1989) Effect of aldose reductase inhibition and insulin treatment on retinal capillary basement membrane thickening in BB rats. Diabetes 38(9):1181–1186
48. Chakrabarti S, Sima AA (1987) Pathogenetic heterogeneity in retinal capillary basement membrane thickening in the diabetic BB-rat. Diabetologia 30(12):966–968
49. Robison WG Jr, McCaleb ML, Feld LG, OEt M, Laver N, Mercandetti M (1991) Degenerated intramural pericytes ('ghost cells') in the retinal capillaries of diabetic rats. Curr Eye Res 10(4):339–350
50. Ellis EA, Guberski DL, Somogyi-Mann M, Grant MB (2000) Increased H2O2, vascular endothelial growth factor and receptors in the retina of the BBZ/Wor diabetic rat. Free Radic Biol Med 28(1):91–101
51. Koletsky S (1975) Pathologic findings and laboratory data in a new strain of obese hypertensive rats. Am J Pathol 80(1):129–142
52. Huang SS, Khosrof SA, Koletsky RJ, Benetz BA, Ernsberger P (1995) Characterization of retinal vascular abnormalities in lean and obese spontaneously hypertensive rats. Clin Exp Pharmacol Physiol Suppl 22(1):S129–S131
53. Michaelis OE IV, Patrick DH, Hansen CT, Canary JJ, Werner RM, Carswell N (1986) Insulin-independent diabetes mellitus (type II). Spontaneous hypertensive/NIH-corpulent rat. Am J Pathol 123(2):398–400
54. Michaelis OE IV, Ellwood KC, Judge JM, Schoene NW, Hansen CT (1984) Effect of dietary sucrose on the SHR/N-corpulent rat: a new model for insulin-independent diabetes. Am J Clin Nutr 39(4):612–618
55. Matsui K, Ohta T, Oda T et al (2008) Diabetes-associated complications in Spontaneously Diabetic Torii fatty rats. Exp Anim 57(2):111–121
56. Feit-Leichman RA, Kinouchi R, Takeda M et al (2005) Vascular damage in a mouse model of diabetic retinopathy: relation to neuronal and glial changes. Invest Ophthalmol Vis Sci 46(11):4281–4287
57. Tamura H, Miyamoto K, Kiryu J et al (2005) Intravitreal injection of corticosteroid attenuates leukostasis and vascular leakage in experimental diabetic retina. Invest Ophthalmol Vis Sci 46(4):1440–1444
58. Barber AJ, Antonetti DA, Kern TS et al (2005) The Ins2Akita mouse as a model of early retinal complications in diabetes. Invest Ophthalmol Vis Sci 46(6):2210–2218
59. Hunt CE, Lindsey JR, Walkley SU (1976) Animal models of diabetes and obesity, including the PBB/Ld mouse. Fed Proc 35(5):1206–1217
60. Herberg L, Major E, Hennigs U, Gruneklee D, Freytag G, Gries FA (1970) Differences in the development of the obese-hyperglycemic syndrome in obob and NZO mice. Diabetologia 6(3):292–299
61. Larabi Y, Dahmani Y, Gernigon T, Nguyen-Legros J (1991) Tyrosine hydroxylase immunoreactivity in the retina of the diabetic sand rat *Psammomys obesus*. J Hirnforsch 32(4):525–531
62. Larabi Y, Nguyen-Legros J (1991) Morphology, density and distribution of tyrosine hydroxylase immunoreactive cells in the retina in the gerbil *Meriones shawi*. Relationships with horizontal cells. J Hirnforsch 32(3):387–395
63. Strasser H (1968) A breeding program for spontaneously diabetic experimental animals:

Psammomys obesus (sand rat) and *Acomys cahirinus* (spiny mouse). Lab Anim Care 18(3):328–338

64. Makino S, Kunimoto K, Muraoka Y, Mizushima Y, Katagiri K, Tochino Y (1980) Breeding of a non-obese, diabetic strain of mice. Jikken Dobutsu 29(1):1–13
65. Coleman DL (1982) Diabetes-obesity syndromes in mice. Diabetes 31(Suppl 1 Pt 2): 1–6
66. Beauchemin ML, Leuenberger PM, Babel J (1975) Retinal capillary basement membrane thickness in spiny mice (*Acomys cahirinus*) with induced and spontaneous diabetes. Invest Ophthalmol 14(7):560–562
67. Ning X, Baoyu Q, Yuzhen L, Shuli S, Reed E, Li QQ (2004) Neuro-optic cell apoptosis and microangiopathy in KKAY mouse retina. Int J Mol Med 13(1):87–92
68. Midena E, Segato T, Radin S et al (1989) Studies on the retina of the diabetic db/db mouse. I. Endothelial cell-pericyte ratio. Ophthalmic Res 21(2):106–111
69. Clements RS Jr, Robison WG Jr, Cohen MP (1998) Anti-glycated albumin therapy ameliorates early retinal microvascular pathology in db/db mice. J Diabetes Complications 12(1):28–33
70. Cheung AK, Fung MK, Lo AC et al (2005) Aldose reductase deficiency prevents diabetes-induced blood-retinal barrier breakdown, apoptosis, and glial reactivation in the retina of db/db mice. Diabetes 54(11): 3119–3125
71. Engerman RL, Kern TS (1984) Experimental galactosemia produces diabetic-like retinopathy. Diabetes 33(1):97–100
72. Engerman RL, Kern TS (1986) Hyperglycemia as a cause of diabetic retinopathy. Metabolism 35(4 Suppl. 1):20–23
73. Kern TS, Tang J, Mizutani M et al (2000) Response of capillary cell death to aminoguanidine predicts the development of retinopathy: comparison of diabetes and galactosemia. Invest Ophthalmol Vis Sci 41(12):3972–3978
74. Robison WG Jr, Jacot JL, Glover JP, Basso MD, Hohman TC (1998) Diabetic-like retinopathy: early and late intervention therapies in galactose-fed rats. Invest Ophthalmol Vis Sci 39(10):1933–1941
75. Berkowitz BA, Ito Y, Kern TS, McDonald C, Hawkins R (2001) Correction of early subnormal superior hemiretinal DeltaPO(2) predicts therapeutic efficacy in experimental diabetic retinopathy. Invest Ophthalmol Vis Sci 42(12):2964–2969
76. Kern TS, Engerman RL (1995) Galactose-induced retinal microangiopathy in rats. Invest Ophthalmol Vis Sci 36(2):490–496
77. Kern TS, Engerman RL (1996) A mouse model of diabetic retinopathy. Arch Ophthalmol 114(8):986–990
78. Vlassara H (1994) Recent progress on the biologic and clinical significance of advanced glycosylation end products. J Lab Clin Med 124(1):19–30
79. Zimmerman GA, Meistrell M III, Bloom O et al (1995) Neurotoxicity of advanced glycation endproducts during focal stroke and neuroprotective effects of aminoguanidine. Proc Natl Acad Sci USA 92(9):3744–3748
80. Giardino I, Fard AK, Hatchell DL, Brownlee M (1998) Aminoguanidine inhibits reactive oxygen species formation, lipid peroxidation, and oxidant-induced apoptosis. Diabetes 47(7):1114–1120
81. Kowluru RA, Engerman RL, Kern TS (2000) Abnormalities of retinal metabolism in diabetes or experimental galactosemia VIII. Prevention by aminoguanidine. Curr Eye Res 21(4):814–819
82. Du Y, Sarthy VP, Kern TS (2004) Interaction between NO and COX pathways in retinal cells exposed to elevated glucose and retina of diabetic rats. Am J Physiol Regul Integr Comp Physiol 287(4):R735–R741
83. Frank RN, Amin R, Kennedy A, Hohman TC (1997) An aldose reductase inhibitor and aminoguanidine prevent vascular endothelial growth factor expression in rats with long-term galactosemia. Arch Ophthalmol 115(8):1036–1047
84. Rakieten N, Rakieten ML, Nadkarni MR (1963) Studies on the diabetogenic action of streptozotocin (NSC-37917). Cancer Chemother Rep 29:91–98
85. Herr RR, Jahnke JK, Argoudelis AD (1967) The structure of streptozotocin. J Am Chem Soc 89(18):4808–4809
86. Elsner M, Guldbakke B, Tiedge M, Munday R, Lenzen S (2000) Relative importance of transport and alkylation for pancreatic beta-cell toxicity of streptozotocin. Diabetologia 43(12):1528–1533
87. Ledoux SP, Wilson GL (1984) Effects of streptozotocin on a clonal isolate of rat insulinoma cells. Biochim Biophys Acta 804(4):387–392
88. Yamamoto H, Uchigata Y, Okamoto H (1981) Streptozotocin and alloxan induce DNA strand breaks and poly(ADP-ribose) synthetase in pancreatic islets. Nature 294(5838):284–286

89. Uchigata Y, Yamamoto H, Kawamura A, Okamoto H (1982) Protection by superoxide dismutase, catalase, and poly(ADP-ribose) synthetase inhibitors against alloxan- and streptozotocin-induced islet DNA strand breaks and against the inhibition of proinsulin synthesis. J Biol Chem 257(11):6084–6088
90. Gandy SE, Buse MG, Crouch RK (1982) Protective role of superoxide dismutase against diabetogenic drugs. J Clin Invest 70(3):650–658
91. Papaccio G, Pisanti FA, Frascatore S (1986) Acetyl-homocysteine-thiolactone-induced increase of superoxide dismutase counteracts the effect of subdiabetogenic doses of streptozocin. Diabetes 35(4):470–474
92. Kwon NS, Lee SH, Choi CS, Kho T, Lee HS (1994) Nitric oxide generation from streptozotocin. FASEB J 8(8):529–533
93. Turk J, Corbett JA, Ramanadham S, Bohrer A, McDaniel ML (1993) Biochemical evidence for nitric oxide formation from streptozotocin in isolated pancreatic islets. Biochem Biophys Res Commun 197(3):1458–1464
94. Tanaka Y, Shimizu H, Sato N, Mori M, Shimomura Y (1995) Involvement of spontaneous nitric oxide production in the diabetogenic action of streptozotocin. Pharmacology 50(2):69–73
95. Junod A, Lambert AE, Stauffacher W, Renold AE (1969) Diabetogenic action of streptozotocin: relationship of dose to metabolic response. J Clin Invest 48(11):2129–2139
96. Ar'Rajab A, Ahren B (1993) Long-term diabetogenic effect of streptozotocin in rats. Pancreas 8(1):50–57
97. Giorgino F, Chen JH, Smith RJ (1992) Changes in tyrosine phosphorylation of insulin receptors and a 170, 000 molecular weight nonreceptor protein in vivo in skeletal muscle of streptozotocin-induced diabetic rats: effects of insulin and glucose. Endocrinology 130(3):1433–1444
98. Whiting PH, Bowley M, Sturton RG, Pritchard PH, Brindley DN, Hawthorne JN (1977) The effect of chronic diabetes, induced by streptozotocin, on the activities of some enzymes of glycerolipid synthesis in rat liver. Biochem J 168(2):147–153
99. Abiko T, Abiko A, Clermont AC et al (2003) Characterization of retinal leukostasis and hemodynamics in insulin resistance and diabetes: role of oxidants and protein kinase-C activation. Diabetes 52(3): 829–837
100. Antonetti DA, Barber AJ, Khin S, Lieth E, Tarbell JM, Gardner TW (1998) Vascular permeability in experimental diabetes is associated with reduced endothelial occludin content: vascular endothelial growth factor decreases occludin in retinal endothelial cells. Penn State Retina Research Group. Diabetes 47(12):1953–1959
101. Kern TS, Miller CM, Du Y et al (2007) Topical administration of nepafenac inhibits diabetes-induced retinal microvascular disease and underlying abnormalities of retinal metabolism and physiology. Diabetes 56(2):373–379
102. Nakhoda A, Wong HA (1979) The induction of diabetes in rats by intramuscular administration of streptozotocin. Experientia 35(12):1679–1680
103. Bar-On H, Roheim PS, Eder HA (1976) Hyperlipoproteinemia in streptozotocin-treated rats. Diabetes 25(6):509–515
104. Vinores SA, Gadegbeku C, Campochiaro PA, Green WR (1989) Immunohistochemical localization of blood-retinal barrier breakdown in human diabetics. Am J Pathol 134(2):231–235
105. McLeod DS, Lefer DJ, Merges C, Lutty GA (1995) Enhanced expression of intracellular adhesion molecule-1 and P-selectin in the diabetic human retina and choroid. Am J Pathol 147(3):642–653
106. Aiello LP, Avery RL, Arrigg PG et al (1994) Vascular endothelial growth factor in ocular fluid of patients with diabetic retinopathy and other retinal disorders. N Engl J Med 331(22):1480–1487
107. Makino N, Maeda T, Sugano M, Satoh S, Watanabe R, Abe N (2005) High serum TNF-alpha level in Type 2 diabetic patients with microangiopathy is associated with eNOS down-regulation and apoptosis in endothelial cells. J Diabetes Complications 19(6):347–355
108. Matsuoka M, Ogata N, Minamino K, Matsumura M (2006) Expression of pigment epithelium-derived factor and vascular endothelial growth factor in fibrovascular membranes from patients with proliferative diabetic retinopathy. Jpn J Ophthalmol 50(2):116–120
109. Nowak M, Wielkoszynski T, Marek B et al (2008) Blood serum levels of vascular cell adhesion molecule (sVCAM-1), intercellular adhesion molecule (sICAM-1) and endothelial leucocyte adhesion molecule-1 (ELAM-1) in diabetic retinopathy. Clin Exp Med 8(3):159–164

110. Yuuki T, Kanda T, Kimura Y et al (2001) Inflammatory cytokines in vitreous fluid and serum of patients with diabetic vitreoretinopathy. J Diabetes Complications 15(5):257–259
111. Bolton SJ, Anthony DC, Perry VH (1998) Loss of the tight junction proteins occludin and zonula occludens-1 from cerebral vascular endothelium during neutrophil-induced blood-brain barrier breakdown in vivo. Neuroscience 86(4):1245–1257
112. Del Maschio A, Zanetti A, Corada M et al (1996) Polymorphonuclear leukocyte adhesion triggers the disorganization of endothelial cell-to-cell adherens junctions. J Cell Biol 135(2):497–510
113. Kurose I, Anderson DC, Miyasaka M et al (1994) Molecular determinants of reperfusion-induced leukocyte adhesion and vascular protein leakage. Circ Res 74(2):336–343
114. Leal EC, Manivannan A, Hosoya KI et al (2007) Inducible nitric oxide synthase isoform is a key mediator of leukostasis and blood-retinal barrier breakdown in diabetic retinopathy. Invest Ophthalmol Vis Sci 48(11):5257–5265
115. Joussen AM, Murata T, Tsujikawa A, Kirchhof B, Bursell SE, Adamis AP (2001) Leukocyte-mediated endothelial cell injury and death in the diabetic retina. Am J Pathol 158(1):147–152
116. Lu M, Perez VL, Ma N et al (1999) VEGF increases retinal vascular ICAM-1 expression in vivo. Invest Ophthalmol Vis Sci 40(8):1808–1812
117. Miyamoto K, Khosrof S, Bursell SE et al (1999) Prevention of leukostasis and vascular leakage in streptozotocin-induced diabetic retinopathy via intercellular adhesion molecule-1 inhibition. Proc Natl Acad Sci USA 96(19):10836–10841
118. Ishida S, Usui T, Yamashiro K et al (2003) VEGF164 is proinflammatory in the diabetic retina. Invest Ophthalmol Vis Sci 44(5):2155–2162
119. Al-Shabrawey M, Rojas M, Sanders T et al (2008) Role of NADPH oxidase in retinal vascular inflammation. Invest Ophthalmol Vis Sci 49(7):3239–3244
120. Gubitosi-Klug RA, Talahalli R, Du Y, Nadler JL, Kern TS (2008) 5-Lipoxygenase, but not 12/15-lipoxygenase, contributes to degeneration of retinal capillaries in a mouse model of diabetic retinopathy. Diabetes 57(5):1387–1393
121. Chen P, Scicli GM, Guo M et al (2006) Role of angiotensin II in retinal leukostasis in the diabetic rat. Exp Eye Res 83(5):1041–1051
122. Yamagishi S, Matsui T, Nakamura K, Takeuchi M, Imaizumi T (2006) Pigment epithelium-derived factor (PEDF) prevents diabetes- or advanced glycation end products (AGE)-elicited retinal leukostasis. Microvasc Res 72(1–2):86–90
123. Muranaka K, Yanagi Y, Tamaki Y et al (2006) Effects of peroxisome proliferator-activated receptor gamma and its ligand on blood-retinal barrier in a streptozotocin-induced diabetic model. Invest Ophthalmol Vis Sci 47(10):4547–4552
124. Ono R, Kakehashi A, Ito Y et al (2006) Effect of topical nipradilol on retinal microvascular leukocyte adhesion in diabetic rats. Ophthalmic Res 38(5):270–273
125. Miyamoto K, Khosrof S, Bursell SE et al (2000) Vascular endothelial growth factor (VEGF)-induced retinal vascular permeability is mediated by intercellular adhesion molecule-1 (ICAM-1). Am J Pathol 156(5): 1733–1739
126. Sun W, Oates PJ, Coutcher JB, Gerhardinger C, Lorenzi M (2006) A selective aldose reductase inhibitor of a new structural class prevents or reverses early retinal abnormalities in experimental diabetic retinopathy. Diabetes 55(10):2757–2762
127. Joussen AM, Poulaki V, Le ML et al (2004) A central role for inflammation in the pathogenesis of diabetic retinopathy. FASEB J 18(12):1450–1452
128. Joussen AM, Poulaki V, Qin W et al (2002) Retinal vascular endothelial growth factor induces intercellular adhesion molecule-1 and endothelial nitric oxide synthase expression and initiates early diabetic retinal leukocyte adhesion in vivo. Am J Pathol 160(2):501–509
129. Kowluru RA, Menon B, Gierhart DL (2008) Beneficial effect of zeaxanthin on retinal metabolic abnormalities in diabetic rats. Invest Ophthalmol Vis Sci 49(4):1645–1651
130. Cukiernik M, Hileeto D, Evans T, Mukherjee S, Downey D, Chakrabarti S (2004) Vascular endothelial growth factor in diabetes induced early retinal abnormalities. Diabetes Res Clin Pract 65(3):197–208
131. Mrudula T, Suryanarayana P, Srinivas PN, Reddy GB (2007) Effect of curcumin on hyperglycemia-induced vascular endothelial growth factor expression in streptozotocin-induced diabetic rat retina. Biochem Biophys Res Commun 361(2):528–532
132. Murata T, Nakagawa K, Khalil A, Ishibashi T, Inomata H, Sueishi K (1996) The relation between expression of vascular endothelial growth factor and breakdown of the

blood-retinal barrier in diabetic rat retinas. Lab Invest 74(4):819–825

133. Qaum T, Xu QW, Joussen AM et al (2001) VEGF-initiated blood-retinal barrier breakdown in early diabetes. Invest Ophthalmol Vis Sci 42(10):2408–2413
134. El-Remessy AB, Behzadian MA, Abou-Mohamed G, Franklin T, Caldwell RW, Caldwell RB (2003) Experimental diabetes causes breakdown of the blood-retina barrier by a mechanism involving tyrosine nitration and increases in expression of vascular endothelial growth factor and urokinase plasminogen activator receptor. Am J Pathol 162(6):1995–2004
135. Zhang SX, Ma JX, Sima J et al (2005) Genetic difference in susceptibility to the blood-retina barrier breakdown in diabetes and oxygen-induced retinopathy. Am J Pathol 166(1):313–321
136. Xu Q, Qaum T, Adamis AP (2001) Sensitive blood-retinal barrier breakdown quantitation using Evans blue. Invest Ophthalmol Vis Sci 42(3):789–794
137. Warboys CM, Toh HB, Fraser PA (2009) Role of NADPH oxidase in retinal microvascular permeability increase by RAGE activation. Invest Ophthalmol Vis Sci 50(3):1319–1328
138. Skondra D, Noda K, Almulki L et al (2008) Characterization of azurocidin as a permeability factor in the retina: involvement in VEGF-induced and early diabetic blood-retinal barrier breakdown. Invest Ophthalmol Vis Sci 49(2):726–731
139. Erickson KK, Sundstrom JM, Antonetti DA (2007) Vascular permeability in ocular disease and the role of tight junctions. Angiogenesis 10(2):103–117
140. Antonetti DA, Barber AJ, Hollinger LA, Wolpert EB, Gardner TW (1999) Vascular endothelial growth factor induces rapid phosphorylation of tight junction proteins occludin and zonula occluden 1. A potential mechanism for vascular permeability in diabetic retinopathy and tumors. J Biol Chem 274(33):23463–23467
141. Kaur C, Foulds WS, Ling EA (2008) Blood-retinal barrier in hypoxic ischaemic conditions: basic concepts, clinical features and management. Prog Retin Eye Res 27(6):622–647
142. Sundstrom JM, Tash BR, Murakami T et al (2009) Identification and analysis of occludin phosphosites: a combined mass spectrometry and bioinformatics approach. J Proteome Res 8(2):808–817
143. Bloodworth JM Jr (1962) Diabetic retinopathy. Diabetes 11:1–22
144. Eddleston M, Mucke L (1993) Molecular profile of reactive astrocytes – implications for their role in neurologic disease. Neuroscience 54(1):15–36
145. Mizutani M, Gerhardinger C, Lorenzi M (1998) Muller cell changes in human diabetic retinopathy. Diabetes 47(3):445–449
146. Lieth E, Barber AJ, Xu B et al (1998) Glial reactivity and impaired glutamate metabolism in short-term experimental diabetic retinopathy. Penn State Retina Research Group. Diabetes 47(5):815–820
147. Rungger-Brandle E, Dosso AA, Leuenberger PM (2000) Glial reactivity, an early feature of diabetic retinopathy. Invest Ophthalmol Vis Sci 41(7):1971–1980
148. Bocker-Meffert S, Rosenstiel P, Rohl C et al (2002) Erythropoietin and VEGF promote neural outgrowth from retinal explants in postnatal rats. Invest Ophthalmol Vis Sci 43(6):2021–2026
149. Zeng K, Xu H, Mi M et al (2009) Dietary taurine supplementation prevents glial alterations in retina of diabetic rats. Neurochem Res 34(2):244–254
150. Newman E, Reichenbach A (1996) The Muller cell: a functional element of the retina. Trends Neurosci 19(8):307–312
151. Allen NJ, Barres BA (2009) Neuroscience: glia – more than just brain glue. Nature 457(7230):675–677
152. Romano C, Price MT, Olney JW (1995) Delayed excitotoxic neurodegeneration induced by excitatory amino acid agonists in isolated retina. J Neurochem 65(1):59–67
153. Barnett NL, Pow DV, Robinson SR (2000) Inhibition of Muller cell glutamine synthetase rapidly impairs the retinal response to light. Glia 30(1):64–73
154. Ambati J, Chalam KV, Chawla DK et al (1997) Elevated gamma-aminobutyric acid, glutamate, and vascular endothelial growth factor levels in the vitreous of patients with proliferative diabetic retinopathy. Arch Ophthalmol 115(9):1161–1166
155. Dyer MA, Cepko CL (2000) Control of Muller glial cell proliferation and activation following retinal injury. Nat Neurosci 3(9):873–880
156. Brooks SE, Gu XL, Kaufmann PM, Marcus DM, Caldwell RB (1998) Modulation of VEGF production by pH and glucose in retinal Muller cells. Curr Eye Res 17(9): 875–882

157. Mizutani M, Kern TS, Lorenzi M (1996) Accelerated death of retinal microvascular cells in human and experimental diabetic retinopathy. J Clin Invest 97(12): 2883–2890
158. Barber AJ, Lieth E, Khin SA, Antonetti DA, Buchanan AG, Gardner TW (1998) Neural apoptosis in the retina during experimental and human diabetes – early onset and effect of insulin. J Clin Invest 102(4):783–791
159. Adams JM, Cory S (1998) The Bcl-2 protein family: arbiters of cell survival. Science 281(5381):1322–1326
160. Podesta F, Romeo G, Liu WH et al (2000) Bax is increased in the retina of diabetic subjects and is associated with pericyte apoptosis in vivo and in vitro. Am J Pathol 156(3):1025–1032
161. Oshitari T, Dezawa M, Okada S et al (2002) The role of c-fos in cell death and regeneration of retinal ganglion cells. Invest Ophthalmol Vis Sci 43(7):2442–2449
162. Levin LA, Schlamp CL, Spieldoch RL, Geszvain KM, Nickells RW (1997) Identification of the bcl-2 family of genes in the rat retina. Invest Ophthalmol Vis Sci 38(12):2545–2553
163. Oshitari T, Roy S (2005) Diabetes: a potential enhancer of retinal injury in rat retinas. Neurosci Lett 390(1):25–30
164. Oshitari T, Yamamoto S, Hata N, Roy S (2008) Mitochondria- and caspase-dependent cell death pathway involved in neuronal degeneration in diabetic retinopathy. Br J Ophthalmol 92(4):552–556
165. Mohr S, Xi X, Tang J, Kern TS (2002) Caspase activation in retinas of diabetic and galactosentic mice and diabetic patients. Diabetes 51(4):1172–1179
166. Caballero S, Sengupta N, Afzal A et al (2007) Ischemic vascular damage can be repaired by healthy, but not diabetic, endothelial progenitor cells. Diabetes 56(4):960–967
167. Martin PM, Roon P, Van Ells TK, Ganapathy V, Smith SB (2004) Death of retinal neurons in streptozotocin-induced diabetic mice. Invest Ophthalmol Vis Sci 45(9): 3330–3336
168. Zhang J, Wu Y, Jin Y et al (2008) Intravitreal injection of erythropoietin protects both retinal vascular and neuronal cells in early diabetes. Invest Ophthalmol Vis Sci 49(2): 732–742
169. Park SH, Park JW, Park SJ et al (2003) Apoptotic death of photoreceptors in the streptozotocin-induced diabetic rat retina. Diabetologia 46(9):1260–1268
170. Zheng L, Du Y, Miller C et al (2007) Critical role of inducible nitric oxide synthase in degeneration of retinal capillaries in mice with streptozotocin-induced diabetes. Diabetologia 50(9):1987–1996
171. Tzekov R, Arden GB (1999) The electroretinogram in diabetic retinopathy. Surv Ophthalmol 44(1):53–60
172. Holopigian K, Seiple W, Lorenzo M, Carr R (1992) A comparison of photopic and scotopic electroretinographic changes in early diabetic retinopathy. Invest Ophthalmol Vis Sci 33(10):2773–2780
173. Palmowski AM, Sutter EE, Bearse MA, Fung W (1997) Mapping of retinal function in diabetic retinopathy using the multifocal electroretinogram. Invest Ophthalmol Vis Sci 38(12):2586–2596
174. Shimada Y, Li Y, Bearse MA Jr, Sutter EE, Fung W (2001) Assessment of early retinal changes in diabetes using a new multifocal ERG protocol. Br J Ophthalmol 85(4): 414–419
175. Fortune B, Schneck ME, Adams AJ (1999) Multifocal electroretinogram delays reveal local retinal dysfunction in early diabetic retinopathy. Invest Ophthalmol Vis Sci 40(11):2638–2651
176. Kizawa J, Machida S, Kobayashi T, Gotoh Y, Kurosaka D (2006) Changes of oscillatory potentials and photopic negative response in patients with early diabetic retinopathy. Jpn J Ophthalmol 50(4):367–373
177. Ghirlanda G, Di Leo MA, Caputo S et al (1991) Detection of inner retina dysfunction by steady-state focal electroretinogram pattern and flicker in early IDDM. Diabetes 40(9):1122–1127
178. Hood DC (2000) Assessing retinal function with the multifocal technique. Prog Retin Eye Res 19(5):607–646
179. Ng JS, Bearse MA Jr, Schneck ME, Barez S, Adams AJ (2008) Local diabetic retinopathy prediction by multifocal ERG delays over 3 years. Prog Retin Eye Res 49(4):1622–1628
180. Bearse MA Jr, Adams AJ, Han Y et al (2006) A multifocal electroretinogram model predicting the development of diabetic retinopathy. Prog Retin Eye Res 25(5):425–448
181. Simonsen SE (1980) The value of the oscillatory potential in selecting juvenile diabetics at risk of developing proliferative retinopathy. Acta ophthalmol 58(6):865–878
182. Frost-Larsen K, Larsen HW, Simonsen SE (1980) Oscillatory potential and nyctometry in insulin-dependent diabetics. Acta ophthalmol 58(6):879–888

183. Parisi V, Uccioli L, Monticone G et al (1997) Electrophysiological assessment of visual function in IDDM patients. Electroencephalogr Clin Neurophysiol 104(2):171–179
184. Li Q, Zemel E, Miller B, Perlman I (2002) Early retinal damage in experimental diabetes: electroretinographical and morphological observations. Exp Eye Res 74(5):615–625
185. Kusari J, Zhou S, Padillo E, Clarke KG, Gil DW (2007) Effect of memantine on neuroretinal function and retinal vascular changes of streptozotocin-induced diabetic rats. Invest Ophthalmol Vis Sci 48(11): 5152–5159
186. Hancock HA, Kraft TW (2004) Oscillatory potential analysis and ERGs of normal and diabetic rats. Invest Ophthalmol Vis Sci 45(3):1002–1008
187. Ramsey DJ, Ripps H, Qian H (2006) An electrophysiological study of retinal function in the diabetic female rat. Invest Ophthalmol Vis Sci 47(11):5116–5124
188. Layton CJ, Chidlow G, Casson RJ, Wood JP, Graham M, Osborne NN (2005) Monocarboxylate transporter expression remains unchanged during the development of diabetic retinal neuropathy in the rat. Invest Ophthalmol Vis Sci 46(8):2878–2885
189. Aizu Y, Oyanagi K, Hu J, Nakagawa H (2002) Degeneration of retinal neuronal processes and pigment epithelium in the early stage of the streptozotocin-diabetic rats. Neuropathology 22(3):161–170
190. Kohzaki K, Vingrys AJ, Bui BV (2008) Early inner retinal dysfunction in streptozotocin-induced diabetic rats. Invest Ophthalmol Vis Sci 49(8):3595–3604
191. Layton CJ, Safa R, Osborne NN (2007) Oscillatory potentials and the b-Wave: partial masking and interdependence in dark adaptation and diabetes in the rat. Graefes Arch Clin Exp Ophthalmol 245(9):1335–1345
192. Phipps JA, Fletcher EL, Vingrys AJ (2004) Paired-flash identification of rod and cone dysfunction in the diabetic rat. Invest Ophthalmol Vis Sci 45(12):4592–4600
193. Phipps JA, Yee P, Fletcher EL, Vingrys AJ (2006) Rod photoreceptor dysfunction in diabetes: activation, deactivation, and dark adaptation. Invest Ophthalmol Vis Sci 47(7):3187–3194
194. Kern TS, Miller CM, Tang J, Du Y, Ball S, Berti-Matera L (2009) Comparison of 3 rat strains with respect to rate of development of diabetic retinopathy and tactile allodynia. Molecular Vision. submitted
195. Lorenzi M (2007) The polyol pathway as a mechanism for diabetic retinopathy: attractive, elusive, and resilient. Exp Diabetes Res 2007:61038
196. Yamagishi S, Ueda S, Matsui T, Nakamura K, Okuda S (2008) Role of advanced glycation end products (AGEs) and oxidative stress in diabetic retinopathy. Curr Pharm Des 14(10):962–968
197. Das Evcimen N, King GL (2007) The role of protein kinase C activation and the vascular complications of diabetes. Pharmacol Res 55(6):498–510
198. Antonetti DA, Barber AJ, Bronson SK et al (2006) Diabetic retinopathy: seeing beyond glucose-induced microvascular disease. Diabetes 55(9):2401–2411
199. Adamis AP, Miller JW, Bernal MT et al (1994) Increased vascular endothelial growth factor levels in the vitreous of eyes with proliferative diabetic retinopathy. Am J Ophthalmol 118(4):445–450
200. Limb GA, Chignell AH, Green W, LeRoy F, Dumonde DC (1996) Distribution of TNF alpha and its reactive vascular adhesion molecules in fibrovascular membranes of proliferative diabetic retinopathy. Br J Ophthalmol 80(2):168–173
201. Funatsu H, Yamashita H, Shimizu E, Kojima R, Hori S (2001) Relationship between vascular endothelial growth factor and interleukin-6 in diabetic retinopathy. Retina 21(5):469–477
202. Funatsu H, Yamashita H, Ikeda T, Nakanishi Y, Kitano S, Hori S (2002) Angiotensin II and vascular endothelial growth factor in the vitreous fluid of patients with diabetic macular edema and other retinal disorders. Am J Ophthalmol 133(4):537–543
203. Brooks HL Jr, Caballero S Jr, Newell CK et al (2004) Vitreous levels of vascular endothelial growth factor and stromal-derived factor 1 in patients with diabetic retinopathy and cystoid macular edema before and after intraocular injection of triamcinolone. Arch Ophthalmol 122(12): 1801–1807
204. Meleth AD, Agron E, Chan CC et al (2005) Serum inflammatory markers in diabetic retinopathy. Invest Ophthalmol Vis Sci 46(11):4295–4301
205. Patel JI, Hykin PG, Gregor ZJ, Boulton M, Cree IA (2005) Angiopoietin concentrations in diabetic retinopathy. Br J Ophthalmol 89(4):480–483

206. Watanabe D, Suzuma K, Matsui S et al (2005) Erythropoietin as a retinal angiogenic factor in proliferative diabetic retinopathy. N Engl J Med 353(8):782–792
207. Watanabe D, Suzuma K, Suzuma I et al (2005) Vitreous levels of angiopoietin 2 and vascular endothelial growth factor in patients with proliferative diabetic retinopathy. Am J Ophthalmol 139(3):476–481
208. Demircan N, Safran BG, Soylu M, Ozcan AA, Sizmaz S (2006) Determination of vitreous interleukin-1 (IL-1) and tumour necrosis factor (TNF) levels in proliferative diabetic retinopathy. Eye 20(12):1366–1369
209. Hernandez C, Fonollosa A, Garcia-Ramirez M et al (2006) Erythropoietin is expressed in the human retina and it is highly elevated in the vitreous fluid of patients with diabetic macular edema. Diabetes Care 29(9): 2028–2033
210. Esser P, Heimann K, Wiedemann P (1993) Macrophages in proliferative vitreoretinopathy and proliferative diabetic retinopathy: differentiation of subpopulations. Br J Ophthalmol 77(11):731–733
211. Canataroglu H, Varinli I, Ozcan AA, Canataroglu A, Doran F, Varinli S (2005) Interleukin (IL)-6, interleukin (IL)-8 levels and cellular composition of the vitreous humor in proliferative diabetic retinopathy, proliferative vitreoretinopathy, and traumatic proliferative vitreoretinopathy. Ocul Immunol Inflamm 13(5):375–381
212. Zeng XX, Ng YK, Ling EA (2000) Neuronal and microglial response in the retina of streptozotocin-induced diabetic rats. Vis Neurosci 17(3):463–471
213. Krady JK, Basu A, Allen CM et al (2005) Minocycline reduces proinflammatory cytokine expression, microglial activation, and caspase-3 activation in a rodent model of diabetic retinopathy. Diabetes 54(5): 1559–1565
214. Zeng HY, Green WR, Tso MO (2008) Microglial activation in human diabetic retinopathy. Arch Ophthalmol 126(2):227–232
215. Zhang J, Gerhardinger C, Lorenzi M (2002) Early complement activation and decreased levels of glycosylphosphatidylinositol-anchored complement inhibitors in human and experimental diabetic retinopathy. Diabetes 51(12):3499–3504
216. Guillot R, Bringuier AF, Porokhov B, Guillausseau PJ, Feldmann G (2001) Increased levels of soluble Fas in serum from diabetic patients with neuropathy. Diabetes Metab 27(3):315–321
217. Okamura N, Ito Y, Shibata MA, Ikeda T, Otsuki Y (2002) Fas-mediated apoptosis in human lens epithelial cells of cataracts associated with diabetic retinopathy. Med Electron Microsc 35(4):234–241
218. Joussen AM, Poulaki V, Mitsiades N et al (2003) Suppression of Fas-FasL-induced endothelial cell apoptosis prevents diabetic blood-retinal barrier breakdown in a model of streptozotocin-induced diabetes. FASEB J 17(1):76–78
219. Ferguson TA, Apte RS (2008) Angiogenesis in eye disease: immunity gained or immunity lost? Semin Immunopathol 30(2):111–119
220. Slepova OS, Gerasimenko VL, Zakharova G, Novikova-Bilak TI (2001) Comparative study of the role of cytokines in various eye diseases. 2. Diabetic retinopathy. Vestn Oftalmol 117(3):35–37
221. Zorena K, Mysliwska J, Mysliwiec M et al (2007) Serum TNF-alpha level predicts nonproliferative diabetic retinopathy in children. Mediators Inflamm 2007:92196
222. Engerman RL (1989) Pathogenesis of diabetic retinopathy. Diabetes 38(10):1203–1206
223. Esser P, Bresgen M, Fischbach R, Heimann K, Wiedemann P (1995) Intercellular adhesion molecule-1 levels in plasma and vitreous from patients with vitreoretinal disorders. Ger J Ophthalmol 4(5):269–274
224. Ben-Mahmud BM, Mann GE, Datti A, Orlacchio A, Kohner EM, Chibber R (2004) Tumor necrosis factor-alpha in diabetic plasma increases the activity of core 2 GlcNAc-T and adherence of human leukocytes to retinal endothelial cells: significance of core 2 GlcNAc-T in diabetic retinopathy. Diabetes 53(11):2968–2976
225. Ben-Mahmud BM, Chan WH, Abdulahad RM et al (2006) Clinical validation of a link between TNF-alpha and the glycosylation enzyme core 2 GlcNAc-T and the relationship of this link to diabetic retinopathy. Diabetologia 49(9):2185–2191
226. Gustavsson C, Agardh E, Bengtsson B, Agardh CD (2008) TNF-alpha is an independent serum marker for proliferative retinopathy in type 1 diabetic patients. J Diabetes Complications 22(5):309–316
227. Mysliwiec M, Balcerska A, Zorena K, Mysliwska J, Lipowski P, Raczynska K (2008) The role of vascular endothelial growth factor, tumor necrosis factor alpha and interleukin-6 in pathogenesis of diabetic retinopathy. Diabetes Res Clin Pract 79(1):141–146
228. Joussen AM, Poulaki V, Mitsiades N et al (2002) Nonsteroidal anti-inflammatory

drugs prevent early diabetic retinopathy via TNF-alpha suppression. FASEB J 16(3): 438–440

229. Wang AL, Yu AC, He QH, Zhu X, Tso MO (2007) AGEs mediated expression and secretion of TNF alpha in rat retinal microglia. Exp Eye Res 84(5):905–913
230. Limb GA, Webster L, Soomro H, Janikoun S, Shilling J (1999) Platelet expression of tumour necrosis factor-alpha (TNF-alpha), TNF receptors and intercellular adhesion molecule-1 (ICAM-1) in patients with proliferative diabetic retinopathy. Clin Exp Immunol 118(2):213–218
231. Le LM, Poulaki V, Koizumi K, Fauser S, Kirchhof B, Joussen AM (2003) Reduced histopathological alterations in long-term diabetic TNF-R deficient mice. Invest Ophthalmol Vis Sci 44:3894
232. McQuillan LP, Leung GK, Marsden PA, Kostyk SK, Kourembanas S (1994) Hypoxia inhibits expression of eNOS via transcriptional and posttranscriptional mechanisms. Am J Physiol 267(5 Pt 2):H1921–H1927
233. Flowers MA, Wang Y, Stewart RJ, Patel B, Marsden PA (1995) Reciprocal regulation of endothelin-1 and endothelial constitutive NOS in proliferating endothelial cells. Am J Physiol 269(6 Pt 2):H1988–H1997
234. Moncada S, Higgs A (1993) The L-arginine-nitric oxide pathway. N Engl J Med 329(27):2002–2012
235. Sennlaub F, Courtois Y, Goureau O (2002) Inducible nitric oxide synthase mediates retinal apoptosis in ischemic proliferative retinopathy. J Neurosci 22(10):3987–3993
236. Berkowitz BA, Luan H, Gupta RR et al (2004) Regulation of the early subnormal retinal oxygenation response in experimental diabetes by inducible nitric oxide synthase. Diabetes 53(1):173–178
237. Ellis EA, Sengupta N, Caballero S, Guthrie SM, Mames RN, Grant MB (2005) Nitric oxide synthases modulate progenitor and resident endothelial cell behavior in galactosemia. Antioxid Redox Signal 7(11–12): 1413–1422
238. Carmo A, Cunha-Vaz JG, Carvalho AP, Lopes MC (1999) L-arginine transport in retinas from streptozotocin diabetic rats: correlation with the level of IL-1 beta and NO synthase activity. Vis Res 39(23):3817–3823
239. Kern TS, Engerman RL (2001) Pharmacological inhibition of diabetic retinopathy: aminoguanidine and aspirin. Diabetes 50(7):1636–1642
240. Vincent JA, Mohr S (2007) Inhibition of caspase-1/interleukin-1beta signaling prevents degeneration of retinal capillaries in diabetes and galactosemia. Diabetes 56(1):224–230
241. Lazovic J, Basu A, Lin HW et al (2005) Neuroinflammation and both cytotoxic and vasogenic edema are reduced in interleukin-1 type 1 receptor-deficient mice conferring neuroprotection. Stroke 36(10):2226–2231
242. Romeo G, Liu WH, Asnaghi V, Kern TS, Lorenzi M (2002) Activation of nuclear factor-kappaB induced by diabetes and high glucose regulates a proapoptotic program in retinal pericytes. Diabetes 51(7):2241–2248
243. Nagai N, Izumi-Nagai K, Oike Y et al (2007) Suppression of diabetes-induced retinal inflammation by blocking the angiotensin II type 1 receptor or its downstream nuclear factor-kappaB pathway. Invest Ophthalmol Vis Sci 48(9):4342–4350
244. Sone H, Kawakami Y, Okuda Y et al (1997) Ocular vascular endothelial growth factor levels in diabetic rats are elevated before observable retinal proliferative changes. Diabetologia 40(6):726–730
245. Gerhardinger C, Brown LF, Roy S, Mizutani M, Zucker CL, Lorenzi M (1998) Expression of vascular endothelial growth factor in the human retina and in nonproliferative diabetic retinopathy. Am J Pathol 152(6): 1453–1462
246. Segawa Y, Shirao Y, Yamagishi S et al (1998) Upregulation of retinal vascular endothelial growth factor mRNAs in spontaneously diabetic rats without ophthalmoscopic retinopathy. A possible participation of advanced glycation end products in the development of the early phase of diabetic retinopathy. Ophthalmic Res 30(6):333–339
247. Buchi ER, Kurosawa A, Tso MO (1996) Retinopathy in diabetic hypertensive monkeys: a pathologic study. Graefes Arch Clin Exp Ophthalmol 234(6):388–398
248. Hofman P, Blaauwgeers HGT, Vrensen G, Schlingemann RO (2001) Role of VEGF-A in endothelial phenotypic shift in human diabetic retinopathy and VEGF-A-induced retinopathy in monkeys. Ophthalmic Res 33(3):156–162
249. Ogata N, Nishikawa M, Nishimura T, Mitsuma Y, Matsumura M (2002) Unbalanced vitreous levels of pigment epithelium-derived factor and vascular endothelial growth factor in diabetic retinopathy. Am J Ophthalmol 134(3):348–353
250. Zhang SX, Wang JJ, Gao G, Shao C, Mott R, Ma JX (2006) Pigment epithelium-derived factor (PEDF) is an endogenous antiinflammatory factor. FASEB J 20(2):323–325

251. Matsuyama K, Ogata N, Matsuoka M et al (2008) Relationship between pigment epithelium-derived factor (PEDF) and renal function in patients with diabetic retinopathy. Mol Vis 14:992–996
252. Ogata N, Tombran-Tink J, Nishikawa M et al (2001) Pigment epithelium-derived factor in the vitreous is low in diabetic retinopathy and high in rhegmatogenous retinal detachment. Am J Ophthalmol 132(3): 378–382
253. Spranger J, Osterhoff M, Reimann M et al (2001) Loss of the antiangiogenic pigment epithelium-derived factor in patients with angiogenic eye disease. Diabetes 50(12):2641–2645
254. Boehm BO, Lang G, Volpert O et al (2003) Low content of the natural ocular anti-angiogenic agent pigment epithelium-derived factor (PEDF) in aqueous humor predicts progression of diabetic retinopathy. Diabetologia 46(3):394–400
255. Hammes HP, Du X, Edelstein D et al (2003) Benfotiamine blocks three major pathways of hyperglycemic damage and prevents experimental diabetic retinopathy. Nat Med 9(3):294–299
256. Nakajima M, Cooney MJ, Tu AH et al (2001) Normalization of retinal vascular permeability in experimental diabetes with genistein. Invest Ophthalmol Vis Sci 42(9):2110–2114
257. Cao J, Majji A, Jensen T, Grebe R, Cooney MM, de Juan E (1999) Effects of genistein on retinal changes in the Zucker fatty diabetic rats. Invest Ophthalmol Vis Sci 40:S472
258. (2007) Ruboxistaurin: LY 333531. Drugs R D 8(3):193–199
259. Zheng L, Szabo C, Kern TS (2004) Poly(ADP-ribose) polymerase is involved in the development of diabetic retinopathy via regulation of nuclear factor-kappaB. Diabetes 53(11):2960–2967
260. Gills JP Jr, Anderson WB Jr (1969) Photocoagulation and local steroid-induced ocular hypertension in the treatment of diabetic retinopathy. Arch Intern Med 123(6):626–631
261. Zheng L, Howell SJ, Hatala DA, Huang K, Kern TS (2007) Salicylate-based anti-inflammatory drugs inhibit the early lesion of diabetic retinopathy. Diabetes 56(2): 337–345
262. Kowluru RA, Engerman RL, Case GL, Kern TS (2001) Retinal glutamate in diabetes and effect of antioxidants. Neurochem Int 38(5):385–390
263. Dene BA, Maritim AC, Sanders RA, Watkins JB 3 rd (2005) Effects of antioxidant treatment on normal and diabetic rat retinal enzyme activities. J Ocul Pharmacol Ther 21(1):28–35
264. Roberts R, Luan H, Berkowitz BA (2006) {alpha}-lipoic acid corrects late-phase supernormal retinal oxygenation response in experimental diabetic retinopathy. Invest Ophthalmol Vis Sci 47(9):4077–4082
265. Kowluru RA, Kanwar M, Chan PS, Zhang JP (2008) Inhibition of retinopathy and retinal metabolic abnormalities in diabetic rats with AREDS-based micronutrients. Arch Ophthalmol 126(9):1266–1272
266. Arita R, Hata Y, Nakao S et al (2009) Rho kinase inhibition by fasudil ameliorates diabetes-induced microvascular damage. Diabetes 58(1):215–226
267. Steinle JJ, Kern TS, Thomas SA, McFadyen-Ketchum LS, Smith CP (2009) Increased basement membrane thickness, pericyte ghosts, and loss of retinal thickness and cells in dopamine beta hydroxylase knockout mice. Exp Eye Res 88(6):1014–1019
268. Spyridopoulos I, Brogi E, Kearney M et al (1997) Vascular endothelial growth factor inhibits endothelial cell apoptosis induced by tumor necrosis factor-alpha: balance between growth and death signals. J Mol Cell Cardiol 29(5):1321–1330
269. Rosenkranz-Weiss P, Sessa WC, Milstien S, Kaufman S, Watson CA, Pober JS (1994) Regulation of nitric oxide synthesis by proinflammatory cytokines in human umbilical vein endothelial cells. Elevations in tetrahydrobiopterin levels enhance endothelial nitric oxide synthase specific activity. J Clin Invest 93(5):2236–2243
270. Kourembanas S, McQuillan LP, Leung GK, Faller DV (1993) Nitric oxide regulates the expression of vasoconstrictors and growth factors by vascular endothelium under both normoxia and hypoxia. J Clin Invest 92(1): 99–104
271. Greenberg DA, Jin K (2005) From angiogenesis to neuropathology. Nature 438(7070): 954–959
272. Nakagawa T (2008) Uncoupling of VEGF with NO as a mechanism for diabetic nephropathy. Diabetes Res Clin Pract 82(Suppl. 1):S67–S69
273. Zheng Z, Chen H, Ke G et al (2009) The protective effect of perindopril on diabetic retinopathy is associated with decreased VEGF/PEDF ratio: involvement of a mitochondria-ROS pathway. Diabetes 58(4): 954–964

274. Yue DK, McLennan S, Handelsman DJ, Delbridge L, Reeve T, Turtle JR (1984) The effect of salicylates on nonenzymatic glycosylation and thermal stability of collagen in diabetic rats. Diabetes 33(8):745–751
275. Oxlund H, Andreassen TT (1992) Aminoguanidine treatment reduces the increase in collagen stability of rats with experimental diabetes mellitus. Diabetologia 35(1):19–25
276. Shaw SG, Boden JP, Biecker E, Reichen J, Rothen B. (2006)Endothelin antagonism prevents diabetic retinopathy in NOD mice: a potential role of the angiogenic factor adrenomedullin. Exp Biol Med (Maywood) 231(6):1101–5

Chapter 8

Nonprimate Models for Glaucoma Retinopathy and Optic Neuropathy

Iok-Hou Pang and Abbot F. Clark

Abstract

The generation and advancement of animal models have contributed significantly to the advancement of glaucoma research. This chapter describes and summarizes major nonprimate animal models useful for the study of this disease. Rodent models, both rats and mice, have been popular for glaucoma studies, because of the relatively better-developed genetic and genomic tools and the similarity of the relevant ocular structures between human and these animals. The larger animals, e.g., rabbit, feline, canine, bovine, ovine, and porcine models, have also been successfully used and provided valuable information on various aspects of the disease. Some of the models depicted in this chapter involve a transient or chronic ocular hypertension. Others do not affect intraocular pressure, but instead address certain specific mechanisms of the disease and serve as surrogate models.

Key words: Glaucoma models, Glaucomatous retinopathy, Glaucomatous optic neuropathy, Rodent, Rabbit, Feline, Canine, Bovine, Ovine, Porcine

1. Introduction

Glaucoma is a leading cause of irreversible visual impairment and blindness, affecting approximately 60–70 million individuals worldwide. Glaucoma is actually a heterogeneous group of diseases resulting in a very similar optic neuropathy and retinopathy. The risk factors for glaucoma include age, elevated intraocular pressure (IOP), race, family history, and steroid (glucocorticoid) responsiveness. Elevated IOP is a causative risk factor for both the development and progression of glaucoma in man, and many animal models of glaucoma utilize this insult to cause glaucomatous damage to the retina and optic nerve. Glaucoma is clinically characterized by thinning of the retinal nerve fiber layer, cupping and excavation of the optic disk, and a characteristic and progressive

Iok-Hou Pang and Abbot F. Clark (eds.), *Animal Models for Retinal Diseases*, Neuromethods, vol. 46
DOI 10.1007/978-1-60761-541-5_8, © Springer Science+Business Media, LLC 2010

change in the visual field. The vision loss in glaucoma is due to a progressive axonopathy and apoptotic death to the retinal ganglion cells (RGCs), with relative sparing of other regions of the retina.

A number of pathogenic pathways have been proposed for glaucomatous damage to the retina, optic nerve head, and optic nerve (Table 1). Increased hydrostatic pressure can damage RGCs and activate optic nerve head astrocytes (ONHAs). Elevated IOP also inhibits axonal transport in the nonmyelinated RGC axons at the lamina cribrosa. Expression and transport of neurotrophins and their receptors in RGCs, RGC axons, and the optic nerve head are altered in glaucoma suggesting that neurotrophic deprivation may be involved in glaucomatous damage. Stretch of ONH cells, mimicking glaucomatous distension of the lamina cribrosa, alters gene and protein expression. There is evidence of retinal and optic nerve vascular dysregulation and ischemia in both normal tension and hypertensive glaucoma. Glaucomatous damage is also associated with the generation and release of wide variety of toxic molecules in the retina and optic nerve head, including glutamate, nitric oxide, endothelin-1 (ET-1), and tumor necrosis factor alpha (TNFα). Oxidative stress and mitochondrial dysfunction have also been reported in the glaucomatous retina and optic nerve head. Glaucoma also is associated with "activation" of optic nerve head astrocytes and microglia as well as with remodeling of the optic nerve head extracellular matrix. And finally, a role for autoimmunity and immune dysregulation has been proposed to cause glaucomatous damage to the retina.

Generation and characterization of animal models of glaucoma are important to test various theories and to better characterize the cellular and molecular mechanisms responsible for glaucomatous damage to the retina and optic nerve. These models

Table 1
Pathogenic pathways for glaucomatous damage to retina and optic nerve

• Pressure-induced damage to retinal ganglion cells and optic nerve head
• Retinal and optic nerve head ischemia
• Loss of neurotrophic support
• Remodeling of the optic nerve head (TGFβ2)
• Generation of cytotoxic molecules (glutamate, NO, TNFα, ET-1)
• Activation of optic nerve head astrocytes and microglia
• Oxidative stress and mitochondrial dysfunction
• Autoimmunity and/or immune dysregulation

are also important to test new glaucoma therapeutic interventions. However, it is very important to recognize the strengths and weaknesses of each model and to determine the relevance of each aspect of the model to human glaucoma.

2. Rodent Glaucoma Models

Among nonprimate glaucoma models, rodent models are the most popular (1). A majority of them involves transient or chronic ocular hypertension to address glaucomatous changes associated with elevated IOP, whereas others are nonpressure related, designed to address specific theoretical mechanisms of the disease progress.

2.1. Pressure Models

2.1.1. Episcleral Injection of Hypertonic Saline

Injection of hypertonic saline ([NaCl]≈1.75 M) into one of the episcleral veins in the rat induced sclerosis and blockade of the aqueous humor outflow pathway, which led to an elevation of IOP (2, 3). Typically, the majority of, but not all, eyes developed ocular hypertension within 10 days after injection. And not all affected eyes acquired the same level of IOP change. Nonetheless, once induced, the IOP increase was long-lasting with prolonged ocular hypertension for up to 200 days (2).

Ocular hypertension produced by this technique generated morphological changes in the retina, optic nerve head, and optic nerve similar to those observed in glaucoma patients. For example, in the injected eyes, selective RGC apoptosis and RGC loss were detected, which correlated positively with the severity of IOP elevation (2, 4–6). In the optic nerve head, pressure-dependent progressive cupping of the optic disk (7) and loss of nerve fibers (6) were evident. Optic nerves of injected eyes also showed axonal degeneration, as indicated by disappearance of axoplasm, axonal swelling, and occasional collapses of myelin sheath. The severity of optic nerve damage correlated well with the degree of ocular hypertension when evaluated by quantifying damaged areas in the nerve cross-sections (2, 8), by axon counts (7, 9), or by the optic nerve injury scores (3, 4, 10–13). Notably, the optic nerve damage was significantly reduced by glaucoma medications, betaxolol and apraclonidine (14).

Positive scotopic ERG threshold responses were selectively reduced in rats with mild IOP elevation induced by hypertonic saline injection, indicative of a selective RGC injury, whereas amplitudes of other ERG components, such as a-wave, b-wave, and oscillatory potential, remained unchanged (12). In contrast, sufficiently high IOP eventually decreased amplitudes of these other ERG components (12).

The hypertonic saline injection model has been successfully established and adopted by many researchers. It is a useful model for glaucoma research, producing impeded aqueous outflow and retina and optic nerve damage as a result of IOP elevation. Unfortunately, the injection does not always produce the same degree of IOP elevation. Consequently, large numbers of animals are usually necessary for most studies to allow sufficient sample sizes in each severity category. Furthermore, the episcleral vein injection procedure is technically challenging and labor intensive, thus limiting the throughput of studies.

2.1.2. Laser Photocoagulation of Outflow Pathway

Elevation of IOP by laser photocoagulation of the aqueous outflow pathway can be achieved by several methods. Ueda and colleagues demonstrated the ocular hypertensive effect of lasering the rat trabecular meshwork (TM) after injecting India ink into the anterior chamber (15). The procedure caused apoptotic RGC death (16) and reduction of axons in the optic nerve (17). Unfortunately, frequent and repeated laser treatments are required to maintain a sustained increase in IOP. Instead of the TM, direct laser photocoagulation of the episcleral and limbal veins also produced long-termed increase in rat IOP, lasting more than a year (18, 19). It caused a significant RGC loss (18) without outer retina damage (20) and a significant reduction in the amplitudes of Pattern ERG (20).

Levkovitch-Verbin and colleagues showed that translimbal laser burns directed at the TM alone or TM plus episcleral veins generated consistent IOP elevation in the rat (21), with subsequent RGC death and optic neuropathy (21–23). Nevertheless, the damage was not limited to the RGC: all retinal layers shrank in thickness (24). The lasered eyes also revealed prominent reduction in a- and b-wave amplitudes, as well as a complete disappearance of oscillatory potentials, confirming injuries on neurons other than the RGC (24).

In the mouse, laser damage to the limbus also produced ocular hypertension (25–27). Similar to the rat, the treatment caused extensive retinal damages: RGC loss, thinning of all retinal layers, optic nerve degeneration, reduction in a- and b-wave amplitudes and oscillatory potentials (26–29).

Considering the small size of the rodent eyes, laser photocoagulation is technically challenging to perform. It requires steady, trained hands to carefully deliver approximately 80 laser burns around the limbus in order to produce a meaningful increase in IOP. Furthermore, IOP fluctuations induced by this method have been reported. The IOP often could not be stably maintained over a long period of time. High IOP spikes were also observed immediately after the procedure, which may explain the outer retina damage described in some studies. Despite these disadvantages, this animal model displays ocular damages that share many characteristics of clinical glaucoma, therefore it is a useful investigative tool.

2.1.3. Intracameral Injection of Foreign Substances

In addition to laser photocoagulation, physical blockade of the outflow pathway can also be accomplished by injecting foreign substances into the anterior chamber. For example, a single intracameral injection of hyaluronic acid produced ocular hypertension lasting for more than a week in the rat (30). With weekly injections, IOP elevation reached a plateau for at least 10 weeks (30). The chronic IOP increase caused losses of cells in the ganglion cell layer and axons in the optic nerve, as well as decreases in a- and b-wave amplitudes and oscillatory potentials of the scotopic ERG (31). Similarly, weekly intracameral injections of latex microspheres with or without hydroxypropylmethylcelluose raised rat IOP that could be maintained for at least 30 weeks, together with RGC loss (32). This preliminary characterization suggests that these study models mimic certain features of glaucomatous retinopathy. However, the repeated weekly intracameral injections are undesirable: it is labor intensive and likely produces adverse health effects in many animals.

2.1.4. Occlusion of Extraocular Veins

Cauterization of extraocular veins of the rat was shown to increase IOP (33), as well as induce apoptotic RGC death, optic disk excavation, and optic nerve degeneration (34–39). However, the damage in retina induced by this technique was not specific to the RGC, as shown by the decline of ERG a- and b-waves (37, 39, 40), indicating involvement of other retinal neurons. A recent report demonstrated that ligation of extraocular veins induced mild ocular hypertension lasting for 7 months (41). Rats in the late-stage IOP increase had selective RGC apoptosis and cupping of the optic disk.

Cauterization of extraocular veins raises IOP consistently, and is a relatively easy technique to perform. However, in some studies, the IOP elevation returned to the baseline level within 2–3 months after surgery (37, 39, 42), which was likely due to the growth of new blood vessels subsequent to the disruption of blood flow. Most importantly, this technique obstructs ocular blood flow and can trigger ischemia and intraocular vascular congestion, thus producing injuries that are unrelated to ocular pressure change. Researchers need to be cautious in interpreting the study results.

2.1.5. Ocular Expression of Transgenes

Delivery of glaucoma-associated transgenes by viral vector to the rodent eyes has successfully induced ocular hypertension. For example, MYOC, encoding myocilin, was the first glaucoma gene identified (43). Transduction of mouse eyes with the human MYOC containing glaucomatous mutations significantly elevated IOP. The critical mutations were associated with exposure of a cryptic signal that mislocates the myocilin protein within cells and consequently causes deleterious TM cell functions (44). This cryptic signal is absent from mouse myocilin, which may explain the lack of effect on IOP with glaucomatous mutations in mouse Myoc.

In addition to mutant myocilin, ocular delivery of other transgenes was also effective in raising IOP. We recently showed that adenoviral gene transfer of active human transforming growth factor β2 (TGFβ2) to the rat or mouse eye caused significant ocular hypertension (45), concomitant with a reduction in outflow facility (unpublished observation). TGFβ2 is a cytokine shown to be higher in the aqueous humor of POAG patients (46), and intravitreal injection of TGFβ2 raised IOP in volunteers (47).

We have also used genomics and various methods to evaluate altered expression of other genes in glaucoma and developed glaucoma mouse models by ocular gene transfer. The Wnt-signaling pathway plays important roles in embryogenesis and morphogenesis, including development of the eye. Increased expression of the Wnt antagonist, secreted frizzled-related protein-1 (sFRP1), was found in the glaucomatous TM (48). Increased expression of sFRP1 by transduction of mouse eyes with an adenoviral vector caused elevated IOP, and the degree of IOP elevation correlated with aqueous humor levels of sFRP1 (48). Similarly, expression of serum amyloid A (SAA), an acute phase response apolipoprotein, was shown to increase in glaucomatous TM cells and tissues (49). The addition of recombinant SAA to the medium of perfusion cultured human eyes elevated IOP (49), and transduction of mouse eyes with an SAA viral expression vector also elevated IOP (unpublished observation). In the mouse eye, adenoviral overexpression of CD44, a cell surface receptor, whose ectodomain fragment is elevated in aqueous humor of POAG patients and toxic to the TM, also significantly increases IOP (50).

Clearly, ocular expression of transgenes appears to be a promising approach to develop rodent models that more closely mimic human glaucoma. For ocular transgene delivery, viral vectors, such as adenovirus, adeno-associated virus, herpes simplex virus, and lentivirus have been used extensively (51). Among these, adenovirus is very popular for transduction of the TM, because of its good tropism, relatively strong transgene expression, and minimal untoward effects. Unfortunately, however, the adenovirus-mediated transgene expression in the mouse eye was transient (52), most probably a result of an immune response generated against viral proteins. This shortcoming can be overcome by immunomodulation by treating the animals with the anti-CD40L antibody (52). Long-term gene delivery to the TM was also achieved by lentiviral vectors (53, 54).

2.1.6. Spontaneous Ocular Hypertensive Rodents

Mutant rats, inbred and transgenic mice have been shown to develop ocular hypertension. They are valuable tools for glaucoma research. A colony of mutant RCS rats has been reported to have an elevated IOP ranging from 25 to 45 mmHg in one or both eyes. The IOP increase correlates with cupping of the optic nerve head and a severe RGC loss: more than 90% of the RGC disappear at 18 months of age. The etiology of glaucoma in these animals is not clear, but was speculated to be a result of an overproduction of aqueous humor (55). The potential usefulness of these animals as a glaucoma model awaits future characterization and the general availability of the animal.

The $Col1a1^{r/r}$ transgenic mouse expressed mutations blocking the degradation of the $\alpha 1$ subunit of collagen type I. Reduction of collagen metabolism led to a gradual, excessive accumulation of collagen in the aqueous outflow pathway and eventually age-dependent ocular hypertension (56). Mean IOP of $Col1a1^{r/r}$ mice was significantly higher than that of the wild-type control animals at 16 weeks of age or older. At 54 weeks of age, these mice developed a significant reduction in the number of axons in the optic nerve (57). With further investigations, this mouse may be a useful model for glaucoma research.

Among the spontaneous ocular hypertensive rodent models, two substrains of mice, DBA/2J and DBA/2NNia, attracted considerable attention as potentially important animal models for glaucoma research (58). The DBA/2J mouse spontaneously develops ocular hypertension starting at 4–8 months of age (59, 60). The increase in IOP is a result of mutations in two genes, glycosylated protein nmb (Gpnmb) and tyrosinase-related protein I (Tyrp1) (61–63), which causes iris stroma atrophy and pigment dispersion (62, 64–66). The iris abnormalities disperse pigment in the anterior segment of the eye and eventually cause an excessive accumulation of pigment and cell debris in the aqueous outflow pathway, leading to an elevation of IOP. Interestingly, the pathological changes are not always uniform bilaterally or in mice of the same age. Thus, IOP of the two eyes of the same animal may vary, and IOP of animals of the same age may differ significantly (59, 60, 67).

Specific apoptotic RGC death, thinning of the nerve fiber layer, cupping of the optic disk, and optic neuropathy were observed in DBA/2J mice (58–60, 66–71). Yet thickness of other retinal layers and morphology of other types of retinal neurons were unaffected (58, 59). Nonetheless, retinal levels of γ-aminobutyric acid and choline acetyltransferase immunoreactivity were reported to be diminished in aged DBA/2J mice, suggesting that certain subtypes of amacrine cells may be somewhat distressed (72).

The DBA/2NNia mouse is similar to the DBA/2J mouse in that both develop age-dependent RGC loss and optic nerve degeneration resulting from IOP elevation (73–76). However, a critical difference between the two substrains is that the DBA/2NNia mouse also showed depletion of neurons in the inner and outer nuclear layers of the retina (77), which correlates with dramatic decreases in the a- and b-wave amplitudes, indicating a widespread damage of the photoreceptor cells (40).

Among rodent models with spontaneous ocular hypertension, the DBA/2J mouse is clearly the best characterized. Its key similarities to the human disease render these mice very popular with glaucoma researchers. However, glaucomatous changes develop over a period of many months in this model, which necessitates long study duration in some experiments. Furthermore, because the rate of pathological change is not equal in all eyes or all mice of the same age, large sample sizes are required to provide sufficient statistical power. Therefore, studies using the DBA/2J mice tend to be resource-intensive. However, until a better model becomes accessible, this mouse is extremely valuable for glaucoma research.

2.2. Nonpressure Models

2.2.1. Excitotoxic Insult in the Retina

Glutamate-related excitotoxicity is implicated in glaucomatous retinopathy. Thus, intravitreal injection of glutamate or its analog, N-methyl-D-aspartate (NMDA), has been exploited as a research model to address specifically the excitotoxicity theory potentially associated with the disease. Injection of glutamate or NMDA (20–200 nmol) produced dose-dependent apoptotic RGC loss without affecting IOP in both the rat and mouse. In the rat, several days after intravitreal injection, NMDA induced apoptosis (78–80) and death of more than 80% of RGC (68, 81), together with similar losses of Thy-1 mRNA and immunoreactivity in the retina (82–84). Equivalent levels of damage to the optic nerve axons were also observed (85). However, the injurious effects of NMDA are not specific to the RGC. The excitotoxic amino acid caused a dose-dependent depletion of choline acetyltransferase, suggesting damages to amacrine cells (86, 87). Indeed, apoptosis of retinal neurons in the inner nuclear layer was reported after intravitreal injection of NMDA (88).

Intravitreal injection of NMDA significantly attenuated electrophysiological indicators associated with the functional integrity of the RGC and optic nerve. Amplitudes of the negative scotopic threshold response (89) and the second harmonic of the pattern visual evoked potential (VEP) (90) were reduced by NMDA treatment in the rat. Moreover, NMDA also suppressed the amplitude of b-wave considerably (87, 89, 91), confirming its effects on retinal cells other than the RGC. Glutamate and NMDA also interfere with visual function. Intravitreal injection of glutamate or NMDA caused a complete loss of the rat's ability to discriminate visual patterns. Interestingly, this loss was transient. The visual function gradually recovered to its normal level even though 50% or more of RGC had been destroyed by the insult (81, 92–94).

Glutamate and NMDA induce comparable damages in the mouse retina. RGC apoptosis was detectable as early as 1 h after intravitreal NMDA injection (95), and reached its peak level 2 days later (96), with a parallel drop in Thy-1 mRNA in the insulted retina (4, 96). Similar to the rat, NMDA also induced apoptotic death of neurons in the inner nuclear layer of the mouse (96, 97).

Intravitreal injection of glutamate or NMDA has been very popular as a study model to delineate molecular and cellular pathways related to excitotoxicity in the retina and to appraise potential neuroprotection strategies. The technique is relatively easy to perform and requires only a short study duration of a few days. In spite of these advantages, a key shortcoming of this model is that it relies solely on excitotoxicity as the insult, whose involvement in glaucomatous retinopathy is still controversial.

Recently, Harada and colleagues showed that knockout mice deficient in glutamate transporters, GLAST and EAAC1, developed spontaneous RGC death and optic nerve damage (98). Glutamate transporters are responsible for the clearance of glutamate after it is released from nerve endings. Paucity of transporters allows the released glutamate to accumulate and inflict excitotoxicity in neuronal tissues, including the retina. Hence, significant RGC loss was observable in GLAST$^{-/-}$ and EAAC1$^{-/-}$ mice after 2 and 8 weeks of age, respectively. Consistent with RGC loss, optic disk cupping, optic nerve degeneration, and multifocal ERG abnormality were apparent in the GLAST$^{-/-}$ mutants. Similar to intravitreal injection of glutamate or NMDA, pathological changes of the GLAST$^{-/-}$ mice were not limited to the RGC. Substantial loss of neurons in the inner nuclear layer was also detected (98).

2.2.2. Optic Nerve Injury

Optic nerve injury, induced either by optic nerve transection or optic nerve crush, is another popular rodent model to study RGC death. These procedures impose mechanical damage to the optic nerve, which mimics certain aspects of pathological changes related to glaucomatous optic neuropathy.

Optic nerve transection injured all axons in the rat optic nerve (99), and generated an almost total loss of RGC within 2 weeks (100–103). The RGC death was mediated by apoptosis, which can be delayed by inhibition of components of the apoptosis pathway (104–106). The damage by optic nerve transection was very specific to the RGC. It obliterated RGC without affecting the survival of amacrine cells (99).

In the rat, optic nerve transection completely eliminated the pattern ERG (102, 107), and significantly reduced both positive and negative components of the scotopic threshold response (89).

The a- and b-waves, however, were not affected (89). Optic nerve transection also completely abolished the visual function (102).

In the mouse, similar pathological changes were observed. Most RGC (85%) were lost 2 weeks after the insult (108), and 2 months after transection, RGC, optic nerve axons, and pattern ERG response were all eliminated (109).

Since optic nerve transection produces severe and total damage to the optic nerve, it is very dissimilar to optic neuropathy seen in glaucoma. Its relevance in glaucoma research is frequently debated. A few years ago, a rat model of partial optic nerve transection was reported (110). In this model, only the superior one-third of the optic nerve was transected, which induced a direct axotomy insult to RGCs of the superior retina, and an indirect, secondary insult to RGCs of the inferior half of the retina (110). The secondary insult is a useful research tool to study the secondary RGC degeneration suggested in glaucoma.

Optic nerve damage can also be induced by crushing the optic nerve with devices such as forceps and clamps. Instead of the total optic nerve axotomy by transection, optic nerve crush produces injuries of different severities to different populations of axons in the optic nerve. In both the rat and mouse, optic nerve crush damaged optic nerve axons (111–113) and caused RGC death. The extents of RGC damage, however, could span from practically no change in cell density to almost complete loss (114, 115). Different levels of cell loss due to apoptosis and sometimes necrosis several days to several weeks after the procedure were reported (4, 58, 116–127). These variances were likely a result of differences in the force and techniques used during the crush. Functionally, various crushing forces caused various magnitudes in reduction of VEP and pattern VEP amplitudes (114).

A major concern of the optic nerve crush as an insult is its reproducibility. The procedural variability makes precise quantitative comparisons among studies very difficult. Furthermore, the typical optic nerve crush procedure involves the whole nerve, and generates pan-retina damages that are very different from glaucomatous retinopathy. A partial optic nerve crush (128) may overcome this limitation.

2.2.3. Retinal Ischemia

Vascular insufficiency has been proposed as an important contributory factor in glaucomatous retinopathy. Many researchers have used retinal ischemia to identify related disease changes and test potential neuroprotectants. There are several means to achieve this insult. The most common method is to raise IOP transiently above the systolic blood pressure, which stops blood flow in the eye (129–132). Another method uses optic nerve ligation, in which a suture is looped around the optic nerve and tightened until blood flow in the central retinal artery and posterior ciliary arteries is stopped (133). Photothrombosis is a noninvasive method to induce retinal ischemia. This procedure requires an intravenous injection of rose bengal, followed by focused green light irradiation on the central retinal artery. The procedure releases singlet molecular oxygen, activates platelets, and leads to the formation of a localized thrombus (134). Although many researchers attempted to use these models to understand glaucoma, the biological significance of the insult in this disease is still being debated. Furthermore, retinal ischemia damages ocular tissues unrelated to glaucomatous retinopathy as well. Hence, the relevance of these experimental models for glaucoma research is questionable. However, it is clearly a meaningful model for retinal ischemic diseases. It will be described and discussed in more details in the "Retinal Ischemia" Chapter of this book.

In addition to these techniques, endothelin-1 (ET-1), a potent vasoactive peptide, was also used initially to induce chronic optic nerve ischemia (135). Surprisingly, it was subsequently found to have direct toxicity to the RGC and optic nerve, independent of its effect on blood flow (136–140). Intravitreal injection or local application of ET-1 to the rat optic nerve caused a reversible reduction in axonal transport in the optic nerve (136, 139). These findings are intriguing because elevated ET-1 levels have been found in the aqueous humor of POAG patients (141, 142), and blockade of axoplasmic transport has been demonstrated in human glaucoma and experimental glaucoma in monkeys and rodents (143–145). Rat optic neuropathy induced by ET-1 may be an interesting model for glaucoma research.

3. Rabbit Models

Rabbits have been used for decades to study aqueous humor dynamics and to determine the effects of various agents on IOP. However, it is important to realize that there are important differences in the rabbit retina and optic nerve. For example, unlike the primate and rodent RGC axons, which are unmyelinated until they exit the eye and join the optic nerve, the rabbit RGC axons are myelinated within the retina.

There have been several reports of congenital glaucoma in rabbits, in which the elevated IOP causes buphthalmia. This ocular hypertension led to a loss in RGCs and was associated with increased retinal expression of semaphorin (146). In contrast, Bunt and colleagues did not find any optic nerve damage in their strain of congenital glaucomatous rabbits (147).

The IOP of rabbits can be artificially elevated by a variety of mechanisms. Acute changes in IOP can block optic nerve retrograde axonal transport in rabbits (148). Anterior segment injection of α-chymotrypsin caused prolonged IOP elevation, resulting in retinal nerve fiber layer and RGC loss as well as cupping of the optic disk (149). Multiple anterior segment injections of methylcellulose have also elevated IOP in rabbits, with the loss of RGCs and retinal glial cell activation (150). In addition, a number of groups have generated glucocorticoid-induced ocular hypertension in rabbits (151–153). Although the glucocorticoid is topically delivered, there are significant systemic side effects (e.g. inhibition of growth, loss in body weight) and to date there has been no report of glucocorticoid-mediated ocular hypertension resulting in retinal or optic nerve damage in the rabbit.

The rabbit has been used to look at ischemic and ischemia/reperfusion damage to the rabbit retina. Pigmented rabbits recovered from acute IOP elevation-induced retinal ischemia, assessed by recovery of the ERG b-wave and optic nerve damage, while albino rabbits developed more severe damage with this insult (154). Acute IOP elevation increased the levels of the excitotoxic amino acid glutamate in rabbit optic nerve head (155). Optic nerve head ischemia has been induced by repeated intravitreal injections of the potent vasoconstrictor ET-1. Optic nerve head microcirculation decreased, visual evoked potential latency decreased, and there was enlargement of the optic disk and axonal loss in the ET-1-injected eyes (156). Another study also examined intraviterally administered ET-1, which caused hypoperfusion of the optic nerve head and retina as well as a loss of myelinated axons and gliosis in the optic nerve (157).

4. Feline Models

Cats have also been used to model certain aspects of glaucoma. The anatomy of the cat retina and optic nerve is more similar to primates than is the rabbit. Although relatively rare, cats can develop primary open angle glaucoma with loss of RGCs and optic nerve head cupping and gliosis (158). Injection of red blood cell ghosts into the anterior segment induced ocular hypertension, which caused RGC dendritic changes that preceded RGC cell death (159). Topical administration of the potent glucocorticoid dexamethasone caused steroid-induced ocular hypertension in cats, although glaucomatous damages to the retina and optic nerve were not studied (160). RGC soma and dendritic field sizes were decreased after optic nerve crush in cats, and intravitreal administration of BDNF partially prevented these RGC changes (161, 162). Intravitreal administration of BDNF plus CNTF or a caspase inhibitor, or electric stimulation of the cut end of the optic nerve increased RGC survival after axotomy of the optic nerve (163).

In an attempt to develop a cat model of human glaucoma, a lentivirus expression vector encoding the glaucoma gene myocilin was used to transduce the TM of cat eyes (164). There was prolonged expression of the reporter gene GFP for up to 2 years in the transduced eyes. However, prolonged overexpression of normal (wild-type) cat myocilin or introduction of a severe myocilin glaucomatous mutation did not elevate IOP in these cats. This inability to induce ocular hypertension may be due to the lack of a PTS1 signal at the end of cat myocilin. A functional PTS1 signal, which is found in human Myocilin, was required for mutant myocilin induced ocular hypertension in mice (44).

5. Canine Models

There have been a number of reports characterizing glaucoma in dogs, particularly the glaucomatous beagle. Breed related primary glaucoma prevalence is similar to that seen in humans (165). The inheritance of primary glaucoma in the beagle is autosomal recessive (166) which is initially an open angle glaucoma. This glaucoma is associated with anterior segment morphological changes including increased basement membrane and extracellular material as well as compressed and disorganized trabeculae (167). There are also glaucomatous changes to the optic nerve, including the loss of larger diameter optic nerve axons (168). Glaucomatous changes to the lamina cribrosa are seen early in disease and increase in severity as the disease progresses (169). In addition, there are ultrastructural differences in optic nerve head capillaries of glaucomatous beagles (169). These beagles also develop glucocorticoid-induced ocular hypertension (170), which may be similar to the increased steroid sensitivity present in glaucomatous humans.

Several studies have been done in attempt to identify the pathogenic mechanisms responsible for glaucoma in dogs. Aqueous humor and vitreous levels of ET-1 and nitric oxide are higher in the glaucomatous beagle compared to age-matched normal beagles (171). A potential role for the human glaucoma gene myocilin has also been examined in glaucomatous dogs. Aqueous humor levels of myocilin are higher in glaucomatous dogs, with higher levels associated with secondary glaucomas (172). Elevated aqueous humor levels of myocilin in the glaucomatous beagle appeared to correlate with the presence and severity of glaucoma (173). In addition, there were increased amounts of myocilin in the anterior segment (especially the TM) of glaucomatous dog eyes (174). However, the elevated levels of myocilin do not necessarily equate with pathogenesis and may be indirectly regulated by elevated IOP, as has been shown experimentally (175).

6. Porcine Models

Pigs and pig eyes have also been used to study and characterize glaucomatous damage to the retina and optic nerve. Episcleral vein cauterization chronically elevated IOP and caused RGC loss in the peripheral and mid-peripheral retina of pig eyes, with a greater loss of the larger RGCs (176). Acute and modest IOP elevation (40–45 mmHg for 6 h) decreased axonal transport, which was associated with optic nerve head axonal cytoskeletal changes (177). Acutely elevated IOP also affected intravascular pO2 in the pig retina, optic nerve head, and choroid, with greater pressure induced sensitivity in the retina and optic nerve head (178). Multifocal ERG studies have been conducted in pig eyes with and without optic nerve transection to identify RGC contributions to the mfERG signal (179).

Pig eyes have also been used for in vitro and ex vivo glaucoma studies. Porcine eye globes were used to evaluate the effects of scleral collagen cross-linking on IOP-mediated biomechanical stress to the optic nerve head (180). Glutamate-induced excitotoxicity has been studied in isolated adult pig RGC cultures (181, 182).

7. Bovine and Ovine Models

Topical ocular administration of the glucocorticoid prednisolone acetate elevated IOP in both cattle (183) and sheep (184), suggesting that these species could be used as models of steroid glaucoma. There are exciting preliminary data showing that transducing sheep eyes with a viral vector that expresses the extracellular proteinase MMP1 under the control of a glucocorticoid response element prevents and even reverses this steroid-induced ocular hypertension in sheep (185). However, as with other animal models of steroid-induced IOP elevation, retinal and optic nerve effects have not yet been studied.

A bovine eye cup model has been used to show that increased hydrostatic pressure caused the release of ATP from the retina, where the elevated ATP may play a pathological role in glaucomatous damage to the RGCs (186).

Table 2
Summary of rodent glaucoma models

		Key characteristics						
		Morphological changes			Functional changes			
Model	Mechanism	RGC loss	Damage specific to RGC	Optic nerve injury	Reduction in a- or b-wave	Reduction in STR	Reduction in VEP	Technical difficulty
Pressure models								
• Episcleral hypertonic saline injection	Sclerosis of outflow pathway	Yes	Yes	Yes	No	Yes		High
• Outflow pathway photocoagulation	Photo-occlusion of outflow pathway	Yes	No	Yes	Yes			Medium
• Intracameral injection of foreign substances								
– Intracameral hyaluronate	Chemical blockade of outflow pathway	Yes	No	Yes	Yes			Low
– Intracameral microspheres	Physical blockade of outflow pathway	Yes						Low
• Extraocular vein cautery	Occlusion of outflow pathway	Yes	No	Yes	Yes			Low
• Ocular transgene expression	Transgene-induced outflow pathway abnormality							Low
• Spontaneous ocular hypertensive rodents								

(continued)

Table 2 (continued)

Model	Mechanism	Key characteristics						
		Morphological changes			Functional changes			
		RGC loss	Damage specific to RGC	Optic nerve injury	Reduction in a- or b-wave	Reduction in STR	Reduction in VEP	Technical difficulty
– Mutant RCS rat	Unknown (excessive aqueous production?)	Yes						N/A
– Col1a1$^{r/r}$ mouse	Mutation predisposed outflow blockade	Yes		Yes				N/A
– DBA/2J mouse	Gene predisposed outflow blockade	Yes	Yes	Yes				N/A
– DBA/2NNia mouse	Gene predisposed outflow blockade	Yes	No	Yes	Yes			N/A
Non-pressure models								
• Excitotoxicity								
– Intravitreal excitatory amino acid	Excitotoxicity	Yes	No	Yes	Yes	Yes	Yes	Low
– GLAST$^{-/-}$ and EAAC1$^{-/-}$ mice	Mutation predisposed excitotoxicity	Yes	No	Yes				N/A
• Optic nerve injury	Optic nerve injury-induced pathology	Yes	Yes	Yes	No	Yes	Yes	Medium
• Retinal ischemia	Ischemia-induced retinal damage	Yes	No	Yes	Yes			Medium

8. Conclusion

Important progress has been made over the past years on the discovery, development, and characterization of various rodent (Table 2) and larger animal (Table 3) models in glaucoma research.

Table 3
Summary of other nonprimate glaucoma models

Species	Model	Mechanism	Key characteristics
Rabbit	Congenital glaucoma	Naturally occuring	Ocular hypertension & buthalmia; loss of RGCs
	Acute IOP elevation	Intracameral cannula	Blockage of retrograde axonal transport
	Chronic IOP elevation	α-Chymotrypsin injection	Prolonged IOP elevation; RNFL & RGC loss; cupping of optic disk
		Methylcellulose injections	Loss of RGCs & retinal glial activation
	Steroid ocular hypertension	Glucocorticoid administration	Effects on retina and optic nerve not studied
	Ischemia	Acute IOP elevation	Loss in ERG b-wave; increased ONH glutamate levels
		Intravitreal ET-1 injection	Increased VEP latency; optic disk enlargement; ON axon loss
Cat	POAG	Naturally occuring (rare)	Loss of RGCs; ONH cupping & gliosis
	Ocular hypertension	Injection of RBC ghosts	RGC dendritic changes preceed RGC death
	Optic nerve crush	Optic nerve crush	Loss of RGCs partially protected by BDNF & CNTF
Dog	Primary glaucoma (beagle)	Naturally occuring (AR)	Loss of ON axons; changes in lamina cribrosa; increased ET-1 & NO
Pig	Chronic IOP elevation	Episcleral vein cauterization	Peripheral & mid-peripheral RGC loss
	Acute IOP elevation	Intracameral cannula	Decreased axonal transport
Cow & Sheep	Steroid ocular hypertension	Glucocorticoid administration	Effects on retina and optic nerve not studied

These models include spontaneous, genetic, transgenic, as well as pharmacologically and mechanically induced changes in the eye, leading to ocular hypertension or certain specific aspects of pathology associated with the disease. They are useful in addressing many areas in glaucoma research. New mouse models of glaucoma will allow glaucoma researchers to molecularly and genetically dissect the various pathogenic pathways responsible for glaucomatous damage to the eye. The power of mouse genetics will also allow researchers to identify and characterize modifier genes that control the susceptibility and severity of glaucoma pathology. New tools, including the ability to increase or knockdown the expression of specific genes in select regions of the eye will be used not only to better understand glaucoma pathogenesis, but also to identify new therapeutic targets. Each model has its own characteristic advantages, its own disadvantages, and its own limitations. The availability of these animal models has contributed significantly to the understanding of the disease, its etiology, pathology, genetic, molecular, and cellular mechanisms, as well as the continuously development and improvement of its intervention and therapy.

References

1. Pang I-H, clark AF (2007) Rodent models for glaucoma retinopathy and optic neuropathy. J Glaucoma 16:483–505
2. Morrison JC, Moore CG, Deppmeier LM, Gold BG, Meshul CK, Johnson EC (1997) A rat model of chronic pressure-induced optic nerve damage. Exp Eye Res 64(1):85–96
3. Morrison JC, Johnson EC, Cepurna W, Jia L (2005) Understanding mechanisms of pressure-induced optic nerve damage. Prog Retin Eye Res 24(2):217–240
4. Schlamp CL, Johnson EC, Li Y, Morrison JC, Nickells RW (2001) Changes in Thyl gene expression associated with damaged retinal ganglion cells. Mol Vis 7:192–201
5. Hanninen VA, Pantcheva MB, Freeman EE, Poulin NR, Grosskreutz CL (2002) Activation of caspase 9 in a rat model of experimental glaucoma. Curr Eye Res 25(6):389–395
6. Guo L, Moss SE, Alexander RA, Ali RR, Fitzke FW, Cordeiro MF (2005) Retinal ganglion cell apoptosis in glaucoma is related to intraocular pressure and IOP-induced effects on extracellular matrix. Invest Ophthalmol Vis Sci 46(1):175–182
7. Chauhan BC, Pan J, Archibald ML, LeVatte TL, Kelly ME, Tremblay F (2002) Effect of intraocular pressure on optic disc topography, electroretinography, and axonal loss in a chronic pressure-induced rat model of optic nerve damage. Invest Ophthalmol Vis Sci 43(9):2969–2976
8. Johnson EC, Deppmeier LM, Wentzien SK, Hsu I, Morrison JC (2000) Chronology of optic nerve head and retinal responses to elevated intraocular pressure. Invest Ophthalmol Vis Sci 41(2):431–442
9. Tezel G, Yang X, Cai J (2005) Proteomic identification of oxidatively modified retinal proteins in a chronic pressure-induced rat model of glaucoma. Invest Ophthalmol Vis Sci 46(9):3177–3187
10. Jia L, Cepurna WO, Johnson EC, Morrison JC (2000) Patterns of intraocular pressure elevation after aqueous humor outflow obstruction in rats. Invest Ophthalmol Vis Sci 41(6):1380–1385
11. Ahmed F, Brown KM, Stephan DA, Morrison JC, Johnson EC, Tomarev SI (2004) Microarray analysis of changes in mRNA levels in the rat retina after experimental elevation of intraocular pressure. Invest Ophthalmol Vis Sci 45(4):1247–1258
12. Fortune B, Bui BV, Morrison JC et al (2004) Selective ganglion cell functional loss in rats with experimental glaucoma. Invest Ophthalmol Vis Sci 45(6):1854–1862

13. Pang I-H, Johnson EC, Jia L et al (2005) Evaluation of inducible nitric oxide synthase in glaucomatous optic neuropathy and pressure-induced optic nerve damage. Invest Ophthalmol Vis Sci 46(4):1313–1321
14. Morrison JC, Nylander KB, Lauer AK, Cepurna WO, Johnson E (1998) Glaucoma drops control intraocular pressure and protect optic nerves in a rat model of glaucoma. Invest Ophthalmol Vis Sci 39(3):526–531
15. Ueda J, Sawaguchi S, Hanyu T et al (1998) Experimental glaucoma model in the rat induced by laser trabecular photocoagulation after an intracameral injection of India ink. Jpn J Ophthalmol 42(5):337–344
16. Lam TT, Kwong JM, Tso MO (2003) Early glial responses after acute elevated intraocular pressure in rats. Invest Ophthalmol Vis Sci 44(2):638–645
17. Ishii Y, Kwong JM, Caprioli J (2003) Retinal ganglion cell protection with geranylgeranylacetone, a heat shock protein inducer, in a rat glaucoma model. Invest Ophthalmol Vis Sci 44(5):1982–1992
18. WoldeMussie E, Ruiz G, Wijono M, Wheeler LA (2001) Neuroprotection of retinal ganglion cells by brimonidine in rats with laser-induced chronic ocular hypertension. Invest Ophthalmol Vis Sci 42(12):2849–2855
19. Hare W, WoldeMussie E, Lai R et al (2001) Efficacy and safety of memantine, an NMDA-type open-channel blocker, for reduction of retinal injury associated with experimental glaucoma in rat and monkey. Surv Ophthalmol 45(Suppl 3):S284–S289
20. Ben Shlomo G, Bakalash S, Lambrou GN et al (2005) Pattern electroretinography in a rat model of ocular hypertension: functional evidence for early detection of inner retinal damage. Exp Eye Res 81(3):340–349
21. Levkovitch-Verbin H, Quigley HA, Martin KR, Valenta D, Baumrind LA, Pease ME (2002) Translimbal laser photocoagulation to the trabecular meshwork as a model of glaucoma in rats. Invest Ophthalmol Vis Sci 43(2):402–410
22. Martin KR, Levkovitch Verbin H, Valenta D, Baumrind L, Pease ME, Quigley HA (2002) Retinal glutamate transporter changes in experimental glaucoma and after optic nerve transection in the rat. Invest Ophthalmol Vis Sci 43(7):2236–2243
23. Martin KR, Quigley HA, Zack DJ et al (2003) Gene therapy with brain-derived neurotrophic factor as a protection: retinal ganglion cells in a rat glaucoma model. Invest Ophthalmol Vis Sci 44(10):4357–4365
24. Grozdanic SD, Kwon YH, Sakaguchi DS, Kardon RH, Sonea IM (2004) Functional evaluation of retina and optic nerve in the rat model of chronic ocular hypertension. Exp Eye Res 79(1):75–83
25. Aihara M, Lindsey JD, Weinreb RN (2003) Experimental mouse ocular hypertension: establishment of the model. Invest Ophthalmol Vis Sci 44(10):4314–4320
26. Grozdanic SD, Betts DM, Sakaguchi DS, Allbaugh RA, Kwon YH, Kardon RH (2003) Laser-induced mouse model of chronic ocular hypertension. Invest Ophthalmol Vis Sci 44(10):4337–4346
27. Gross RL, Ji J, Chang P et al (2003) A mouse model of elevated intraocular pressure: retina and optic nerve findings. Trans Am Ophthalmol Soc 101:163–169
28. Mabuchi F, Aihara M, Mackey MR, Lindsey JD, Weinreb RN (2003) Optic nerve damage in experimental mouse ocular hypertension. Invest Ophthalmol Vis Sci 44(10): 4321–4330
29. Ji J, Chang P, Pennesi ME et al (2005) Effects of elevated intraocular pressure on mouse retinal ganglion cells. Vision Res 45(2):169–179
30. Benozzi J, Nahum LP, Campanelli JL, Rosenstein RE (2002) Effect of hyaluronic acid on intraocular pressure in rats. Invest Ophthalmol Vis Sci 43(7):2196–2200
31. Moreno MC, Marcos HJ, Oscar Croxatto J et al (2005) A new experimental model of glaucoma in rats through intracameral injections of hyaluronic acid. Exp Eye Res 81(1):71–80
32. Urcola JH, Hernandez M, Vecino E (2006) Three experimental glaucoma models in rats: comparison of the effects of intraocular pressure elevation on retinal ganglion cell size and death. Exp Eye Res 83(2):429–437
33. Shareef SR, Garcia-Valenzuela E, Salierno A, Walsh J, Sharma SC (1995) Chronic ocular hypertension following episcleral venous occlusion in rats. Exp Eye Res 61(3): 379–382
34. Garcia-Valenzuela E, Shareef S, Walsh J, Sharma SC (1995) Programmed cell death of retinal ganglion cells during experimental glaucoma. Exp Eye Res 61(1):33–44
35. Sawada A, Neufeld AH (1999) Confirmation of the rat model of chronic, moderately elevated intraocular pressure. Exp Eye Res 69(5):525–531
36. Neufeld AH, Sawada A, Becker B (1999) Inhibition of nitric-oxide synthase 2 by aminoguanidine provides neuroprotection

of retinal ganglion cells in a rat model of chronic glaucoma. Proc Natl Acad Sci USA 96(17):9944–9948
37. Mittag TW, Danias J, Pohorenec G et al (2000) Retinal damage after 3 to 4 months of elevated intraocular pressure in a rat glaucoma model. Invest Ophthalmol Vis Sci 41(11):3451–3459
38. Ko ML, Hu DN, Ritch R, Sharma SC (2000) The combined effect of brain-derived neurotrophic factor and a free radical scavenger in experimental glaucoma. Invest Ophthalmol Vis Sci 41(10):2967–2971
39. Grozdanic SD, Betts DM, Sakaguchi DS, Kwon YH, Kardon RH, Sonea IM (2003) Temporary elevation of the intraocular pressure by cauterization of vortex and episcleral veins in rats causes functional deficits in the retina and optic nerve. Exp Eye Res 77(1):27–33
40. Bayer AU, Danias J, Brodie S et al (2001) Electroretinographic abnormalities in a rat glaucoma model with chronic elevated intraocular pressure. Exp Eye Res 72(6):667–677
41. Yu S, Tanabe T, Yoshimura N (2006) A rat model of glaucoma induced by episcleral vein ligation. Exp Eye Res 83(4):758–770
42. Kanamori A, Nakamura M, Nakanishi Y, Yamada Y, Negi A (2005) Long-term glial reactivity in rat retinas ipsilateral and contralateral to experimental glaucoma. Exp Eye Res 81(1):48–56
43. Stone EM, Fingert JH, Alward WLM et al (1997) Identification of a gene that causes primary open angle glaucoma. Science 275(5300):668–670
44. Shepard AR, Jacobson N, Millar JC et al (2007) Glaucoma-causing myocilin mutants require he Peroxisomal Targeting Signal-1 Receptor (PTS1R) to elevate intraocular pressure. Hum Mol Genet 16:609–617
45. Clark AF, Millar JC, Pang I-H, Jacobson N, Shepard A (2006) Adenoviral gene transfer of active human transforming growth factor-β2 induces elevated intraocular pressure in rats. ARVO Abstract 2006:4771.
46. Tripathi RC, Li J, Chan WF, Tripathi BJ (1994) Aqueous humor in glaucomatous eyes contains an increased level of TGF-beta 2. Exp Eye Res 59(6):723–727
47. Thompson JT, Smiddy WE, Williams GA et al (1998) Comparison of recombinant transforming growth factor-beta-2 and placebo as an adjunctive agent for macular hole surgery. Ophthalmology 105:700–706
48. Wang W-H, McNatt LG, Pang I-H et al (2008) Increased expression of Wnt antagonist sFRP1 in glaucoma and its regulation of intraocular pressure. J Clin Invest 118:1056–1064
49. Wang W-H, McNatt LG, Pang I-H et al (2008) Increased expression of serum amyloid A in glaucoma and its effect on intraocular pressure. Invest Ophthalmol Vis Sci 49:1916–1923
50. Shepard AR, Nolan MJ, Millar JC, et al (2008) CD44 overexpression causes ocular hypertension in the mouse. ARVO Abstract 2008:2880.
51. Borras T, Brandt CR, Nickells R, Ritch R (2002) Gene therapy for glaucoma: treating a multifaceted, chronic disease. Invest Ophthalmol Vis Sci 43(8):2513–2518
52. Millar JC, Wang Y, Pang I-H, Wang W-H, Clark AF (2008) Effect of immunosuppression with anti-CD40L antibody on adenoviral-mediated transgene expression in mouse anterior segment. Mol Vis 14:10–19
53. Challa P, Luna C, Liton PB et al (2005) Lentiviral mediated gene delivery to the anterior chamber of rodent eyes. Mol Vis 11(49):425–430
54. Balaggan KS, Binley K, Esapa M et al (2006) Stable and efficient intraocular gene transfer using pseudotyped EIAV lentiviral vectors. J Gene Med 8(3):275–285
55. Thanos S, Naskar R (2004) Correlation between retinal ganglion cell death and chronically developing inherited glaucoma in a new rat mutant. Exp Eye Res 79(1): 119–129
56. Aihara M, Lindsey JD, Weinreb RN (2003) Ocular hypertension in mice with a targeted type I collagen mutation. Invest Ophthalmol Vis Sci 44(4):1581–1585
57. Mabuchi F, Aihara M, Mackey MR, Lindsey JD, Weinreb RN (2004) Regional optic nerve damage in experimental mouse glaucoma. Invest Ophthalmol Vis Sci 45(12): 4352–4358
58. Libby RT, Li Y, Savinova OV et al (2005) Susceptibility to neurodegeneration in a glaucoma is modified by bax gene dosage. PLoS Genet 1(1):e4
59. John SW, Smith RS, Savinova OV et al (1998) Essential iris atrophy, pigment dispersion, and glaucoma in DBA/2J mice. Invest Ophthalmol Vis Sci 39(6):951–962
60. Zhou X, Li F, Kong L, Tomita H, Li C, Cao W (2005) Involvement of inflammation, degradation, and apoptosis in a mouse model of glaucoma. J Biol Chem 280(35): 31240–31248
61. Chang B, Smith RS, Hawes NL et al (1999) Interacting loci cause severe iris atrophy and

glaucoma in DBA/2J mice. Nat Genet 21(4):405–409

62. Anderson MG, Smith RS, Hawes NL et al (2002) Mutations in genes encoding melanosomal proteins cause pigmentary glaucoma in DBA/2J mice. Nat Genet 30(1): 81–85
63. Anderson MG, Libby RT, Mao M et al (2006) Genetic context determines susceptibility to intraocular pressure elevation in a mouse pigmentary glaucoma. BMC Biol 4:20
64. Mo JS, Anderson MG, Gregory M et al (2003) By altering ocular immune privilege, bone marrow-derived cells pathogenically contribute to DBA/2J pigmentary glaucoma. J Exp Med 197(10):1335–1344
65. John SW (2005) Mechanistic insights into glaucoma provided by experimental genetics the cogan lecture. Invest Ophthalmol Vis Sci 46(8):2649–2661
66. Libby RT, Anderson MG, Pang I-H et al (2005) Inherited glaucoma in DBA/2J mice: pertinent disease features for studying the neurodegeneration. Vis Neurosci 22(5): 637–648
67. Libby RT, Gould DB, Anderson MG, John SW (2005) Complex genetics of glaucoma susceptibility. Annu Rev Genomics Hum Genet 6:15–44
68. Schuettauf F, Vorwerk C, Naskar R et al (2004) Adeno-associated viruses containing bFGF or BDNF are neuroprotective against excitotoxicity. Curr Eye Res 29(6):379–386
69. Schlamp CL, Li Y, Dietz JA, Janssen KT, Nickells RW (2006) Progressive ganglion cell loss and optic nerve degeneration in DBA/2J mice is variable and asymmetric. BMC Neurosci 7:66
70. Reichstein D, Ren L, Filippopoulos T, Mittag T, Danias J (2007) Apoptotic retinal ganglion cell death in the DBA/2 mouse model of glaucoma. Exp Eye Res 84(1):13–21
71. Howell GR, Libby RT, Jakobs TC et al (2007) Axons of retinal ganglion cells are insulted in the optic nerve early in DBA/2J glaucoma. J Cell Biol 179:1523–1537
72. Moon JI, Kim IB, Gwon JS et al (2005) Changes in retinal neuronal populations in the DBA/2J mouse. Cell Tissue Res 320(1): 51–59
73. Sheldon WG, Warbritton AR, Bucci TJ, Turturro A (1995) Glaucoma in food-restricted and ad libitum-fed DBA/2NNia mice. Lab Anim Sci 45(5):508–518
74. Danias J, Kontiola AI, Filippopoulos T, Mittag T (2003) Method for the noninvasive measurement of intraocular pressure in mice. Invest Ophthalmol Vis Sci 44(3): 1138–1141
75. Filippopoulos T, Danias J, Chen B, Podos SM, Mittag TW (2006) Topographic and morphologic analyses of retinal ganglion cell loss in old DBA/2NNia mice. Invest Ophthalmol Vis Sci 47(5):1968–1974
76. May CA, Mittag T (2006) Optic nerve degeneration in the DBA/2NNia mouse: is the lamina cribrosa important in the development of glaucomatous optic neuropathy? Acta Neuropathol 111(2):158–167
77. Bayer AU, Neuhardt T, May AC et al (2001) Retinal morphology and ERG response in the DBA/2NNia mouse model of angle-closure glaucoma. Invest Ophthalmol Vis Sci 42(6):1258–1265
78. Sun Q, Ooi VE, Chan SO (2001) N-methyl-D-aspartate-induced excitotoxicity in adult rat retina is antagonized by single systemic injection of MK-801. Exp Brain Res 138(1):37–45
79. Kwong JM, Lam TT, Caprioli J (2003) Hyperthermic pre-conditioning protects retinal neurons from N-methyl-D-aspartate (NMDA)-induced apoptosis in rat. Brain Res 970(1):119–130
80. El Remessy AB, Khalil IE, Matragoon S et al (2003) Neuroprotective effect of (-)Delta9-tetrahydrocannabinol and cannabidiol in N-methyl-D-aspartate-induced retinal neurotoxicity: involvement of peroxynitrite. Am J Pathol 163(5):1997–2008
81. Vorwerk CK, Kreutz MR, Dreyer EB, Sabel BA (1996) Systemic L-kynurenine administration partially protects against NMDA, but not kainate-induced degeneration of retinal ganglion cells, and reduces visual discrimination deficits in adults rats. Invest Ophthalmol Vis Sci 37(12):2382–2392
82. Osborne NN, DeSantis L, Bae JH et al (1999) Topically applied betaxolol attenuates NMDA-induced toxicity to ganglion cells and the effects of ischaemia to the retina. Exp Eye Res 69(3):331–342
83. Nash MS, Osborne NN (1999) Assessment of Thy-1 mRNA levels as an index of retinal ganglion cell damage. Invest Ophthalmol Vis Sci 40(6):1293–1298
84. Chidlow G, Osborne NN (2003) Rat retinal ganglion cell loss caused by kainate, NMDA and ischemia correlates with a reduction in mRNA and protein of Thy-1 and neurofilament light. Brain Res 963(1–2):298–306
85. Shi JM, Jiang YQ, Liu XY (2004) Morphological changes of retina after

N-methyl-D-aspartate induced damage in rats. Zhong Nan Da Xue Xue Bao Yi Xue Ban 29(3):287–291
86. Siliprandi R, Canella R, Carmignoto G et al (1992) N-methyl-D-aspartate-induced neurotoxicity in the adult rat retina. Vis Neurosci 8(6):567–573
87. Casson RJ, Chidlow G, Wood JP, Vidal Sanz M, Osborne NN (2004) The effect of retinal ganglion cell injury on light-induced photoreceptor degeneration. Invest Ophthalmol Vis Sci 45(2):685–693
88. Lam TT, Abler AS, Kwong JM, Tso MO (1999) N-methyl-D-aspartate (NMDA)–induced apoptosis in rat retina. Invest Ophthalmol Vis Sci 40(10):2391–2397
89. Bui BV, Fortune B (2003) Ganglion cell contributions to the rat full-field electroretinogram. J Physiol 555(Pt 1):153–173
90. Kermer P, Klocker N, Bahr M (2001) Modulation of metabotropic glutamate receptors fails to prevent the loss of adult rat retinal ganglion cells following axotomy or N-methyl-D-aspartate lesion in vivo. Neurosci Lett 315(3):117–120
91. Maruyama I, Maeda T, Okisaka S, Mizukawa A, Nakazawa M, Ohguro H (2002) Autoantibody against neuron-specific enolase found in glaucoma patients causes retinal dysfunction in vivo. Jpn J Ophthalmol 46(1): 1–12
92. Sisk DR, Kuwabara T, Kirsch AD (1984) Behavioral recovery in albino rats with glutamate-damaged retinas. Invest Ophthalmol Vis Sci 25(10):1124–1128
93. Sisk DR, Kuwabara T (1985) Histologic changes in the inner retina of albino rats following intravitreal injection of monosodium L-glutamate. Graefes Arch Clin Exp Ophthalmol 223(5):250–258
94. Sabel BA, Sautter J, Stoehr T, Siliprandi R (1995) A behavioral model of excitotoxicity: retinal degeneration, loss of vision, and subsequent recovery after intraocular NMDA administration in adult rats. Exp Brain Res 106(1):93–105
95. Li Y, Schlamp CL, Poulsen GL, Jackson MW, Griep AE, Nickells RW (2002) p53 regulates apoptotic retinal ganglion cell death induced by N-methyl-D-aspartate. Mol Vis 8: 341–350
96. Li Y, Schlamp CL, Nickells RW (1999) Experimental induction of retinal ganglion cell death in adult mice. Invest Ophthalmol Vis Sci 40(5):1004–1008
97. Kumada M, Niwa M, Hara A et al (2005) Tissue type plasminogen activator facilitates NMDA-receptor-mediated retinal apoptosis through an independent fibrinolytic cascade. Invest Ophthalmol Vis Sci 46(4): 1504–1507
98. Harada T, Harada C, Nakamura K et al (2007) The potential role of glutamate transporters in the pathogenesis of normal tension glaucoma. J Clin Invest 117:1763–1770
99. Kielczewski JL, Pease ME, Quigley HA (2005) The effect of experimental glaucoma and optic nerve transection on amacrine cells in the rat retina. Invest Ophthalmol Vis Sci 46(9):3188–3196
100. Isenmann S, Engel S, Gillardon F, Bahr M (1999) Bax antisense oligonucleotides reduce axotomy-induced retinal ganglion cell death in vivo by reduction of Bax protein expression. Cell Death Differ 6(7):673–682
101. Kermer P, Ankerhold R, Klocker N, Krajewski S, Reed JC, Bahr M (2000) Caspase-9: involvement in secondary death of axotomized rat retinal ganglion cells in vivo. Brain Res Mol Brain Res 85(1–2):144–150
102. Kittlerova P, Valouskova V (2000) Retinal ganglion cells regenerating through the peripheral nerve graft retain their electroretinographic responses and mediate light-induced behavior. Behav Brain Res 112(1–2):187–194
103. Ju WK, Kim KY, Lee MY et al (2000) Up-regulated CNTF plays a protective role for retrograde degeneration in the axotomized rat retina. NeuroReport 11(17): 3893–3896
104. Chaudhary P, Ahmed F, Quebada P, Sharma SC (1999) Caspase inhibitors block the retinal ganglion cell death following optic nerve transection. Brain Res Mol Brain Res 67(1): 36–45
105. Heiduschka P, Thanos S (2000) Aurintricarboxylic acid promotes survival and regeneration of axotomised retinal ganglion cells in vivo. Neuropharmacology 39(5):889–902
106. Lingor P, Koeberle P, Kugler S, Bahr M (2005) Down-regulation of apoptosis mediators by RNAi inhibits axotomy-induced retinal ganglion cell death in vivo. Brain 128(Pt 3):550–558
107. Domenici L, Gravina A, Berardi N, Maffei L (1991) Different effects of intracranial and intraorbital section of the optic nerve on the functional responses of rat retinal ganglion cells. Exp Brain Res 86(3):579–584
108. Kilic E, Hermann DM, Isenmann S, Bahr M (2002) Effects of pinealectomy and melatonin on the retrograde degeneration of retinal

ganglion cells in a novel model of intraorbital optic nerve transection in mice. J Pineal Res 32(2):106–111

109. Chierzi S, Cenni MC, Maffei L et al (1998) Protection of retinal ganglion cells and preservation of function after optic nerve lesion in bcl-2 transgenic mice. Vision Res 38(10):1537–1543
110. Levkovitch Verbin H, Quigley HA, Martin KR, Zack DJ, Pease ME, Valenta DF (2003) A model to study differences between primary and secondary degeneration of retinal ganglion cells in rats by partial optic nerve transection. Invest Ophthalmol Vis Sci 44(8):3388–3393
111. Zalish M, Lavie V, Duvdevani R, Yoles E, Schwartz M (1993) Gangliosides attenuate axonal loss after optic nerve injury. Retina 13(2):145–147
112. Minzenberg M, Berkelaar M, Bray G, McKerracher L (1995) Changes in retinal ganglion cell axons after optic nerve crush: neurofilament expression is not the sole determinant of calibre. Biochem Cell Biol 73(9–10):599–604
113. Yoles E, Schwartz M (1998) Elevation of intraocular glutamate levels in rats with partial lesion of the optic nerve. Arch Ophthalmol 116(7):906–910
114. Klocker N, Zerfowski M, Gellrich NC, Bahr M (2001) Morphological and functional analysis of an incomplete CNS fiber tract lesion: graded crush of the rat optic nerve. J Neurosci Meth 110(1):147–153
115. Gellrich NC, Schimming R, Zerfowski M, Eysel UT (2002) Quantification of histological changes after calibrated crush of the intraorbital optic nerve in rats. Br J Ophthalmol 86(2):233–237
116. Allcutt D, Berry M, Sievers J (1984) A qualitative comparison of the reactions of retinal ganglion cell axons to optic nerve crush in neonatal and adult mice. Brain Res 318(2):231–240
117. Misantone LJ, Gershenbaum M, Murray M (1984) Viability of retinal ganglion cells after optic nerve crush in adult rats. J Neurocytol 13(3):449–465
118. Barron KD, Dentinger MP, Krohel G, Easton SK, Mankes R (1986) Qualitative and quantitative ultrastructural observations on retinal ganglion cell layer of rat after intraorbital optic nerve crush. J Neurocytol 15(3):345–362
119. Buys YM, Trope GE, Tatton WG (1995) (-)-Deprenyl increases the survival of rat retinal ganglion cells after optic nerve crush. Curr Eye Res 14(2):119–126
120. Schmitt U, Sabel BA (1996) MK-801 reduces retinal ganglion cell survival but improves visual performance after controlled optic nerve crush. J Neurotrauma 13(12): 791–800
121. Schuettauf F, Naskar R, Vorwerk CK, Zurakowski D, Dreyer EB (2000) Ganglion cell loss after optic nerve crush mediated through AMPA-kainate and NMDA receptors. Invest Ophthalmol Vis Sci 41(13):4313–4316
122. Levkovitch Verbin H, Harris Cerruti C, Groner Y, Wheeler LA, Schwartz M, Yoles E (2000) RGC death in mice after optic nerve crush injury: oxidative stress and neuroprotection. Invest Ophthalmol Vis Sci 41(13):4169–4174
123. Freeman EE, Grosskreutz CL (2000) The effects of FK506 on retinal ganglion cells after optic nerve crush. Invest Ophthalmol Vis Sci 41(5):1111–1115
124. Naskar R, Quinto K, Romann I, Schuettauf F, Zurakowski D (2002) Phenytoin blocks retinal ganglion cell death after partial optic nerve crush. Exp Eye Res 74(6):747–752
125. Tezel G, Yang X, Yang J, Wax MB (2004) Role of tumor necrosis factor receptor-1 in the death of retinal ganglion cells following optic nerve crush injury in mice. Brain Res 996(2):202–212
126. Maeda K, Sawada A, Matsubara M, Nakai Y, Hara A, Yamamoto T (2004) A novel neuroprotectant against retinal ganglion cell damage in a glaucoma model and an optic nerve crush model in the rat. Invest Ophthalmol Vis Sci 45(3):851–856
127. Swanson KI, Schlieve CR, Lieven CJ, Levin LA (2005) Neuroprotective effect of sulfhydryl reduction in a rat optic nerve crush model. Invest Ophthalmol Vis Sci 46(10):3737–3741
128. Dieterich DC, Trivedi N, Engelmann R, Gundelfinger ED, Gordon-Weeks PR, Kreutz MR (2002) Partial regeneration and long-term survival of rat retinal ganglion cells after optic nerve crush is accompanied by altered expression, phosphorylation and distribution of cytoskeletal proteins. Eur J NeuroSci 15:1433–1443
129. Smith GG, Baird CD (1952) Survival time of retinal cells when deprived of their blood supply by increased intraocular pressure. Am J Ophthalmol 35(5:2):133–136
130. Buchi ER, Suivaizdis I, Fu J (1991) Pressure-induced retinal ischemia in rats: an experimental model for quantitative study. Ophthalmologica 203(3):138–147

131. Hughes WF (1991) Quantitation of ischemic damage in the rat retina. Exp Eye Res 53(5):573–582
132. Li B, Pang I-H, Barnes G, McLaughlin M, Holt W (2002) A new method and device to induce transient retinal ischemia in the rat. Curr Eye Res 24(6):458–464
133. Stefansson E, Wilson CA, Schoen T, Kuwabara T (1988) Experimental ischemia induces cell mitosis in the adult rat retina. Invest Ophthalmol Vis Sci 29(7): 1050–1055
134. Mosinger JL, Olney JW (1989) Photothrombosis-induced ischemic neuronal degeneration in the rat retina. Exp Neurol 105(1):110–113
135. Cioffi GA, Orgul S, Onda E, Bacon DR, Van Buskirk EM (1995) An in vivo model of chronic optic nerve ischemia: the dose-dependent effects of endothelin-1 on the optic nerve microvasculature. Curr Eye Res 14(12):1147–1153
136. Stokely ME, Brady ST, Yorio T (2002) Effects of endothelin-1 on components of anterograde axonal transport in optic nerve. Invest Ophthalmol Vis Sci 43(10): 3223–3230
137. Chauhan BC, LeVatte TL, Jollimore CA et al (2004) Model of endothelin-1-induced chronic optic neuropathy in rat. Invest Ophthalmol Vis Sci 45(1):144–152
138. Lau J, Dang M, Hockmann K, Ball AK (2006) Effects of acute delivery of endothelin-1 on retinal ganglion cell loss in the rat. Exp Eye Res 82:132–145
139. Wang X, baldridge WH, Chauhan BC (2008) Acute endothelin-1 application induces reversible fast axonal transport blockade in adult rat optic nerve. Invest Ophthalmol Vis Sci 49:961–967
140. Krishnamoorthy RR, Rao VR, Dauphin R, Prasanna G, Johnson C, Yorio T (2008) Role of the ETB receptor in retinal ganglion cell death in glaucoma. Can J Physiol Pharmacol 86:380–393
141. Noske W, Hensen J, Wiederholt M (1997) Endothelin-like immunoreactivity in aqueous humor of patients with primary open-angle glaucoma and cataract. Graefes Arch Clin Exp Ophthalmol 235(9):551–552
142. Tezel G, Kass MA, Kolker AE, Becker B, Wax MB (1997) Plasma and aqueous humor endothelin levels in primary open-angle glaucoma. J Glaucoma 6(2):83–89
143. Quigley HA, Guy J, Anderson DR (1979) Blockade of rapid axonal transport. Effect of intraocular pressure elevation in primate optic nerve. Arch Ophthalmol 97(3): 525–531
144. Johansson JO (1988) Inhibition and recovery of retrograde axoplasmic transport in rat optic nerve during and after elevated IOP in vivo. Exp Eye Res 46:223–227
145. Knox DL, Eagle RCJ, Green WR (2007) Optic nerve hydropic axonal degeneration and blocked retrograde axoplasmic transport: histopathologic features in human high-pressure secondary glaucoma. Arch Ophthalmol 125:347–353
146. Solomon AS, Kimron M, Holdengreber V et al (2003) Up-regulation of semaphorin expression in retina of glaucomatous rabbits. Graefes Arch Clin Exp Ophthalmol 241:673–681
147. Bunt-Milam AH, Dennis MB, Bensinger RE (1987) Optic nerve head axonal transport in rabbits with hereditary glaucoma. Exp Eye Res 44:537–551
148. Chihara E, Honda Y (1981) Analysis of orthograde fast axonal transport and nonaxonal transport along the optic pathway of albino rabbits during increased and decreased intraocular pressure. Exp Eye Res 32: 229–239
149. Best M, Rabinovitz AZ, Masket S (1975) Experimental alpha-chymotrypsin glaucoma. Ann Ophthalmol 7:803–810
150. Manni G, Lambiase A, Centofanti M et al (1996) Histopathological evaluation of retinal damage during intraocular hypertension in rabbit: involvement of ganglion cells and nerve fiber layer. Graefes Arch Clin Exp Ophthalmol 234(Suppl 1):S209–S213
151. Lorenzetti OJ (1970) Effects of corticosteroids on ocular dynamics in rabbits. J Pharmacol Exp Ther 175:763–772
152. Knepper PA, Breen M, Weinstein HG, Blacik JL (1978) Intraocular pressure and glycosaminoglycan distribution in the rabbit eye: effect of age and dexamethasone. Exp Eye Res 27(5):567–575
153. Ticho U, Lahav M, Berkowitz S, Yoffe P (1979) Ocular changes in rabbits with corticosteroid-induced ocular hypertension. Br J Ophthalmol 63:646–650
154. Muller A, Villain M, Favreau B, Sandillon F, Privat A, Bonne C (1996) Differential effect of ischemia/reperfusion on pigmented and albino rabbit retina. J Ocul Pharmacol Ther 12:337–342
155. Okuno T, Oku H, Sugiyama T, Ikeda T (2006) Glutamate level in optic nerve head is increased by artificial elevation of intraocular pressure in rabbits. Exp Eye Res 82: 465–470

156. Oku H, Sugiyama T, Kojima S, Watanabe T, Azuma I (1999) Experimental optic cup enlargement caused by endothelin-1-induced chronic optic nerve head ischemia. Surv Ophthalmol 44(Suppl 1):S74–S84
157. Sasaoka M, Taniguchi T, Shimazawa M, Ishida N, Shimazaki A, Hara H (2006) Intravitreal injection of endothelin-1 caused optic nerve damage following to ocular hypoperfusion in rabbits. Exp Eye Res 83:629–637
158. Jacobi S, Dubielzig RR (2008) Feline primary open angle glaucoma. Vet Ophthalmol 11:162–165
159. Shou T, Liu J, Wang W, Zhou Y, Zhao K (2003) Differential dendritic shrinkage of alpha and beta retinal ganglion cells in cats with chronic glaucoma. Invest Ophthalmol Vis Sci 44:3005–3010
160. Zhan GL, Miranda OC, Bito LZ (1992) Steroid glaucoma: corticosteroid-induced ocular hypertension in cats. Exp Eye Res 54:211–218
161. Chen H, Weber AJ (2001) BDNF enhances retinal ganglion cell survival in cats with optic nerve damage. Invest Ophthalmol Vis Sci 42(5):966–974
162. Weber AJ, Harman CD (2008) BDNF preserves the dendritic morphology of alpha and beta ganglion cells in the cat retina after optic nerve injury. Invest Ophthalmol Vis Sci 49:2456–2463
163. Watanabe M, Fukuda Y (2002) Survival and axonal regeneration of retinal ganglion cells in adult cats. Prog Retin Eye Res 21(6):529–553
164. Khare PD, Loewen N, Teo W et al (2008) Durable, safe, multi-gene lentiviral vector expression in feline trabecular meshwork. Mol Ther 16:97–106
165. Gelatt KN, MacKay EO (2004) Prevalence of the breed-related glaucomas in pure-bred dogs in North America. Vet Ophthalmol 7:97–111
166. Gelatt KN, Gum GG (1981) Inheritance of primary glaucoma in the beagle. Am J Vet Res 42:1691–1693
167. Samuelson DA, Gum GG, Gelatt KN (1989) Ultrastructural changes in the aqueous outflow apparatus of beagles with inherited glaucoma. Invest Ophthalmol Vis Sci 30:550–561
168. Brooks DE, Strubbe DT, Kubilis PS, MacKay EO, Samuelson DA, Gelatt KN (1995) Histomorphometry of the optic nerves of normal dogs and dogs with hereditary glaucoma. Exp Eye Res 60:71–89
169. Brooks DE, Samuelson DA, Gelatt KN, Smith PJ (1989) Morphologic changes in the lamina cribrosa of beagles with primary open-angle glaucoma. Am J Vet Res 50:936–941
170. Gelatt KN, Mackay EO (1998) The ocular hypertensive effects of topical 0.1% dexamethasone in beagles with inherited glaucoma. J Ocul Pharmacol Ther 14:57–66
171. Källberg ME, Brooks DE, Gelatt KN, Garcia-Sanchez GA, Szabo NJ, Lambrou GN (2007) Endothelin-1, nitric oxide, and glutamate in the normal and glaucomatous dog eye. Vet Ophthalmol 10(Suppl 1):46–52
172. MacKay EO, Kallberg ME, Barrie KP et al (2008) Myocilin protein levels in the aqueous humor of the glaucomas in selected canine breeds. Vet Ophthalmol 11:234–241
173. MacKay EO, Kallberg ME, Gelatt KN (2008) Aqueous humor myocilin protein levels in normal, genetic carriers, and glaucoma Beagles. Vet Ophthalmol 11:177–185
174. Hart H, Samuelson DA, Tajwar H et al (2007) Immunolocalization of myocilin protein in the anterior eye of normal and primary open-angle glaucomatous dogs. Vet Ophthalmol 10(Suppl 1):28–37
175. Borras T, Rowlette LL, Tamm ER, Gottanka J, Epstein DL (2002) Effects of elevated intraocular pressure on outflow facility and TIGR/MYOC expression in perfused human anterior segments. Invest Ophthalmol Vis Sci 43:33–40
176. Ruiz-Ederra J, García M, Hernández M et al (2005) The pig eye as a novel model of glaucoma. Exp Eye Res 81:561–569
177. Balaratnasingam C, Morgan WH, Bass L, Matich G, Cringle SJ, Yu DY (2007) Axonal transport and cytoskeletal changes in the laminar regions after elevated intraocular pressure. Invest Ophthalmol Vis Sci 48:3632–3644
178. Blumenröder S, Augustin AJ, Koch FH (1997) The influence of intraocular pressure and systemic oxygen tension on the intravascular pO2 of the pig retina as measured with phosphorescence imaging. Surv Ophthalmol 42(suppl 1):S118–S126
179. Lalonde MR, Chauhan BC, Tremblay F (2006) Retinal ganglion cell activity from the multifocal electroretinogram in pig: optic nerve section, anaesthesia and intravitreal tetrodotoxin. J Physiol 570:325–338
180. Thornton IL, Dupps WJ, Roy AS, Krueger RR (2009) Biomechanical effects of intraocular pressure elevation on optic nerve/lamina cribrosa before and after peripapillary scleral collagen cross-linking. Invest Ophthalmol Vis Sci 50:1227–1233

181. Luo X, Heidinger V, Picaud S et al (2001) Selective excitotoxic degeneration of adult pig retinal ganglion cells in vitro. Invest Ophthalmol Vis Sci 42(5):1096–1106
182. Wehrwein E, Thompson SA, Coulibaly SF, Linn DM, Linn CL (2004) Acetylcholine protection of adult pig retinal ganglion cells from glutamate-induced excitotoxicity. Invest Ophthalmol Vis Sci 45(5): 1531–1543
183. Gerometta R, Podos SM, Candia OA et al (2004) Steroid-induced ocular hypertension in normal cattle. Arch Ophthalmol 122:1 492–1497
184. Gerometta R, Podos SM, Danias J, Candia OA (2009) Steroid-induced ocular hypertension in normal sheep. Invest Ophthalmol Vis Sci 50:669–673
185. Gerometta R, Spiga MG, Candia OA, Borras T (2009) Treatment of steroid-induced ocular hypertension by inducible gene transfer of MMP1 in sheep. ARVO Abstract 2009: 5722
186. Reigada D, Lu W, Zhang M, Mitchell CH (2008) Elevated pressure triggers a physiological release of ATP from the retina: possible role for pannexin hemichannels. Neuroscience 157:396–404

Chapter 9

Primate Glaucoma Models

Ronald S. Harwerth

Abstract

Glaucoma is a relatively common disease in which the pathological death of retinal ganglion cells causes progressive losses of sight, often leading to blindness. The diagnosis of glaucoma and the assessment of progression are based on a clinical quantification of the ocular characteristics of cupping of the optic nerve head, a loss of retinal nerve fiber layer thickness, and associated functional vision defects. Consequently, clinical tests are based on the quantification of these clinical characteristics of glaucomatous optic neuropathy. However, the basic neural and cellular pathophysiology that cause the characteristic signs of glaucoma cannot be studied in clinical patients and, therefore, animal models must be employed for basic research on glaucomatous optic neuropathy. For basic research that is directly applicable to the clinical disease, the primate model of experimental glaucoma is especially appropriate because the visual systems of macaque monkeys and humans are essentially identical, in terms of visual sensitivity, the anatomy and physiology of aqueous humor circulation, and the structure and neurology of the eye and visual pathway.

This chapter attempts to illustrate the scope of basic research on glaucoma, using the primate model to study fundamental aspects of the disease process relating to: (1) the relationship between intraocular pressure and visual field defects, (2) the correlation of visual sensitivity and retinal ganglion cell density, (3) the association between reduced visual sensitivity and retinal nerve fiber layer thinning, (4) the use on noninvasive electrophysiology to objectively assess functional integrity of the retina, and (5) the alterations of the tissue properties and biomechanics of the connective tissue of the optic nerve head that lead to clinical cupping and axonal injury in early glaucoma. In many cases, the results and models developed from data on experimental glaucoma have been applied to clinical patients to illustrate that the primate model offers an approach to clinical research that provides essential data that should be applicable to improving methods for assessing glaucomatous optic neuropathy in patients.

Key words: Glaucoma, Macaque monkeys, Optical coherence tomography, Perimetry, Retinal ganglion cells, Retinal nerve fiber layer

1. Introduction

The glaucomas are a group of diseases that all manifest a progressive optic neuropathy caused by the pathological loss of retinal ganglion cells (RGCs). The glaucomas are the second leading cause

Iok-Hou Pang and Abbot F. Clark (eds.), *Animal Models for Retinal Diseases*, Neuromethods, vol. 46
DOI 10.1007/978-1-60761-541-5_9,

of blindness in the United States, with 2.2–2.5 million people older than 40 years of age suffering from the disease, and as many as 120,000 of these people being blind from glaucoma (1, 2). The diseases are an increasingly important problem of public health because the number of patients is expected to continue to grow, with 5,400 cases advancing to blindness and more than 300,000 new cases each year. Further, it has been estimated that 3.3 million Americans will have glaucoma by the year 2020 (1). However, on the positive side, treatment is effective. The major clinical trials have demonstrated that treatment to lower intraocular pressure has value in preventing the onset of glaucoma in ocular hypertensive patients and in slowing the progression in glaucoma patients (3–6). It is, therefore, necessary to continue the development of procedures for diagnosis and assessment of glaucoma to optimize clinical decisions about when treatment is needed and when treatment is effective.

The diagnosis and assessment of glaucoma must rely on ophthalmic testing because the biological causes are unknown and there is no known biochemical or genetic marker. Glaucoma is described as a multifactorial disease, or a constellation of diseases, because there is not an identifiable single etiology (7, 8), although there are epidemiological risk factors that are associated with an increased likelihood of having the disease (8, 9) and there are cellular-level risk factors leading to pathological injury and death of retinal ganglion cells (10–13). However, all of the underlying factors lead to the single final manifestation of glaucoma, the death of retinal ganglion cells, which can be observed by ophthalmic examination as a cupping of the optic nerve head (ONH) (14–16), the loss of retinal nerve fiber layer (RNFL) (16–18), and functional vision defects (6, 8, 19, 20). Consequently, clinical procedures for the diagnosis and assessment of the progression of glaucoma are based on quantification of these clinical characteristics of glaucomatous optic neuropathy.

An understanding of the physiological mechanisms underlying alterations of optic nerve head morphology and characteristic loss of visual sensitivity in glaucoma is a likely key to improving diagnosis and treatment. However, glaucoma is a complex disease and the pathophysiology cannot easily be studied in clinical patients or from postmortem patient material and, therefore, animal models must be employed in clinical research on glaucomatous optic neuropathy. For this purpose, the primate model of experimental glaucoma is especially appropriate because the visual systems of macaque monkeys are essentially identical to humans, in terms of visual sensitivity (21, 22), the anatomy and physiology of aqueous humor circulation (23, 24), and the structure and neurology of the eye and visual pathway (25, 26).

The primate model of glaucoma is created by an obstruction of aqueous humor outflow, sometimes by an injection of a substance to block drainage or, more often, by a laser treatment to decrease aqueous outflow (27). Thus, the model is an exploitation of the best-established risk factor for clinical glaucoma, i.e., elevated intraocular pressure (IOP), and because the monkeys are usually young adults, it allows an isolation of a single factor of elevated intraocular pressure (IOP) as a cause of the death of retinal ganglion cells (RGCs). The principal method that has been used to create the model for recent research is a laser treatment of the trabecular meshwork (28, 29), although one lab has used injections of latex beads into the anterior chamber (30, 31). Both methods produce similar levels of IOP elevation and variability over time, but because the model of laser-induced ocular hypertension has been used much most frequently, it will be emphasized here. In most cases the experimental treatment is unilateral with the fellow eye serving as an intrasubject control, but in a few cases the treatment has been bilateral.

2. Laser-Induced Experimental Glaucoma

To create the model, the intraocular pressures of monkeys' eyes are elevated by scarification of their trabecular meshwork by Argon laser burns, using energy levels that destroy the trabecular meshwork and obliterate Schlemm's canal in the vicinity of the burn. The methods are patterned after the procedures initially described by Gaasterland and Kupfer (28) and, subsequently, refined by Quigley and Hohman (29). In the general protocol, laser treatments are performed on anesthetized monkeys using a standard clinical laser and slit lamp system, with a gonio laser lens (designed for monkey eyes) to focus the laser light on the mid-trabecular meshwork. Typically, Argon lasers are used (32–47), but other lasers have been also used (48). With Argon lasers (blue-green mode), the light energies are about 10× higher than for clinical laser trabeculoplasty, with only small variations in the treatment parameters across investigators. The usual nominal laser parameters are: 1.0–1.5 watts of power, 50 μm spot size, and 0.5–1.0 s exposure durations and the burns are closely spaced to produce contiguous tissue blanching. There is, however, some variation across investigations in the portion of the drainage angle that is initially treated and in the number of burns per treatment. In some cases, the initial laser treatment involves the full 360 degrees of the drainage angle (39, 43–46), while others treat only a portion (one-half (37, 40) or three-quarters (33, 34, 44)). The number of burns in each treatment varies across investigations from 30 to 150 spots, with repeated treatments (typically every 2–3 weeks, but sometimes with longer intervals) until a sustained elevation of the IOP has been achieved.

The retreatments involve the previously untreated trabecular meshwork, if the full circumference has not been treated, as well as retreating areas that were previously lasered. The monkeys' intraocular pressures generally do not become elevated until the full trabecular has been treated and because the entire trabecular meshwork is treated and the Argon laser energies are relatively high, the resulting ocular hypertension can be quite elevated (35–50 mm Hg) compared to the IOPs of chronic ocular hypertension patients, but systematic relationships have not been reported between total laser energy and either the maximum IOP or the rate of elevation of IOP (33).

Elevated IOPs can be sustained over several months, but the variability of IOP across time is usually large. In a study of experimental glaucoma in 18 monkeys, the mean and SD of the lasered eyes were 38.5 ± 13.4 mmHg, compared to 14.6 ± 2.7 mmHg for the control eyes (33). These data for IOPs and variances for laser-treated and control eyes are similar to other studies (27, 30, 40, 46) and, greater variance of IOPs throughout the day for laser-treated compared to control eyes also has been reported (41).

3. Experimental Glaucoma for Studies of Cell Biology

The primate model has been of great value in studies of the cellular and molecular mechanisms of glaucomatous neuropathy, primarily, because of the ability to collect tissue immediately on sacrifice with tissue perfusion or other tissue processing for specific assays (10, 11, 36, 44, 47, 49–52). In many ways, the use of the primate model for these studies replicates investigations with rodent models, except for some questions it is desirable to use a model with eyes that closely resemble the anatomy and physiology of the eyes of clinical patients. However, because there is considerable overlap in these areas between primate models and the rodent models described in Chap. 8, as well as several recent reviews (10–13), the present chapter will describe investigations of experimental glaucoma that are more unique to the primate model.

4. Clinical Assessment of Experimental Glaucoma

One form of validation of the primate model is the parallel in the structural and functional effects of laser-induced ocular hypertension and the clinical manifestation of glaucoma in patients. In the original studies, the evaluation of the laser procedure was based on ophthalmological observation of the optic nerve head and assessment of aqueous outflow facility (28). Subsequently, investigations have applied the model to study the basic pathophysiology of glaucoma through histological studies of the retinal ganglion cell layer (29, 32, 45–47, 53) and the optic nerve head (39, 43, 45), visual function studies by behavioral perimetry and contrast sensitivity (22, 38, 46, 54–57), retinal function studies by electroretinography (ERG) using either the pattern ERG (PERG) (32, 58), photopic negative response (PhNR) (59, 60), or multifocal ERG (mfERG) (37, 40, 61–63), and high-resolution imaging of retinal structure by optical coherence tomography (OCT) (48, 56), Heidelberg retinal tomography (HRT) (39, 43, 64, 65), or scanning laser polarimetry (SLP) (45).

On the face of it, there should be a causal relationship in experimental glaucoma in the time or amount of elevated IOP and the degree of loss of RGCs and visual sensitivity. Such a relationship was supported by a recent study that suggested a relationship between IOP and glaucomatous optic nerve head changes (66). In contrast, the susceptibly to pressure-induced neural damage, as measured by behavioral perimetry, has shown considerable variability (33). In the perimetry studies with 18 monkeys, sometimes visual field defects were correlated with the increases in IOPs and the defects progressed monotonically, while for other monkeys the stage of glaucoma (quantified by the mean deviation (MD) perimetric index) was not well-correlated with IOP. Across the subjects, significant functional deficits were measured after very short periods of elevated IOP in some eyes, but others withstood pressures in excess of 35 mmHg for several months before there were significant reductions in visual sensitivity. Three examples are presented (Fig. 1a–c) to illustrate the range of the relationships between laser treatments, intraocular pressures and visual field defects, as well as, the forms of individualistic relationships between MD and IOP over the time-course of experimental glaucoma. For each monkey, intraocular pressures (upper graph) and the associated MD perimetry indices, in standard deviation (*Z*-score) units (lower graph), are plotted as a function of time. Zero on the time axis was defined as the day of the first laser treatment applied to create an elevated IOP. The times of the initial and subsequent laser treatments are indicated by the vertical dashed lines that extend across the height of the coordinate system. As a reference for the expected normal IOP, the horizontal dashed line on the plot of intraocular pressures designates the mean pressure of the control eye over the time-course of the experiment. The reference line for the visual field data is the 95% confidence limit for normal visual fields and, thus, MD defects greater than the reference are statistically significant ($p < 0.05$).

The examples illustrate three major points. (1) There are considerable differences in the number of laser treatments required

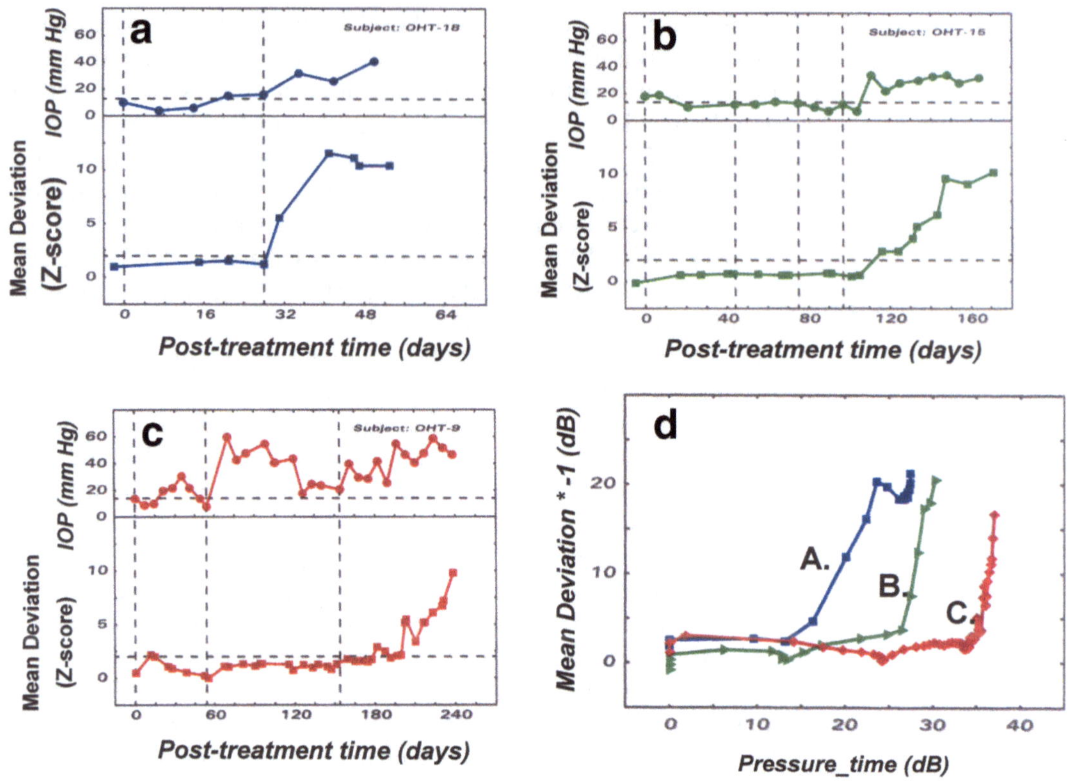

Fig. 1. **(a–c)** The intraocular pressure histories (*upper graphs*) and the time-courses for pressure-induced visual field defects (*lower graphs*) for three monkeys with experimental glaucoma. The times of the Argon laser treatments to create experimental glaucoma are indicated by the *dashed vertical lines* (note that the time scales are different for each subject). The *dashed horizontal line* superimposed on the intraocular pressure data represents the mean IOP for the subject's control eye during the time-course of the experiment. *Dashed lines* are superimposed on the visual field data to represent the upper 95% confidence limit for the mean deviation perimetric index for normal untreated eyes of monkeys. (**d**) The perimetric mean deviation (MD multiplied by −1 to invert the scale) as a function of the product of the magnitude of the intraocular pressure elevation and the amount of time that the pressure was elevated. Data are presented for three monkeys (**a–c**) to demonstrate the form of the function and the range of effects. (Adapted from Harwerth et al., ref. (33))

to cause pressure-induced visual field defects, ranging between two (Fig. 1a) to four treatments (Fig. 1b) for these monkeys. (2) Visual field defects are not well-correlated with increased intraocular pressures for some animals. Although for the two subjects illustrated in Fig. 1a, b, the perimetric defects followed the elevation of IOP, for the monkey presented in Fig. 1c, the initial increase in IOP was episodic and did not produce functional damage. (3) The time between the elevation of IOP and the beginning of progressive field defects for these three monkeys varied from less than 1 week (Fig. 1a) to about 3 months (Fig. 1c).

At least in part, the variability in the results may be explained by nonlinear effects in the relationship between IOP and RGCs and in the relationship between the RGC populations and visual sensitivity. In an attempt to gain a better understanding of pressure-induced neuronal damage, the monkeys' intraocular pressure histories were transformed to corresponding pressure–time (mmHg–days) values from time-course data for intraocular pressure (e.g., upper plots for each monkey). The calculations involved the determination of the time-dependent area under the curve for which the IOP was higher than the mean pressure of the control eye. The pressure–time at a given point in time was then paired with the MD (multiplied by −1 to make increases in effect positive number) as a measure of the stage of glaucoma at the same time, to define the individual points. Because the pressure–time varied greatly across the monkeys, the values were transformed to a logarithmic scale and the multiplied by 10 to make the scale comparable to the MD in decibels (dB).

The time-course data from the three monkeys (Fig. 1a–c) have been used to illustrate the range of results as illustrated in Fig. 1d. Typically, the functions are initially shallow, denoting a period of ocular hypertension that caused only small losses in visual sensitivity. Following the initial period of relatively benign pressure, there was a transition to a much more rapid pressure-induced neural damage. The critical pressure–time product associated with the transition, or knee, of the function demonstrated marked differences in the tolerances of the individual eyes to the effects of high intraocular pressure. For example, the difference between the subject presented in Fig. 1a and the subject for Fig. 1c was more than 3.5 orders of magnitude (>3,000×) in pressure–time preceding a significant effect of experimental glaucoma, i.e., a loss that exceeded the 95% confidence interval for the MD index. The functions accelerated rapidly from a threshold effect and continued until the total losses were severe and, then, the functions leveled off or progressed at a much slower rate, as would be expected.

Further evidence of the idiosyncratic behavior of the laser-induced experimental glaucoma has been demonstrated by the visual field defects of bilaterally treated monkeys. Figure 2 presents the gray-scale plots and summary treatment data for three monkeys with bilateral laser treatment of the trabecular meshwork. In each case, the total energy used and resulting IOPs for the two eyes were similar, but the symmetry of visual field defects varied considerably between the eyes. For example, for subject OHT-37 the experimental glaucoma caused similar neural damage in both eyes, but for the other two subjects the visual fields were obviously asymmetric, with the left eye within normal limits for subject OPT-38 and mild defects for subject OHT-39, while

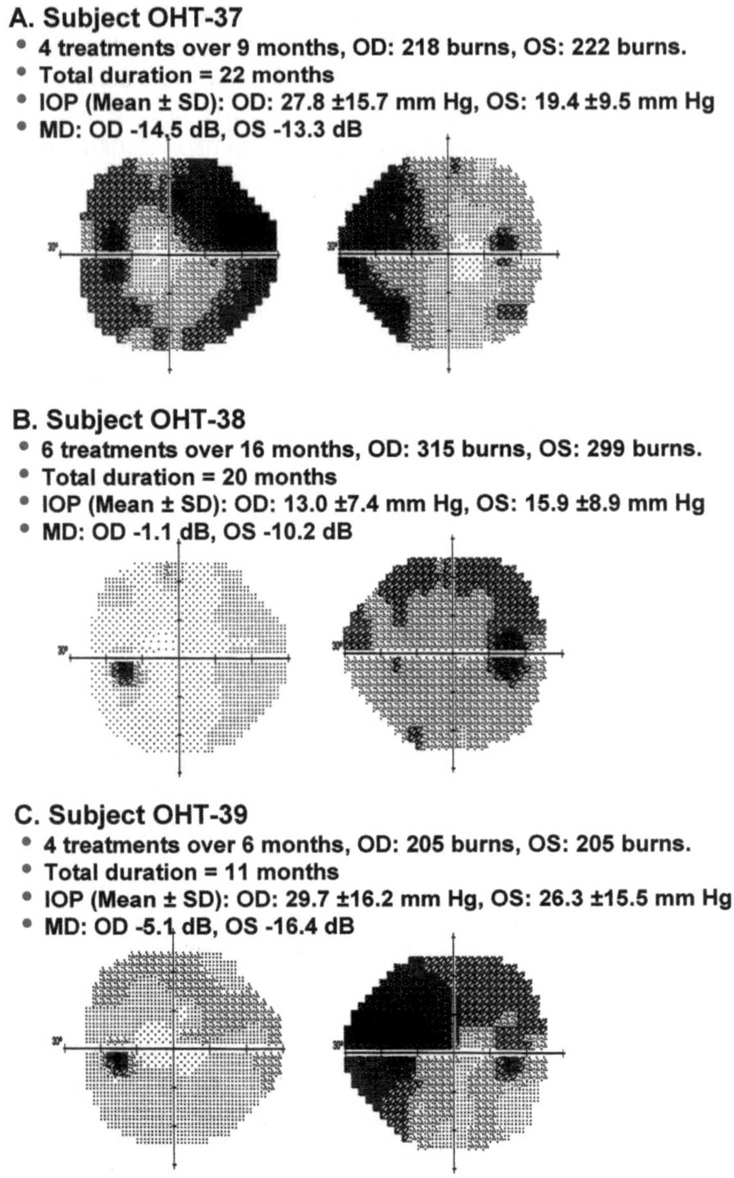

Fig. 2. Perimetry gray-scale plots for three monkeys with bilateral laser-induced experimental glaucoma that illustrate the variability of the relationships between laser energy, IOP elevation, and functional defects in the primate model. Details of the laser treatments and IOPs are presented for each monkey

the visual fields of the right eyes of both monkeys showed moderate-to-advanced defects.

It may be concluded from the evaluation of the relationship between intraocular pressure and the stage of glaucoma that an elevated intraocular pressure is adequate to cause the death of RGCs and other consequent effects of glaucomatous neuropathy, but there is considerable intersubject variability in susceptibility/tolerance to high IOP. The variability of the model may affect the use of the model in prospective studies of potential treatments for glaucoma. Nonetheless, experimental glaucoma at any point should provide a valid model of the pathophysiology of glaucoma at the specific stage that was achieved by the experimental induction of ocular hypertension. Therefore, the primate model has been most useful for investigations of structure-relationships, optic nerve physiology and biomechanics, cellular biology, and neurophysiology, where the state at the time of investigation is more important than the time-course prior to the investigation. On the other hand, in spite of the variability, the primate model has been used successfully to identify the potential benefit of a neuro-protection agent (40).

5. Visual Losses Correlated to Neural Losses in Experimental Glaucoma

The primate model of glaucoma allows point-in-time studies of the structure–function relationships that underlie basic measurements that are used to diagnose and assess progression in glaucoma. For example, monkeys that were behaviorally trained for clinical perimetry have been used to determine the relationship between visual sensitivity and RGC density (34, 38, 46, 55, 57). The measures of visual sensitivity were based on the clinical standard for assessment of visual sensitivity defects, i.e., computer-automated perimetry (SAP), using white-light test targets superimposed on a white-light background (19, 20, 67, 68). The psycho-physiological link between visual sensitivity and RGC density that best described the relationship was a nonlinear pooling of the outputs of neural detectors (38, 55). Response pooling represents an inverse relationship with RGC density, so the visual sensitivity (i.e., the reciprocal of visual threshold) depends on the density of RGCs, which in turn, varies with retinal eccentricity, normal aging, and the stage of glaucoma. The general prediction of the linking proposition, that the degree of vision loss would be proportional to the amount of ganglion cell loss, is reasonable (34, 46, 69, 70), but a more quantitative relationship between visual sensitivity and ganglion cell density has been established, using data from the primate model (55, 71, 72).

The structure–function model that was developed for experimental glaucoma is a nonlinear relationship for a point-wise correlation of retinal ganglion cell density to single measurements of visual sensitivity by SAP. The model becomes linear with logarithmic scaling, of both visual sensitivities and retinal ganglion cell densities, but the parameters for the linear functions vary systematically with eccentricity. Thus, the parameters (*y*-intercept and slope) of the structure–function relationship can be determined for a

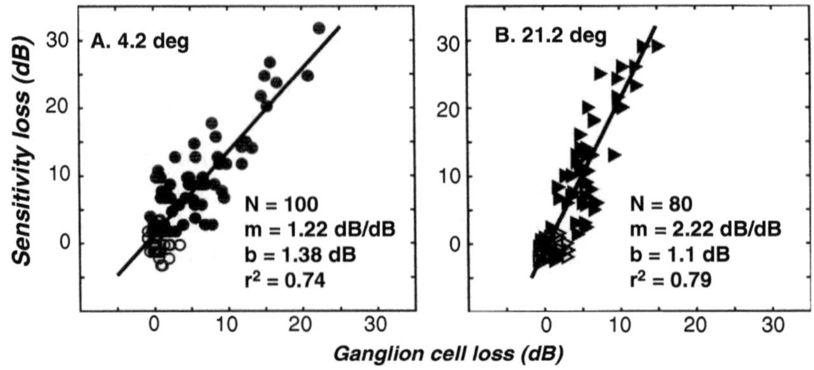

Fig. 3. The relationship between the reduction in visual field sensitivity and the loss of RGC density caused by experimental glaucoma in monkeys for two eccentricities for visual field and retinal locations. The data represent the difference between the expected normal values for SAP sensitivity or RGC density and the empirically determined values for both control eyes (*open symbols*) and laser-treated eyes with various stages of glaucomatous neuropathy (*filled symbols*). The lines superimposed on the data were derived by linear regression analyses of all of the data at each eccentricity with the parameters shown as an inset on each plot. (Adapted from Harwerth et al., ref. (71))

given eccentric location in the visual field and, the ganglion cell density for a given visual sensitivity can be predicted. Examples of the relationship between the losses of visual sensitivity and RGC density are presented in Fig. 3 for two locations, one for a retinal location of a stimulus near fixation (Fig. 3a) and the other at a more peripheral location (Fig. 3b). In both cases the relationship is obviously strong and statistically significant ($p < 0.0001$). Thus, the analysis of the relation between visual sensitivity and ganglion cell density demonstrates that perimetry visual sensitivity provides an accurate and relatively precise quantification of retinal ganglion cell losses caused by experimental glaucoma in monkeys. Further, the relationships derived from experimental glaucoma, with modifications for differences in the axial lengths of the eyes of monkeys and humans, also accurately predicts the RGC losses for glaucoma patients (72) and the age-related losses of RGCs in humans (73, 74).

6. Axonal Losses in the Nerve Fiber Layer in Experimental Glaucoma

During the past couple of decades, methods for objective imaging of retinal structure have been introduced for the diagnosis of glaucoma and the assessment of glaucomatous progression. The Heidelberg retinal tomography (HRT) for measuring optic nerve head parameters (75, 76) has the longest history of clinical use, but measures of the retinal nerve fiber layer (RNFL) thickness can be more directly related to RGC losses. The measurement of RNFL thickness by optical coherence tomography (OCT) is easily accomplished with anesthetized monkeys for assessment of axon losses from experimental glaucoma (56).

OCT of the RNFL in clinical glaucoma has been assessed by spectral domain and time domain technologies but, to date, the only investigations of experimental glaucoma with time domain OCT have been reported. Time domain OCT is a high-resolution cross-sectional imaging technique that provides in vivo measurements of the RNFL thickness with a resolution of 8–10 micrometers (μm) by measuring the echo time delay and intensity of back-reflected and back-scattered light from tissues at different depths (77). The various retinal layers in the OCT image can be differentiated as, for example, the RNFL thickness can be measured because axons of the RNFL are highly optically back-scattering compared to the RGC layer. The OCT methodology for RNFL thickness has been incorporated for clinical application and many studies have documented thinning of RNFL thickness that is correlated with the stage of glaucoma (16–18). In the primate model, the time-course measurements for experimental glaucoma have demonstrated a thinning of RNFL that seems to parallel progressive visual field defects, usually demonstrating a generalized loss of RGC axons that is consistent with the diffuse loss of visual sensitivity that is characteristic of the primate model with high IOP (56).

Measurements of glaucomatous neuropathy by SAP and OCT should be correlated because SAP thresholds from a given location are determined by the number of RGCs in the corresponding retinal area and the RNFL thickness represents the number of RGCs from the same location in the retina. Thus, the number of retinal ganglion cells provides a common factor underlying the clinical procedures and should define a quantitative relationship between the measurements. However, the relationship is complicated by several factors, such as: (1) RNFL thickness is not a localized measure that maps directly to a specific region of the optic nerve head, (2) OCT provides a measure of the total thickness and does not differentiate the neuronal from nonneuronal composition of the RNFL, and (3) there is likely to be remodeling of the RNFL accompanying axonal loss so that RNFL thinning is not directly proportional to neuronal loss.

Experimental glaucoma was used as a model to investigate potential methods of relating the results of different tests by estimating the number of RGC's from subjective or objective measurements, and then using the RGC estimates as an assessment of the state of glaucomatous neuropathy. Initially (56), using normative data for SAP and OCT from the control eyes of rhesus monkeys, procedures were developed (1) to translate visual sensitivity to a corresponding number of RGCs, (2) to derive the number of RGC axons from RNFL thickness measures and, (3) to map the visual field onto the ONH. The subsequent application

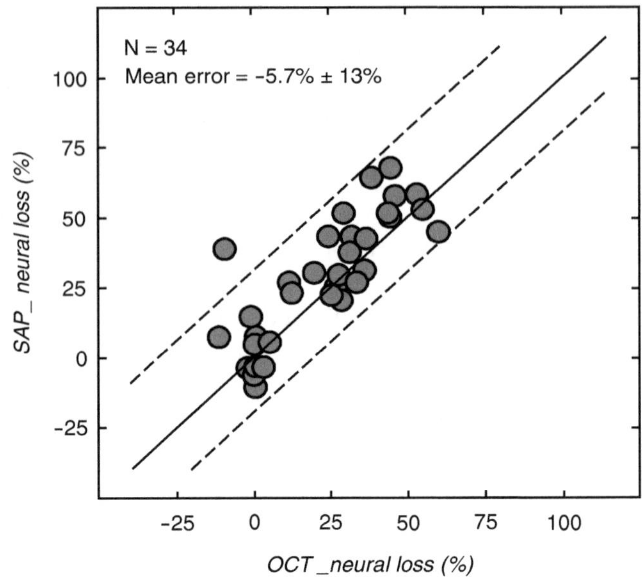

Fig. 4. Comparison of the neural losses that were estimated by SAP and OCT measurements from monkeys with experimental glaucoma. The data represent percentage loss of neurons or axons for the laser-treated eye, with respect to the control eye, based on estimates from SAP visual sensitivities versus OCT data for RNFL thicknesses. The *solid line* represents the unity correlation and the *dashed lines* are the 95% confidence limits from linear regression. (Adapted from Harwerth et al., ref. (56))

of the procedures to relate the neural bases of visual field defects and RNFL defects suggested a common progressive RGC pathophysiology over the time-course of experimental glaucoma. The systematic correlation of neural losses is demonstrated by Fig. 4, which compares the neural losses for SAP and OCT measurements. For this analysis, the neural loss by SAP measurements is the percent difference between a subject's control and laser-treated eyes in the number of RGCs estimated for the entire visual field. Similarly, the corresponding OCT neural loss represents the percent difference between the same subject's eyes in the number of axons estimated from measures of RNFL thickness. For reference, the solid line superimposed on the data is the one-to-one line representing perfect agreement and the dashed lines represent the 95% confidence limits of agreement for neural losses by SAP and OCT. It is apparent that the relationship is linear ($r=0.83$), but the differences are not evenly distributed around the line of unity correlation. The mean difference between SAP and OCT estimates is $5.7 \pm 13\%$, with visual function, on average, producing estimations of greater losses of neurons than the objective measurement of structure. The direction and magnitude of disagreement from the unity relationship implies compensating changes in the glial tissue (78–80) of the RNFL that occur with axonal loss and alters the transformation of RNFL thickness to the number of axons, if a constant value for axonal density is used. Although these aspects of RNFL thinning must be addressed in subsequent histological studies of the retinas of these monkeys, the general results establish that when the measurements are translated to a common factor related to RGCs there is concordance between the structure and function of normal and defective vision from glaucoma. There are, however, differences in experimental glaucoma in monkeys and clinical glaucoma in humans, and the further application to clinical glaucoma required modifications for axial length, aging effects (73, 74), and the stage of glaucoma (81).

7. Electrophysiological Assessment of Experimental Glaucoma

Although subjective measures of visual function seem to be a direct, valid assessment of the clinical stage of the pathology of glaucoma, there is considerable variability in subjective measures, as well as continuing questions about the underlying structure–function relationship for SAP (82–86) and low patient acceptance (87). In addition, for studies of retinal function with correlated histological verification, the use of behavioral testing of primates as a gage of neural dysfunction cannot be implemented in most laboratories. For these reasons, alternative approaches of using objective measures of RGC function that not rely on specific psycho-physiological linking propositions, but which are applicable to both experimental animals and human patients have been advocated. This approach has involved the direct measure of glaucomatous neuropathy from the reduced electrical activity generated by the retinal ganglion cell bodies or axons in the electroretinogram (ERG) (88–93).

The ERG is important as a general clinical tool for assessing the functional integrity of the retina, in vivo, because the electrical response to a change in illumination can be recorded noninvasively, but an application of ERG methods to glaucoma requires an isolation of components that are specifically RGC responses. Several ERG techniques for assessing retinal ganglion cell function have been investigated in the primate model of glaucoma, including the dark-adapted scotopic threshold response (STR) (89) and light-adapted photopic negative response (PhNR) from full-field flash stimulation (59, 60), as well as, pattern ERGs (PERG) (32, 58), multifocal ERGs (mfERG) (61–63), and oscillatory potentials (OPs) (94). All of these ERG techniques have been applied to experimental glaucoma using the same general methods for primate studies as have been used for clinical studies, except that the monkeys are anesthetized during recording.

In an early study of ERGs applied to experimental glaucoma, PERG responses were recorded for counterphased checkerboard patterns (32). The study by Johnson et al. is noteworthy because it demonstrated the potential use of experimental glaucoma to obtain histological validation of neuronal losses underlying a clinical test of retinal function. In that study PERG responses were obtained from macaque monkeys with unilateral laser treatments and the PERG responses of the eyes with experimental glaucoma were reduced in rough proportion to the histological losses of axons in the optic nerve. Across the four monkeys with histologic data, the neuronal losses ranged from 29 to 97% compared to PERG amplitude reductions of 20–69%. These studies provided a structure–function foundation for the use of PERG for the early diagnosis of glaucoma, but most of the subsequent development of PERG methodology has relied on clinical glaucoma, while the later studies of experimental glaucoma have utilized other ERG methods.

One potential alternative is the STR of the dark adapted eye, as a measure of inner retinal integrity, which has shown that reductions in the STR are qualitatively correlated with the stage of glaucomatous visual field defect (89). Although the investigations demonstrated the potential for the STR as an objective quantification of glaucomatous optic atrophy, in practice the procedure seems unlikely to have clinical application because of the length of time for administration, specifically, the successful recording of STRs required the subject to be completely dark-adapted (>30 min) and the very small responses ($<10\,\mu V$) dictated extensive averaging to improve signal-to-noise ratios. Therefore, other methods using the photopic ERG that produce larger signals and do not require dark adaptation are considered to be better for objective measurements of RGC losses from glaucoma.

The cellular origins of the photopic ERG components (a- and b-waves) are neurons that are not affected by glaucoma; but a later negative potential (PhNR) has been described, which originates in the neurons of the inner retina and can be eliminated by pharmacologically blocking spiking neurons or with advanced glaucoma (59, 60). In experimental glaucoma, the PhNRs of experimental eyes were significantly reduced with respect to the control eye when visual field measurements showed mild-to-moderate MD losses and, thereafter, further reductions of PhNR with field loss were quite gradual. Using the MD perimetric index as a measure of the stage of glaucoma, the relationship between PhNR and MD was linear, but with a shallow slope and a nonzero y-intercept (38). Consequently, the data from experimental glaucoma indicated that significantly reduced amplitudes of the PhNR occur in early stages of the optic neuropathy, while at later stages the further reductions in amplitude occur at a low rate with respect to the perimetric MD defect. In fact, the significant reduction of PhNR amplitude precedes clinically significant MD defects and may be closely associated with the early remodeling of the lamina cribrosa in early glaucoma that has been described in studies of the biomechanics of the ONH (15, 43, 64, 65). On the other hand, progression of glaucomatous optic neuropathy may be more difficult to quantify by analysis of PhNR amplitudes.

The sensitivity and specificity of the PhNR for detecting glaucoma in monkeys has been replicated for glaucoma patients and compares favorably with similar analyses of PERG measurements in human patients (95). In addition, the PhNR elicited with red stimuli on a blue background is usually larger than PERG amplitudes, sometimes by a factor of two and large signals are easier to measure (96). Also with respect to clinical application, the PhNR amplitude is relatively unaffected by opacities in the ocular media and it does not require accurate correction of refractive errors. However, the PhNR stimulus is a full-field flash that does not identify localized functional defects, but the focal losses of RGCs in glaucoma are a fundamental characteristic for the diagnosis and assessment of progression and, perhaps, other ERG techniques, such as the multifocal ERG (mfERG), may be more useful in that respect.

The procedure for mfERG utilizes simultaneous recordings of focal responses from more than 100 different retinal regions, and the mfERG can, therefore, provide the possibility for objective testing of localized retinal function (97–99). Studies of the mfERG have been conducted on both clinical and experimental glaucoma with similar results, but the primate model of glaucoma provided a way to determine the extent to which retinal ganglion cell activity contributes to the mfERG responses in localized areas of the retina.

A study of seven monkeys by Hare et al., (40) compared conventional and mfERG responses to histologically determined RGC losses for laser-treated and control eyes. They reported that while the amplitudes of conventional ERG responses were largely unaffected, even in eyes with severe loss of RGCs, the amplitudes of both the first- and second-order mfERG responses were attenuated in the laser-treated ocular hypertensive eyes compared to the normal control eyes. The relative amplitudes of the mfERG peaks were highly correlated with interocular ratio of the densities of surviving RGCs. Based on the strong correlation, it was concluded that both the first- and second-order components of the mfERG reflect a significant contribution from the activity of RGCs within the regions of stimulation.

The use of mfERGs as an objective method of assessing neural losses also has been investigated in monkeys with experimental glaucoma during the progression of their visual field defects (38, 62) to determine whether changes in the mfERG waveform correlate with perimetric measurements. Correlated measurements by objective and subjective methods would have important implications for clinical and experimental work, but the issues are complex. Although some components of the mfERG waveform are systematically reduced with loss of RGCs from experimental glaucoma (37, 40), studies of clinical glaucoma have not shown that mfERG changes are correlated to the localized defects found by perimetry (100, 101). Therefore, investigations of correlation of progressive visual field defects from of experimental glaucoma on mfERG responses were conducted with the MD perimetry index as an indicator of the generalized loss of sensitivity across the visual field (38).

It was found that, in contrast to the PhNR, experimental glaucoma did not cause a reduction in the amplitude of the principal mfERG responses until the visual field defects were highly significant. With more advanced visual defects, the mfERG reductions were linearly related to the MD of the visual fields. On the basis of these data, it appears that the mfERG may be more useful as an objective measure of advanced glaucomatous neuropathy than for its early detection. In this respect, the data from the two ERG procedures may be in agreement with the general concept that the best techniques for early diagnosis may not be the best techniques for determining progression, especially in advanced cases (38).

A second result of the investigation was that the normal nasotemporal variations in the mfERG responses were progressively reduced with deeper losses of visual sensitivity from experimental glaucoma and the first-order responses became progressively smoother (loss of oscillatory potentials) as the nasotemporal differences disappeared. In fact, the most obvious change in the mfERG associated with the advancing visual field defects was the loss of oscillatory potentials (OPs) (94, 102, 103). The OPs in two distinct frequency bands were subsequently analyzed in the monkey mfERG, i.e., fast OPs, with a center frequency of 143 Hz, and slow OPs, with a center frequency at 77 Hz (94). The mean amplitude of the fast OPs in control eyes was significantly larger in temporal than nasal retina and in superior versus inferior retina, while the slow OPs were largest in the foveal region. Experimental glaucoma reduced the amplitudes of fast OPs in all locations studied, even when visual field defects were moderate, whereas the slow OPs were reduced significantly only when field defects were severe. The fast OPs showed a moderate correlation with local visual field sensitivity and with local ganglion cell density (calculated from visual field sensitivity), but the correlation for the slow OPs was low. Thus, it seems that the fast OPs, rely more on the integrity of retinal ganglion cells and axons than the slow OPs, and, potentially, have utility for monitoring the progression of glaucoma and the effects of treatment.

Altogether, the investigations of ERGs in experimental glaucoma support the application of ERG techniques for an objective measure of the stage of pathophysiology of glaucoma. Components of the ERG waveforms that were generated by the retinal ganglion cells have been identified with each method. The photopic full-field flash ERG and mfERG procedures both are viable candidates, but each may be more useful at different stages of glaucoma. The differences on the form of correlation between ERG amplitudes and perimetry MD indices suggest that the structure–function relationships for the responses may be different, with the PhNR amplitude affected more by early losses of ganglion cells and the mfERG affected more by advanced losses of ganglion cells. For example, if it is assumed that the amplitude of the PhNR represents the linear sum of signals from all of the ganglion cells, then it is expected that it would be significantly reduced by a 40–50% loss of ganglion cells, whereas associated losses of visual sensitivity have not obtained statistical significance. On the other hand, the mfERG response to sequential changes in contrast would be influenced by response probability considerations, in the same way as visual sensitivity. However, neither type of ERG produces discrete responses that can be associated with the localized visual field defects that are typical of perimetric measures of visual sensitivity in clinical glaucoma. Experimental glaucoma, on the other hand, is a high IOP model that generally creates diffuse visual loss and the ERG procedures seem to provide accurate and precise reflections of the overall state of ganglion cell loss.

8. Biomechanical Properties of the Optic Nerve Head in Experimental Glaucoma

The biomechanical properties of the optic nerve head (ONH) are particularly important in the pathophysiology of glaucoma because the ONH represents a weak spot in the structure the eyeball (104). Within the ONH, the lamina cribrosa provides structural and functional support for the RGC axons as they exit the globe and much evidence suggests that axonal damage within the lamina cribrosa is the initial insult in the series of events leading to glaucomatous vision loss (15, 39, 105–108). The biomechanical mechanisms underlying these initial events of glaucomatous damage have been investigated through a series of studies of the primate model and much of our current knowledge of the alterations in the properties of the ONH tissues caused by glaucoma have been gained through these investigations of experimental glaucoma (43, 64, 65, 104, 107).

The results of several recent investigations of the alterations of tissue properties at early stages of glaucoma have provided significant new information about glaucomatous neuropathy.

For these studies of early glaucoma, laser treatments were administered to one of the monkey's eyes every 2 weeks until the IOP is statistically elevated above pretreatment pressure, usually to a level of 25–40 mmHg. The onset of glaucomatous damage was defined by the onset of ONH surface change as detected by confocal scanning laser tomography and, then, ONH tissues are collected 3–8 weeks after the initial changes in ONH cupping. The tissues were processed for histological analysis and reconstruction of ONH connective tissues (15, 39).

Three-dimensional analyses of the connective tissues of the ONH were accomplished by high-resolution imaging of the ONH tissue sections in 3 μm increments and, subsequently, creating a reconstruction from several hundred images (109). The analyses the connective tissues demonstrated extensive alterations in anatomy and structure by the time of the onset of observable ONH surface changes. For three monkeys that have been studied extensively, the signs of cupping occurred with axonal losses of only 16–30% (65), which are levels not likely to be detected by clinical SAP or OCT standard clinical measures.

In the investigations of early glaucomatous effects, clinical ONH cupping was divided into two components; laminar (deep) and prelaminar (shallow) pathophysiologic components (65). It was found that all of the three eyes with early experimental glaucoma demonstrated a large laminar component of cupping without a significant prelaminar component, which suggested a pathologic expansion of the connective tissue of the neural canal wall and the lamina cribrosa. Thus, in the young adult monkey eyes with moderate levels of chronic IOP elevation, early glaucomatous cupping was, principally, the result of permanent deformation of the connective tissues of the neural canal and lamina cribrosa, rather than prelaminar neural tissue thinning due to compression or axon loss. In fact, the early ONH surface changes occurred with prelaminar tissue thickening, not thinning, which was an unexpected finding.

In extending these results to clinical glaucoma, the results were interpreted as showing that a pathologic expansion of the tissues can occur under the influence IOP-related stress and strain, if it exceeded the elastic limit of the connective tissues, regardless of whether the IOP was physiologically elevated (65). Consequently, the detection of ONH cupping should suggest the presence of an optic neuropathy, but does not confirm that IOP is an active etiologic agent. In clinical application, if IOP elevation cannot be identified, then clinical cupping without connective tissue deformation should not necessarily indicate a need to lower IOP. Conversely, for patients with robust ONH connective tissues, IOP-related stress and strain could cause axonal degeneration, without damage to the underlying connective tissues. Therefore, direct evidence of ONH connective tissue damage is necessary to implicate IOP in an individual's optic neuropathy.

In a more recent study to characterize the alterations of connective tissue of the lamina cribrosa in early glaucoma, it was found that the volume of connective tissue was increased substantially, but with little change in the total fraction of connective tissue (110). These alterations suggested a remodeling of the lamina cribrosa with significant changes in both connective tissue and nonconnective tissue components in early glaucoma that preserve the normal tissue proportions. An additional form of tissue remodeling was implied by alterations in the viscoelastic material properties of the peripapillary sclera in early glaucoma that had occurred by time that ONH surface change (clinical cupping) could be detected by confocal scanning laser tomography (43).

The quantification of the biomechanical properties of the ONH and posterior sclera can provide important data for modeling of IOP-related stress and strain within the ONH connective tissue (15, 39). From a biomechanical standpoint, the pathophysiology of glaucomatous optic neuropathy is likely to be multifactorial, with the principal mechanisms of axonal insult acting in the region of the ONH and dependent on the stage of underlying connective tissue damage and other cellular involvement. In general, the data obtained from the primate model of experimental glaucoma have led to the postulation that for an individual ONH to demonstrate classical glaucomatous cupping, the connective tissues must have become damaged and undergone mechanical failure in a process that is determined by the distribution of IOP-related stress and strain, regardless of the level of IOP at which it occurs (39). Accordingly, the biomechanical alterations of the ONH that cause axonal damage within the lamina cribrosa, and lead to glaucomatous vision loss, can help to explain the considerable intersubject variability in susceptibility/tolerance to high IOP that was associated with the development and progression of visual field defects in monkeys with experimental glaucoma.

9. Summary and Conclusions

In a general sense, the studies of experimental glaucoma have reinforced the notion that the primate model is a unique model for investigations of the psychophysical, electrophysiological, histochemical, and/or anatomical characteristics of a prevalent ocular disease in humans. In part, the uniqueness comes from the close similarity of the visual function, physiology, and anatomy of monkeys' and humans' eyes. In addition, by using the model, ocular hypertension is isolated as the sole factor in the development of the glaucomatous optic neuropathy and, because the IOP is generally quite high, the full extent of the damage occurs in a period of a few months. Although experimentally induced ocular hypertension eliminates all but one of the well-known epidemiological risk factors for clinical glaucoma, all other aspects of cell death and clinical presentation appear to be identical to the clinical disease. The major drawbacks of the model for clinical research are, (1) the intersubject variability in susceptibility/tolerance to high IOP, which complicates its use in prospective studies of potential treatments for glaucoma and (2) the model cannot be used for studies of the mechanisms by which factors such as age, race, genetics, etc., cause an increase in the probability of acquiring glaucoma and/or affect the rate of progression. Nevertheless, it is a valid model for point-in-time investigations of glaucomatous neuropathy with respect to structure-relationships, optic nerve physiology and biomechanics, cellular biology, and neurophysiology.

The application of the model to studies of the psycho-physiological links of glaucoma through behavioral measurements of visual sensitivity and histologically determined RGC densities defined a quantitative causal relationship, which demonstrated that SAP provides an accurate and relatively precise quantification of local RGC densities. The subsequent extension of these structure–function relationships to human patients with glaucoma validated the utility of the primate model to develop procedures for the interpretation of clinical measurements. Further investigations were undertaken to develop procedures to correlate clinical measurements by SAP and OCT. The data from experimental glaucoma established the general principle that, if the measurements are translated to a common factor related to RGCs, then there is concordance between the visual sensitivity of a local area and the corresponding region of RNFL thickness for both normal and defective vision from glaucoma. In this case, however, the further application to clinical glaucoma required modifications for axial length, aging effects, and the stage of glaucoma.

Although the structure–function relations derived from experimental glaucoma substantiate the validity of perimetry for diagnosis and assessment of progression of glaucomatous neuropathy, there are downsides to SAP, such as measurement variability, insensitivity to early glaucoma and, for clinical use, low patient acceptance. As alternative tests of retinal function, the direct assessment of RGC function by ERG methodology has been advocated and the investigations of ERGs in experimental glaucoma have supported their application for objective measures of the stage of pathophysiology of glaucoma. The PhNR, mfERG, and PERG procedures are all viable candidates with RGC-generated waveforms, but none of these ERG methods produce discrete responses that can be associated with the localized visual field defects that are prototypical of perimetric measures of visual sensitivity in clinical glaucoma. That notwithstanding, the evidence from experimental glaucoma suggested that ERG procedures provide a reflection of the overall state of ganglion cell loss, with the PhNR sensitive to early glaucoma and the mfERG better correlated to more advanced stages of RGC losses.

The measures of RGCs by function (SAP, ERG) or structure (OCT) are generally considered to reflect the stage of glaucomatous neuropathy, but alterations of the ONH surface may reveal the earliest signs of the disease. Clinical cupping of the ONH is a standard diagnostic sign and, because the lamina cribrosa provides structural and functional support for the RGC axons as they exit the globe, axonal damage within the lamina cribrosa may initiate neural damage leading to glaucomatous vision loss. The notable investigations of early experimental glaucoma in monkeys to quantify the biomechanical properties of the ONH and posterior sclera has provided data for modeling of IOP-related stress and strain within the ONH connective tissue. Accordingly, these investigations of the primate model to understand the biomechanical alterations of the ONH that cause axonal damage are instrumental to our understanding of the pathophysiology of the clinical disease.

The overall conclusion from the review of many of the experimental findings from glaucoma in monkeys is that they provide fundamental data that should be applicable to improving methods for understanding and assessing glaucomatous optic neuropathy in patients. Furthermore, because the visual systems of macaque monkeys and humans are practically identical, the results from the array of experiments that have been described can be applied directly to glaucoma in patients.

References

1. American Health Assistance Foundation, National Glaucoma Research web site, http://www.ahaf.org/glaucoma/about/
2. Quigley HA, Vitale S (1997) Models of open-angle glaucoma prevalence and incidence in the United States. Invest Ophthalmol Vis Sci 38:83–91
3. AGIS Investigators (2000) The advanced glaucoma intervention study (AGIS): 7. The relationship between the control of intraocular pressure and visual field deterioration. Am J Ophthalmol 130: 429–440
4. Heijl A, Leske MC, Bengtsson B et al (2002) Reduction of intraocular pressure and glaucoma progression. Arch Ophthalmol 120:1268–1279
5. Kass MA, Heuer DK, Higginbothan EJ et al (2002) The ocular hypertension treatment study: a randomized trial determines that topical ocular hypotensive medication delays or prevents the onset of primary open-angle glaucoma. Arch Ophthalmol 120:701–713
6. Weinreb RN (2004) Primary open-angle glaucoma. Lancet 363:1711–1720
7. Epstein DL (1993) Primary open angle glaucoma. In: Epstein DL, Allingham RR, Schuman JS (eds) Chandler and Grant's glaucoma, 4th edn. Williams & Wilkins, Baltimore, pp 183–198
8. Quigley HA (1993) Open-angle glaucoma. N Engl J Med 328:1097–1106
9. Gramer E, Tausch M (1995) The risk profile of the glaucomatous patient. Curr Opin Ophthalmol 6:78–88
10. Kuehn MH, Fingert JH, Kwon YH (2005) Retinal ganglion cell death in glaucoma:

mechanisms and neuroprotective strategies. Ophthalmol Clin N Am 18:383–395
11. Nickells RW (1996) Retinal ganglion cell death in glaucoma: the how, the why, and the maybe. J Glaucoma 5:345–356
12. Quigley HA (2005) New paradigms in the mechanisms and management of glaucoma. Eye 19:1241–1248
13. Schumer RA, Podos SM (1994) The nerve of glaucoma. Arch Ophthalmol 112:37–44
14. Ernst JT, Potts AM (1968) Pathophysiology of the distal portion of the optic nerve. I. Tissue pressure relationships. Am J Ophthalmol 66:373–387
15. Burgoyne CF, Downs JC, Bellezza AJ et al (2005) The optic nerve head as a biomechanical structure: a new paradigm for understanding the role of IOP-related stress and strain in the pathophysiology of glaucomatous optic nerve head damage. Prog Ret Eye Research 24:39–73
16. Budenz DL, Michael A, Chang RT et al (2005) Sensitivity and specificity of the StratusOCT for perimetric glaucoma. Ophthalmol 112:3–9
17. Greenfeld DS (2002) Optic nerve and retinal nerve fiber analyzers in glaucoma. Curr Opin Ophthalmol 13:68–76
18. Zangwill LM, Bowd C (2006) Retinal nerve fiber layer analysis in the diagnosis of glaucoma. Curr Opin Ophthalmol 17: 120–131
19. Anderson DR (1987) Perimetry, with and without automation, 2nd edn. CV Mosby Co, St. Louis
20. Bengtsson B, Heijl A (2008) Visual field index for calculation of glaucoma rate progression. Am J Ophthalmol 145:343–353
21. Harwerth RS, Smith EL (1985) The rhesus monkey as a model for normal vision of humans. Am J Optom Physiol Optics 62:633–641
22. Harwerth RS, Smith EL, DeSantis L (1993) Behavioral perimetry in monkeys. Invest Ophthalmol Vis Sci 34:31–40
23. Gabelt BT, Kaufman PL (2003) Aqueous humor hydrodynamics. In: Kaufman PL, Alm A (eds) Adler's physiology of the eye: clinical application, 10th edn. Mosby Press, St. Louis, pp 237–289
24. Lutjen-Drecoll E (1998) Functional morphology of the trabecular meshwork in primate eyes. Prog Ret Eye Research 18: 91–119
25. Sharma RK, Ehinger BEJ (2003) Development and structure of the retina. In: Kaufman PL, Alm A (eds) Adler's physiology of the eye: clinical application, 10th edn. Mosby Press, St. Louis, pp 319–347
26. Casagrande VA, Ichida JM (2003) The primary visual cortex. In: Kaufman PL, Alm A (eds) Adler's physiology of the eye: clinical application, 10th edn. Mosby Press, St. Louis, pp 669–685
27. Rasmussen CA, Kaufman PL (2005) Primate glaucoma models. J Glaucoma 14:311–314
28. Gaasterland D, Kupfer C (1974) Experimental glaucoma in the rhesus monkey. Invest Ophthalmol 13:455–457
29. Quigley HA, Hohman RM (1983) Laser energy levels for trabecular meshwork damage in the primate eye. Invest Ophthalmol Vis Sci 24:1305–1307
30. Weber AJ, Zelenak D (2001) Experimental glaucoma in the primate induced by latex microspheres. J Neurosci Meth 111:39–48
31. Weber AJ, Harman CD (2005) Structure-function relations of parasol cells in the normal and glaucomatous primate retina. Invest Ophthalmol Vis Sci 46:3197–3207
32. Johnson MA, Drum BA, Quigley HA et al (1989) Pattern-evoked potentials and optic nerve fiber loss in monocular laser-induced glaucoma. Invest Ophthalmol Vis Sci 30: 897–907
33. Harwerth RS, Smith EL, DeSantis L (1997) Experimental glaucoma; perimetric field defects and intraocular pressure. J Glaucoma 6:390–401
34. Harwerth RS, Carter-Dawson L, Shen F et al (1999) Ganglion cell losses underlying visual field defects from glaucoma. Invest Ophthalmol Vis Sci 40:2242–2250
35. Jonas JB, Hayreh SS (1999) Localised retinal nerve fibre layer defects in chronic experimental high pressure glaucoma in rhesus monkeys. Br J Ophthalmol 83: 1291–1295
36. Morgan JE, Uchida H, Caproli J (2000) Retinal ganglion cell death in experimental glaucoma. Br J Ophthalmol 84:303–310
37. Hare WA, Ton H, Ruitz G et al (2001) Characterization of retinal injury using ERG measures obtained with both conventional and multifocal methods in chronic ocular hypertensive primates. Invest Ophthalmol Vis Sci 42:127–136
38. Harwerth RS, Crawford ML, Frishman LJ et al (2002) Visual field defects and neural losses from experimental glaucoma. Prog Retinal Eye Res 21:91–125
39. Burgoyne CF, Downs JC, Bellezza AJ et al (2004) Three-dimensional reconstruction of normal and early glaucoma monkey optic

nerve head connective tissues. Invest Ophthalmol Vis Sci 45:4388–4399

40. Hare WA, WoldeMussie E, Lai RK et al (2004) Efficacy and safety of memantine treatment for reduction in changes associated with experimental glaucoma in monkey. I. Functional measures. Invest Ophthalmol Vis Sci 45:2625–2639
41. Ollivier FJ, Brooks DE, Kallberg ME et al (2004) Time-specific intraocular pressure curves in Rhesus macaques (*Macaca mulatta*) with laser induced ocular hypertension. Vet Ophthalmol 7:23–27
42. Brooks DE, Kallberg ME, Komaromy AM et al (2005) Optive nerve head neuroretinal rim blood flow differences in monkeys with laser-induced glaucoma. Vet Ophthalmol 8:113–119
43. Downs JC, Suh JKF, Thomas KA et al (2005) Viscoelastic material properties of the peripapillary sclera in normal and early-glaucoma monkey eyes. Invest Ophthalmol Vis Sci 46:540–546
44. Hashimoto K, Parker A, Molone P et al (2005) Long-term activation of c-Fos and c-Jun in optic nerve head astrocytes in experimental ocular hypertension in monkeys and after exposure to elevated pressure in vitro. Brain Res 1054:103–115
45. Shimazawa M, Tomita G, Taniguchi T et al (2006) Morphometric evaluation of changes with time in optic disc structure and thickness of retinal nerve fibre layer in chronic ocular hypertensive monkeys. Exp Eye Res 82:427–440
46. Sasaoka M, Nakamura K, Shimazawa M et al (2008) Changes in visual fields and lateral geniculate nucleus in monkey laser induced high intraocular pressure model. Exp Eye Res 86:770–782
47. Kompass KS, Agapova OA, Li W, et al. Bioinformatic and statistical analysis of the optic nerve head in a primate model of ocular hypertension. http://www.biomedcentral.com/1471-2202/9/93
48. Schuman JS, Pedut-Kloizman T, Pakter H et al (2007) Optical coherence tomography and histologic measurements of nerve fiber layer thickness in normal and glaucomatous monkey eyes. Invest Ophthalmol Vis Sci 48:3645–3654
49. Quigley HA (1998) Neuronal death in glaucoma. Prog Ret Eye Res 18:39–57
50. Hof PR, Lee PY, Yeung G et al (1998) Glutamate receptor subunit GluR2 and NMDARl immunoreactivity in the retina of macaque monkeys with experimental glaucoma does not identify vulnerable neurons. Exp Neurol 153:234–241
51. Carter-Dawson L, Crawford MLJ, Harwerth RS et al (2002) (2002) Vitreal glutamate concentration in monkeys with experimental glaucoma. Invest Ophthalmol Vis Sci 43:2633–2637
52. Kashiwagi K, Ou B, Nakamura S et al (2003) Increase in dephosphorylation of the heavy neurophilament subunit in the monkey chronioc glaucoma model. Invest Ophthalmol Vis Sci 44:154–159
53. Desatnik H, Quigley HA, Glovinsky Y (1996) Study of central ganglion cell loss in experimental glaucoma in monkey eyes. J Glaucoma 5:46–53
54. Harwerth RS, Smith EL, Chandle M (1999) Progressive visual field defects from experimental glaucoma: measurements with white and colored stimuli. Optom Vis Sci 76:558–570
55. Harwerth RS, Carter-Dawson L, Smith EL et al (2004) Neural losses correlated with visual losses in clinical perimetry. Invest Ophthalmol Vis Sci 45:3152–3160
56. Harwerth RS, Vilupuru AS, Rangaswamy NV et al (2007) The relationship between nerve fiber layer and perimetry measurements. Invest Ophthalmol Vis Sci 48:763–773
57. Harwerth RS The, Charles F (2008) Prentice award lecture 2006: a neuron doctrine for glaucoma. Optom Vis Sci 85:436–444
58. Marx MS, Podos SM, Bodis-Wollner I et al (1988) Signs of early damage in glaucomatous monkey eyes: low spatial frequency losses in the pattern ERG and VEP. Exp Eye Res 46:173–184
59. Viswanathan S, Frishman LJ, Robson JG et al (1999) The photopic negative response of the macaque electroretinogram is reduced by experimental glaucoma. Invest Ophthalmol Visl Sci 40:1124–1136
60. Viswanathan S, Frishman LJ, Robson JG (2000) The uniform field and pattern ERG in macaques with experimental glaucoma: removal of spiking activity. Invest Ophthalmol Vis Sci 41:2797–2810
61. Hood DC, Frishman LJ, Viswanathan S et al (1999) Evidence for a substantial ganglion cell contribution to the primate electroretinogram (ERG): the effects of TTX on the multifocal ERG in macaque. Vis Neurosci 16:411–416
62. Frishman LJ, Saszik S, Harwerth RS et al (2000) Effects of experimental glaucoma in macaques on the multifocal ERG. Doc Ophthalmol 100:231–251

63. Hood DC, Bearse MA, Sutter EE et al (2001) The optic nerve head component of the monkey's (*Macaca mulatta*) multifocal electroretinogram (mERG). Vision Res 41:2029–2041
64. Yang H, Downs JC, Girkin C et al (2007) 3-D histomorphometry of the normal and early glaucomatous optic nerve head: lamina cribrosa and peripapillary scleral position and thickness. Invest Ophthalmol Vis Sci 48:4597–4607
65. Yang H, Downs JC, Bellezza A et al (2007) 3-D histomorphometry of the normal and early glaucomatous optic nerve head: prelaminar neural tissues and cupping. Invest Ophthalmol Vis Sci 48:5068–5084
66. Nickells RW, Cassandra L, Schlamp Y et al (2007) Surgical lowering of elevated intraocular pressure in monkeys prevents progression of glaucomatous disease. Exp Eye Res 84:729–736
67. Heijl A (1985) Strategies for detection of glaucoma defects. In: Drance SM, Anderson DR (eds) Automatic perimetry in Glaucoma. A practical guide. Grune & Stratton Inc., Orlando, pp 43–54
68. Johnson CA (1996) Standardizing the measurement of visual fields for clinical research. Ophthalmol 103:186–189
69. Quigley HA, Dunkelberger GR, Green WR (1989) Retinal ganglion cell atrophy correlated with automated perimetry in human eyes with glaucoma. Am J Ophthalmol 107:453–464
70. Kerrigan-Baumrind LA, Quigley HA, Pease ME et al (2000) Number of ganglion cells in glaucoma eyes compared with threshold visual field tests in the same persons. Invest Ophthalmol Vis Sci 41:741–748
71. Harwerth RS, Carter-Dawson L, Smith EL et al (2005) Scaling the structure-function relationship for clinical perimetry. Acta Ophthalmol Scand 83:448–455
72. Harwerth RS, Quigley HA (2006) Visual field defects and retinal ganglion cell losses in human glaucoma patients. Arch Ophthalmol 124:853–859
73. Harwerth RS, Wheat JL (2008) Modeling the effects of aging on retinal ganglion cell density and nerve fiber layer thickness. Graefes Arch Clin Exp Ophthalmol 246:305–314
74. Harwerth RS, Wheat JL, Rangaswamy NV (2008) Age-related losses of retinal ganglion cells and axons. Invest Ophthalmol Vis Sci 49:4437–4443
75. Janknecht P, Funk J (1994) Optic nerve head analyzer and Heidelberg retina tomography: accuracy and reproducibility of topographic measurements in a model eye and in volunteers. Br J Ophthalmol 78:760–768
76. Philippin H, Unsoeld A, Maier P et al (2005) Ten-year results: detection of long-term progressive optic disc changes with confocal laser tomography. Graefes Arch Clin Exp Ophthalmol 25:1–5
77. Fujimoto JG, Hee MR, Huang D et al (2004) Principles of optical coherence tomography. In: Schuman JS, Puliafito CA, Fujimoto JG (eds) Optical Coherence Tomography of Ocular Diseases, 2nd edn. Slack, inc, Thorofare, NJ, pp 3–19
78. Wang L, Cioffi GA, Cull G et al (2002) Immunohistologic evidence for retinal glial cell changes in human glaucoma. Invest Ophthalmol Vis Sci 43:1088–1094
79. Varela HJ, Hernandez MR (1997) Astrocyte responses in human optic nerve hrad with primary open-angle glaucoma. J Glaucoma 6:303–313
80. Grieshaber MC, Orgul S, Schoetzau A et al (2007) Relationship between retinal glial cell activation in glaucoma and vasacular dysregulation. J Glaucoma 16:215–219
81. Wheat JL, Rangaswamy NV, Harwerth RS (under review) Correlating RNFL thickness by OCT with perimetric sensitivity in glaucoma patients. Invest Ophthalmol Vis Sci
82. Garway-Heath DF, Capriolo J, Fitzke FW et al (2000) Scaling the hill of vision: the physiological relationship between light sensitivity and ganglion cell numbers. Invest Ophthalmol Vis Sci 41:1774–1782
83. Swanson WH, Felius J, Pan F (2004) Perimetric defects and ganglion cell damage: Interpreting linear relations using a two-stage linear model. Invest Ophthalmol Vis Sci 45:466–472
84. Ventura LM, Sorokac N, De Los Santos R et al (2006) The relationship between retinal ganglion cell function and retinal nerve fiber layer thickness in early glaucoma. Invest Ophthalmol Vis Sci 47:3904–3911
85. Hood DC, Anderson SC, Wall M et al (2007) Structure versus function in glaucoma: an application of a linear model. Invest Ophthalmol Vis Sci 48:3662–3668
86. Hood DC, Kardon RH (2007) A framework for comparing structural and functional measure of glaucomatous damage. Prog Ret Eye Res 26:688–710
87. Gardiner SK, Demirel S (2008) Assessment of patient opinions of different clinical tests used in the management of glaucoma. Am J Ophthalmol 115:2127–2131

88. Porciatti V, Falsini B, Brunori S et al (1987) Pattern electroretinogram as a function of spatial frequency in ocular hypertension and early glaucoma. Doc Ophthalmol 65: 349–355
89. Frishman LJ, Shen FF, Du L et al (1996) The scotopic electroretinogram of macaque after retinal ganglion cell loss from experimental glaucoma. Invest Ophthalmol Vis Sci 37:125–141
90. Graham SL, Klistorner A (1998) Electrophysiology: a review of signal origins and applications to investigating glaucoma. Aust New Zea J Ophthalmol 26:71–85
91. Hood DC (2000) Assessing retinal function with the multifocal technique. Prog Ret Eye Res 19:607–646
92. Venrura LM, Porciatti V (2006) Pattern electroretinogram in glaucoma. Curr Opin Ophthalmol 17:196–202
93. Bach M, Hoffmann MB (2008) Update on the pattern electroretinogram in glaucoma. Optom Vis Sci 85:386–395
94. Rangaswamy NV, Zhou W, Harwerth RS et al (2006) Effects of experimental glaucoma in primates on oscillatory potentials of the slow-sequence mfERG. Invest Ophthalmol Vis Sci 47:753–767
95. Graham SL, Drance SM, Chauhan BC et al (1996) Comparison of psychophysical and electrophysiological testing in early glaucoma. Investi Ophthalmol Vis Sci 37:2651–2662
96. Viswanathan S, Frishman LJ, Robson JG et al (2001) The photopic negative response of the flash electroretinogram in primary open angle glaucoma. Invest Ophthalmoll Vis Sci 42:514–522
97. Sutter EE (1991) The fast m-transform: a fast computation of cross-correlations with binary m-sequences. Soc Indust Appl Math Comp 20:686–694
98. Sutter EE, Tran D (1992) The field topography of ERG components in man. I. The photopic luminance response. Vision Res 32:433–466
99. Sutter EE, Bearse MA (1999) The optic nerve head component of the human ERG. Vision Res 39:419–436
100. Fortune B, Johnson CA, Cioffi GA (2001) The topographic relationship between multifocal electroretinographic and behavioral perimetric measures of function in glaucoma. Optom Vis Sci 78:206–214
101. Klistorner AI, Grahan SL, Martins A (2000) Multifocal pattern electroretinogram does not demonstrate localized field defects in glaucoma. Doc Ophthalmol 100:155–165
102. Gur M, Zeevi YY, Bielik M et al (1987) Changes in the oscillatory potentials of the electroretinogram in glaucoma. Curr Eye Res 6:457–466
103. Heynen H, Wachtmeister I, von Norren D (1985) Origin of the oscillatory potentials in primate retina. Visio Res 25:1365–1373
104. Downs JC, Roberts MD, Burgoyne CF (2008) Mechanical environment of the optic nerve head in glaucoma. Optom Vis Sci 85:425–435
105. Gaasterland D, Tanishima T, Kuwabara T (1978) Axoplasmic flow during chronic experimental glaucoma I Light and electron microscope studies of the monkey optic nerve head during development of glaucomatous cupping. Invest Ophthalmol Vis Sci 17:838–846
106. Quigley HA, Addicks EM, Green WR et al (1981) Optic nerve damage in human glaucoma. II. The site of injury and susceptibility to damage. Arch Ophthalmol 99: 635–649
107. Bellezza AJ, Rintalan CJ, Thompson HW et al (2003) Deformation of the lamina cribrosa and anterior scleral canal wall in early experimental glaucoma. Invest Ophthalmol Vis Sci 44:623–637
108. Whitmore AV, Libby RT, John SWM (2005) Glaucoma:thinking in new ways – a role for autonomous axonal self-destruction and other compartmentalized processes. Prog Ret Eye Res 24:639–662
109. Yang H, Downs JC, Burgoyne CF (2009) Physiologic inter-eye differences in monkey optic nervehead architecture and their relation to changes in early experimental glaucoma. Invest Ophthalmol Vis Sci 50: 224–234
110. Roberts MD, Grau V, Grimm J. et al (2009) Remodeling of the Connective Tissue Microarchitecture of the Lamina Cribrosa in Early Experimental Glaucoma. Invest Ophthalmol Vis Sci 50:681–690

Chapter 10

Animal Models of Retinal Ischemia

Jacky Man Kwong Kwong and Joseph Caprioli

Abstract

Ischemic disorders of the retina constitute a common cause of blindness and visual impairment worldwide. Retinal ischemia is a disorder initially caused by an imbalance between the supply of metabolic substrates to the retina and its demand for nutrients. If treatment cannot be implemented to correct this imbalance, the result is irreversible ischemic and apoptosis-related cascades leading to cell death. A number of animal models are available for both studying the mechanisms of retinal ischemia and exploring potential treatments to prevent neuronal degeneration. However, the vascular supply and induction procedures of retinal ischemia in animal models must be better understood for application to human disorders.

Key words: Retinal ischemia, Central retinal artery occlusion, Central retinal vein occlusion, Branch retinal vein occlusion, Pressure elevation, Ligation of optic nerve, Ligation of ophthalmic vessels, Photodynamic ablation, Endothelin-1

1. Introduction

Retinal ischemia is a pathological condition that occurs when blood supply is insufficient to meet the metabolic demands of the retina. As retinal tissue has a higher metabolic rate than other tissues, any disturbance in the circulation may affect the supply of oxygen and metabolic substrates, and the removal of metabolic wastes contributing to retinal ischemia (1).

Retinal ischemia is a common cause of visual impairment and blindness worldwide. It is associated with various high impact ocular disorders such as retinal artery and venous occlusion, hypertension, diabetic retinopathy, and glaucoma. The outcomes of retinal ischemia are variable and probably multifactorial. However, the exact location of the vascular occlusion will determine its impact (2). The blood supply and nutrients of the inner retina are generally supplied by the central retinal artery, while the

Iok-Hou Pang and Abbot F. Clark (eds.), *Animal Models for Retinal Diseases*, Neuromethods, vol. 46
DOI 10.1007/978-1-60761-541-5_10, © Springer Science+Business Media, LLC 2010

outer retina is supplied by the choriocapillaris through the retinal pigment epithelium. Hence, the occlusion of the retinal artery results in inner retinal ischemia, but it may not cause ischemia to the outer retina. However, the occlusion of the ophthalmic artery leads to global retinal ischemia, because it supplies the central retinal artery and the posterior ciliary arteries which subdivide into the choriocapillaries. Overall, the most noticeable outcome of strictly retinal ischemia is the loss of retinal ganglion cells (RGCs) and the thinning of the nerve fiber layer.

2. Retinal Ischemia in Humans

The extent of ischemic injury to the retina in humans depends on the number of occluded vessels, the exposure time to vascular disturbance, susceptibility, and specific cell types. Due to the complex anatomy of retinal vasculature such as anastomoses of vessels, the roles of retinal ischemia in ocular diseases vary.

Acute retinal vascular occlusive disorders, found to be one of the major causes of blindness and severe visual impairment, is a collection of disorders. The pathogeneses, clinical features, and management of these disorders vary. Retinal ischemia is found to be a common consequence with variable severity in these disorders (2, 3). In general, retinal arterial occlusions are often embolic in origin, while retinal venous occlusions are nearly always thrombotic.

Central retinal artery occlusion (CRAO) is the second most common acute retinal vascular occlusive disorder with a very poor prognosis. It accounts for approximately 1 per 10,000 outpatient visits in United States and is recognized as an ophthalmic emergency (4, 5). CRAO usually results in severe retinal ischemia which leads to visual acuity of 20/200 or worse and occurs within hours of onset. The dramatic loss of visual function may be painless and only 21–35% of affected eyes retain useful vision. In selected cases (<10%), the central vision may be spared due to the presence of a patent cilioretinal artery, in which the central inner retina is supplied by the choroidal circulation. A number of treatment modalities for CRAO have been practiced but all have limited success (6). In the elderly, CRAO is often associated with the risk factors for arteriosclerosis while the causes in younger patients are blood dyscrasias or systemic inflammatory disease (3). Its cherry-red spot appearance is a major clinical characteristic due to swelling of the inner retina and loss of retinal transparency which increases the pallor of the surrounding macula. Interestingly, in severe retinal ischemia, the inner retinal layers are lost without reactive gliosis, followed by optic atrophy, while the outer layers remain intact. Branch retinal artery occlusion (BRAO), has similar risk factors and pathological features, but the ischemic features are only limited to well-demarcated areas of the retina supplied by the occluded artery.

Retinal venous occlusions are the most common among retinal occlusive disorders. Mitchell et al. (7) reported that retinal venous occlusive disorders increased exponentially with age and were present in 4.6% of those 80 years or older in a population. This result is strongly associated with a cardiovascular risk profile and elevated intraocular pressure. Younger patients usually have underlying hematological or inflammatory diseases (8). Retinal venous occlusions do not represent one single disease entity. Due to a lack of classification standards, the literature is unclear in its etiologies, pathogeneses, clinical features, classifications, prognoses, and managements.

Many studies are based solely on fluorescein angiography to divide central retinal vein occlusion (CRVO) into two subclasses, ischemic and nonischemic. However, Hayreh and co-workers found that a combination of functional tests to examine visual acuity, visual field by Goldmann perimeter, relative afferent papillary defect, and electroretinography, is more reliable to detect ischemic or nonischemic CRVO, especially during the early stages (2). Clinical presentations of ischemic CRVO are similar to CRAO but have the major characteristic of a widespread retinal hemorrhage which corresponds to retinal necrosis with scattered "cotton wool" spots representing nerve fiber layer infarcts. Nonischemic CRVO is less dramatic with normal or moderately reduced visual acuity, no afferent pupillary defect, dilated retinal veins with very few scattered hemorrhages, and a near-absence of "cotton wool" spots. Fluorescein angiography demonstrates capillary obliteration and capillary nonperfusion in the areas of retinal ischemia. In regard to CRVO's etiology, the site of occlusion is still controversial. The histopathological study by Green et al. (9), demonstrated that thrombus was present at the level of lamina cribrosa in all 29 human eyes which were enucleated due to neovascular glaucoma. In contrast, Hayreh et al. argued that those enucleated human eyes represented a select group of ischemic CRVO with neovascular glaucoma (2, 10). Hayreh et al. also demonstrated in anatomical and clinical studies as well as experimental studies in monkeys that the site of occlusion in CRVO is in the optic nerve at a variable distance posterior to the lamina cribrosa (10).

Branch retinal vein occlusion (BRVO) is similar to CRVO but only affects a segment of retina occurring at arteriovenous crossings in which the arteriole crosses over the vein and changes the hemodynamics in the vein. Although the pathophysiology of BRVO remains largely unknown, retinal ischemia is believed to determine the disease outcomes.

Diabetic retinopathy is the most frequent complication of diabetes and a leading cause of blindness in developed countries (11). It is a retinal microangiopathy but the causes of retinal damage and disease progression are not clearly known. Elevation of blood and tissue glucose levels has been suggested to initiate oxidative stress, localized hypoxia and inflammatory reactions. Adhesion of leukocytes to vessel walls, aggregation of platelets, altered blood flow, loss of intramural pericytes, and thickening of the basement membrane are the earliest vascular signs (12). Blockage of the retinal capillaries may stimulate the production of angiogenic factors such as vascular endothelial growth factor (VEGF), a potent survival factor for endothelial cells, which causes the loosening of the vascular cell–cell junction and increases vascular permeability (13). Ophthalmoscopic examination reveals dot hemorrhages, "cotton wool" spots, and retinal edema, indicating vascular leaking, and retinal ischemia leads to macular edema and central vision loss. Microscopically, swelling of the retinal ganglion cells precedes degenerative changes of retinal neurons and gliosis in the inner retinal layers. Advanced ischemia in the retina leads to formation of acellular nonperfused capillaries, blood leakage, proliferative vitreous scarring, and, finally, retinal detachment which may result in blindness (14).

Glaucoma is the second leading cause of blindness worldwide. Visual deterioration in patients with glaucoma is caused by the loss of retinal ganglion cells. Progression of glaucoma is often associated with risk factors including intraocular pressure, age, ethnicity, family history, myopia, and central corneal thickness. Clinically, characteristic optic nerve head excavation and thinning of nerve fiber layer are found corresponding to arcuate scotoma in glaucomatous patients. Current medical and surgical interventions aim to lower intraocular pressure; however, the disease progress cannot be completely arrested. Deprivation of neurotrophic factors due to blockage of axonal transport at the site of lamina cribrosa can lead to RGC degeneration. Episodes of ischemia and reperfusion in the optic nerve head area have been proposed as the basis for retinal pathology in glaucoma (15). Whether retinal ischemia plays a role in glaucomatous degeneration is still controversial, but retinal ischemia has been widely studied to explore new treatments for glaucoma due to loss of retinal ganglion cells and poor visual outcomes (16, 17).

To research effective treatments and comprehensive management of patients, experimental animal models mimicking these disorders became invaluable tools to both understand the pathophysiology of retinal ischemia and evaluate potential new treatments. However, whether the data can be translated to the clinic relies heavily on the experimental design and the selection of proper animal models.

3. Animal Models of Retinal Ischemia

While in vitro ischemic models provide much insight into some of the biochemical and molecular events caused by hypoxia and substrate deprivation, animal models remain essential in understanding the pathophysiology. The entire vascular and associated immune and other support systems provide the opportunity to examine the complex interactions and responses after ischemic insults. In all animal models of retinal ischemia, the retinal circulation is experimentally altered to study the balance between energy substrate supply and metabolic demands, vascular and neuronal changes, and remodeling of retinal tissue.

The transient decreased supply of oxygen and glucose may deplete ATP stores and later may result in functional changes if an adequate blood supply cannot resume within the tolerance time, which is defined as the time within which the measurable changes are reversible (18). When the ischemic insult lasts longer than the tolerance time, the damage becomes irreversible. The result changes ion flows, neurotransmitter release, high metabolic waste levels, and morphological alterations. Reperfusion, the restoration of circulation after blockage, also damages cells because the restoration of oxygen to ischemic regions may amplify the injury. Retinal ischemia causes the leakage of LDH, a marker of ischemic cell death after introduction of oxygen with continued glucose absence (19). In stroke, the ischemic penumbra, a better perfused region surrounding ischemic core, remains intact for several hours and can be saved by reperfusion or by drugs that prevent the spreading of the infarct extending to the penumbral region (20). Therefore, the quality and quantity of ischemia and reperfusion have to be taken into consideration in setting up an animal model.

Monkeys, rabbits, rats, mice, guinea pigs, cats, dogs, and pigs have been used as animal models of retinal ischemia. In choosing a species, several factors are usually considered: (1) anatomy of vascular circulation and retina; (2) relevance to humans in terms of disease outcome; (3) sample size and cost; (4) availability of animals; and (5) ease of handling. There is no species that can perfectly fit all these requirements. In rabbits, the central retinal artery runs superficially and horizontally toward the nasal and temporal sides and most of the retina is avascular while its nutrition is mostly supplied by the choroidal circulation. Cats have no central artery of the retina and the entire thickness of the retina is supplied by the posterior ciliary arterial (PCA) circulation. Conversely, in humans and primates, the CRA supplies only the inner layers of the retina while the outer retina is supplied entirely by the PCA circulation. Rats are most widely used in laboratories

Table 1
Summary of commonly used rat retinal ischemia models

Model	Surgical location	Procedures	Degree of ischemia	Possible effect	Technical difficulty
Pressure elevation	Intraocular	Intraocular pressure > systolic pressure 110 mmHg	Complete	Global ischemia	Medium
Optic nerve ligation	Optic nerve	Suture	Complete	Optic nerve trauma	High
Vessels ligation	Ophthalmic vessels	Suture	Complete	Delayed RGC loss	High
	Unilateral carotid arteries	Suture	Incomplete	Cerebral ischemia	High
	Bilateral carotid arteries	Suture	Complete	Cerebral ischemia	High
Photodynamic ablation	Retinal and choroidal vessels	Intravenous rose Bengal + light illumination	Incomplete	Phototoxicity	Medium
	Central or branch retinal vessels	Intravenous rose Bengal + argon laser + tissue-type plasminogen activator	Incomplete	Photoreceptor degeneration	Medium
Endothelin administration	Anterior optic nerve	Mini-pump implant	Incomplete	Variable response	Medium
	Vitreous	Intravitreal injection	Incomplete	Misregulated anterograde axonal transport	Low
	Conjunctiva	Subconjunctival injection	Incomplete	Systemic vasoconstriction	Low

because their vascular and retinal structures are closer to those of humans, they are less expensive, and easier to procure and handle. This chapter describes several of the most common rat models of retinal ischemia (Table 1).

3.1. Pressure Elevation Model

Elevated intraocular pressure has been most widely used as a model to impede ocular blood flow and consequently to induce transient retinal ischemia. Intraocular pressure above systolic pressure is produced by cannulation of the anterior chamber with a 25-gage needle connected to a raised saline reservoir in rats. Intraocular pressure maintained at about 110–130 mmHg blocks retinal and uveal circulations and leads to ocular ischemia. Details of the procedure was described by Buchi and co-workers (21–23) and elsewhere. Branching of retinal vessels is an indication of successful induction of retina ischemia. At least 20 min of ischemia is required to cause irreversible functional damage, while more than 35 min is needed to produce detectable histological changes, and 45 min for consistent and reliable morphological quantification. Buchi found that there was no morphological difference between 90 and 120 min of ischemia, suggesting that it is the maximal ischemic condition (22). Notably, the degree of ischemic injury may vary with pigmentation (24).

3.2. Ligation of Optic Nerve Bundle

In rats, the ophthalmic artery travels along the optic nerve sheath and trifurates into the central retinal artery and two posterior ciliary arteries (25, 26). A lateral conjunctival peritomy is performed to expose the optic nerve by blunt dissection. A 6–0 silk suture is passed around the optic nerve and the ligature on the optic nerve is tightened until blood flow in the retinal vessels is stopped (27). Immobile clumps of blood cells can be seen in all the large retinal blood vessels with an operating microscope (28). Some researchers confirmed a complete retinal ischemia by obtaining a flat wave in ERG in animals (27). Ligation is removed at different times after tightening. Due to possible mechanical damage by ligation, eyes that fail to reperfuse within 15 min are excluded.

3.3. Ligation of Ophthalmic Vessels

The ophthalmic vessels travel along the optic nerve sheath in rats. To interrupt the ocular blood supply but not directly damage the optic nerve, the optic nerve sheath is opened longitudinally with a microsurgical blade and carefully separated from the optic nerve. A ligature is placed between the optic nerve and the optic nerve sheath, around the latter, and tied (29, 30). The ligature is released and removed after 90 min of ischemia time. Complete retinal reperfusion is assessed and animals that do not show full re-establishment of the retinal blood flow within 15 min are usually excluded.

3.4. Bilateral Occlusion of Carotid Arteries

Blockage of major proximal arteries such as the common carotid and internal carotid arteries, which branch to the ophthalmic artery causes retinal ischemia. The severity of the insult depends on the number of occluded vessels and their anastomoses. Unilateral occlusion of the common carotid artery produces subtle retinal changes, while bilateral occlusion causes ophthalmoscopic evidence of retinal ischemia, retinal ganglion cell loss, optic atrophy, and cerebral damage in rats which have vertebral arteries (31).

However, occlusion of both bilateral carotid and vertebral arteries may cause serious ocular and cerebral ischemia or even death in animals after a long period of occlusion. Suture ligation is commonly used, but, lately, intraluminal suture ligation is used for transient artery occlusion and reperfusion (32).

3.5. Photodynamic Ablation

To induce thrombosis of the retinal vasculature, an iodinated photosensitive dye, rose bengal (40 mg/kg), is injected intravenously through a tail vein. The eyes are immediately exposed to 7 min of intense light from a slide lantern projector fitted with a filter (550 nm) to optimize the wavelength of the light for rose bengal sensitivity (peak absorption = 550 nm) (33). Light illumination might be applied to produce diffuse microthrombi in retinal and choroidal vessels. However, Buchi et al. (34) found that retinal damage was irregular in its extent and character due to this model of phototoxicity.

To specifically occlude the major retinal vessels, argon green laser irradiation at 514 nm wavelength is used to sensitize rose bengal (35). A one-mirror goniolens filled with methylcellulose is placed on a rat's eye for fundus viewing. Immediately after the injection of a rose bengal bolus, the laser with settings of 0.3 s exposure, 50 μm, diameter and 0.1 W power, is used to irradiate the targeted retinal vessels. Thrombolytic therapy with tissue-type plasminogen activator (tPA), is effective for recanalization of the coronary or other vessels obstructed by intraluminal thrombi. Thrombolytic reperfusion of the occluded central retinal artery by tPA is administered through a catheter placed in the femoral vein 60 min after thrombus formation. One-half of the total tPA (0.27 mg/kg body weight), is first injected as a bolus and the remaining half is further infused within 10 min. After tPA administration, animals are included in the study only when reperfusion is confirmed.

3.6. Injection of Endothelin-1

Endothelin-1 (ET-1) is a potent vasoconstrictor peptide produced by vascular endothelial cells. ET-1 and its G-protein-coupled receptors, ET_A and ET_B were demonstrated to be expressed abundantly in ocular tissues (36, 37), while there were increased levels of ET-1 in the plasma and aqueous humor of glaucomatous patients. Intravitreal injection of ET-1 or administration of ET-1 to intraorbital optic nerve with an implanted mini-pump was shown to decrease blood flow in the retina and optic nerve, respectively (38–40). Administration of ET-1 causes retinal ganglion cell loss, activates retinal glial cells with increased immunoreactivity of glial fibrillary acidic protein, and prolongs VEP implicit times. Recently, subconjunctival injection of ET-1 was demonstrated to delay the perfusion of central retinal artery by fluorescein angiography, but its ischemic outcome has not yet been verified (41).

4. Comparison of Different Retinal Ischemia Models

Compared to cerebral ischemia, a retina ischemia model is less invasive, less labor intensive, easily reproduced, and more readily monitored either visually or electrophysiologically. In addition, retinal ischemia models are more readily used for pharmacological manipulation because the vitreous cavity can serve as a reservoir for drug administration to the ischemic retina, bypassing occluded blood vessels and the blood retinal barrier. To eliminate interindividual variation, the finding of the experimental eye can always be used to compare to the contralateral control eye. These advantages suggest that a retinal ischemia model could be a desirable alternative to a cerebral ischemia model to understand the disease process of ischemia and to explore related pharmacological treatments.

For retinal ischemia, the pressure elevation model is commonly used because it is easily reproduced and provides consistent results. It relates to CRAO, acute closed angle glaucoma, and occlusion of the ophthalmic artery in humans. While the procedures are less technically demanding, it has been argued that this procedure causes panretinal ischemia instead of CRAO because of obstruction of both retinal and choroidal circulations (42).

Osborne et al. (43) compared the ischemic effects of high pressure, two-vessel occlusion, and optic nerve ligation. Choline acetyltransferase (ChAT) immunoreactivity, expressed in amacrine cells, was found to be unaffected in the two-vessel occlusion model but completely eliminated in the high pressure model and drastically reduced in the optic nerve ligation model. Because the b-wave in the two-vessel model was reduced by 50%, while it was suppressed almost completely immediately after ischemia in the other two models, the authors suggest that the ischemic damages in these models were different and ranged from mild (oligomia), acute (hypoxia), and chronic (true ischemia). Approximately 30–36 h after reperfusion, neither the αPKC (expressed by on-bipolar cells), nor Thy-1 (expressed by retinal ganglion cells) immunoreactivities were affected in high pressure and optic nerve ligation models while the retinal parvalbumin (localized to amacrine cells) immunoreactivity was greatly reduced in the high pressure model but not in the optic nerve ligation model. Rosenbaum et al. (44) found more cell loss in the retinal ganglion cell layer and inner nuclear layer in the high pressure model than the optic nerve ligation model 1 day after reperfusion, but no histological difference after 7 days of reperfusion. However, Kuroiwa et al. (28) reported that the time courses of TdT-dUTP terminal nick-end labeling (TUNEL) which was used for detecting

DNA fragmentation in apoptosis, in the retinal ganglion cell layer and the inner nuclear layer were similar in both models, but the number of TUNEL positive cells in the high pressure model was less than that in the optic nerve ligation model. Overall, these findings suggested that the differential responses in these models depend on the induction procedures and possibly may relate to incomplete vascular occlusion, hydrostatic pressure on the retina, and mechanical damage of the optic nerve in two-vessel occlusion, high pressure, and optic nerve ligation models, respectively.

The procedure of optic nerve ligation is relatively easy to perform and requires little specialized equipment or surgical manipulation. However, suture ligation may cause direct trauma to the optic nerve and there is some indirect influence or secondary degeneration to the retinal ganglion cell due to disturbed neurotrophic transport through the ligated nerve bundles, in addition to a possible ligation of collateral blood supplies. Similar to the pressure elevation model, a major limitation in interpreting results in these models is that it is unusual to find complete cessation of retinal flow in clinical patients.

Ligation of the ophthalmic artery seems to isolate the ischemic effect in the retina and, more closely, to the experimental CRAO model in monkeys and human patients with CRAO. However, the procedures demand a more skilled dissection to avoid damaging ophthalmic vessels in the optic nerve sheath, as there are anatomical variations in the vascular circulation between individuals and species. Vidal-Sanz et al. (30) noted that the loss of RGCs throughout the retina were diffuse and there were apparently two different phases of RGC losses, which represented 47% of the RGC population in the early phase (7 days) and an additional 15% in delayed (21 days) degenerative processes.

Unilateral and bilateral occlusion of the carotid artery requires more skill in vascular surgery, and the animal may suffer from extensive surgical exposure. Moreover, this model is more relevant to carotid artery disease in humans (1). The procedures may cause incomplete retinal ischemia and the number of proximal vessels being ligated and their anastomoses may vary the degree of ischemic damage. All of these procedures alter the blood flow to the brain and induce cerebral ischemia, which is of concern as a cause or effect of retinal ischemia (17).

The photodynamic ablation method with light illumination is relatively simple, but has substantial disadvantages, such as variable retinal damage due to different degrees of light exposure throughout the rat retina, retinal detachment, and widespread retinal necrosis (34). Since capillary thrombosis occurs, the ischemic damage may be permanent rather than reversible, so that this model is not suitable to study postischemic reperfusion events. For the photothrombic model with green laser irradiation and thrombolytic therapy, investigators should be aware that the parapapillary and central areas of the retina are highly affected, the peripheral area and the outer cell layer rather preserved, while the peripheral area was reported to be equally affected in optic nerve ligation and pressure elevation models (35).

5. Contributions to Understand Disease Mechanisms

Compared to cerebral ischemia, the tolerance time to ischemic damage is longer in the retina than in the brain. Only a few minutes of ischemia may result in permanent neuronal damage in the brains of humans, cats, and dogs. In contrast, a period over 45 min may be required to cause irreversible morphological and functional damage in the retina without a complete recovery (1, 45, 46). Hayreh and co-workers have shown that transient clamping of the CRA for up to 97 min only caused transient visual loss in normal rhesus monkeys while the prolonged circulation disturbance for over 105 min led to irreversible histological and functional loss (18, 42). The significant differences between the ischemic tolerance times of cerebral and retinal neurons may be due to the high content of glucose, glycogen, and oxygen in vitreous and the limited space for buffering the expansion of the tissue during cerebral edema. However, the underlying mechanisms of retinal tolerance are not yet completely understood.

Loss of retinal ganglion cells and thinning of the nerve fiber layer is well-documented as an irreversible outcome after ischemia-reperfusion injury. Because of this outcome, animal models of retinal ischemia are widely used to help understand the mechanism of inner retinal degeneration associated with CRAO, glaucoma, retinal vascular occlusion, diabetic retinopathy, retinopathy of prematurity, and traumatic optic neuropathy. Extreme ischemic conditions cause photoreceptor degeneration; however, the damage is usually much less severe than occurs to inner retinal neurons (23). The question of why photoreceptors appear to be less sensitive to ischemia than inner retinal neurons is unclear, but may be linked to compensatory mechanisms by their high density of inner segment mitochondria, high neuroglobin levels, and lower photoreceptor sensitivity, determined by retinal pigment epithelium or illumination levels. Greater N-methyl-D-aspartate (NMDA) receptor density on the magnocellular retinal ganglion cells may be related to more susceptibility to ischemic damage (47, 48).

In the past decades, necrosis has been believed to be a major cell death mechanism in retinal ischemia. In the 1990s, apoptosis, or programmed cell death, was found to relate not only to developmental but also pathological conditions. Morphological changes and molecular pathways were noted to vary in a spectrum between necrosis and apoptosis, depending on the intensity of ischemic insults (49). Initiation and execution of apoptosis after ischemic injury were shown to involve many apoptotic mediators and signaling cascades in the retina. Vidal-Sanz et al. (29) investigated the effects of intravitreal injection of brain-derived growth factor (BDNF), caspase inhibitor (Z-DEVD), and peptide Bcl-2 on the retinal ganglion cell survival after optic nerve transection and transient ischemia by ligation of ophthalmic vessels and pressure elevation. For transient ischemia and optic nerve transection models, BDNF enhanced the retinal ganglion cell survival in all models. Interestingly, Z-DEVD increased RGC survival 7 days after 90 min of ophthalmic vessel ligation, but not after 60 min, while Bcl-2 increased RGC survival 7 days after 60, but not 90 min of vascular occlusion. The marked differences in amount and pattern of retinal ganglion cell death demonstrate that retinal ganglion cell death may undergo pathways ranging from necrosis, and caspase-dependent and caspase-independent apoptosis.

Neovascularization is a serious complication of ischemic disease, but the process of neovascularization is complex. Dysfunction of the vascular system such as vasodilation, increased permeability, and matrix degradation, may cause visual deterioration and degeneration. Activation of hypoxia inducing factor (HIF)-1 is shown to be a key factor to regulate angiogenesis through the production of VEGF in the retina. Adenosine, a neuromodulator, was also shown to stimulate endothelial cell migration and vascular tube formation (50), via nitric oxide. Antiangiogenesis therapy seems to be a promising therapy (12).

Significant findings demonstrate an important role of glutamate excitotoxicity in neuronal death mechanism after retinal ischemia. Although the exact mechanism of glutamate release during retinal ischemia remains controversial, accumulated evidence supports the hypothesis that glutamate levels rise during retinal ischemia and is also associated with glial dysfunction, which leads to failure of glutamate recycling. Evidence also suggests that both ischemic and excitotoxic cascades share similar pathways which involve metabolic imbalance, ion channel influx, release of neurotransmitters, free radical production, acidosis, and inflammation (1). Since neurodegenerative processes and cellular interactions are very complex, the search for neuroprotective treatment is not straightforward. For example, interleukin (IL)-6, a multifunctional cytokine, was upregulated in microglial cells in retina via modulation of nuclear factor kappa-B. When exposed to ischemia reperfusion and exogenously administered, IL-6 prevents the loss of retinal ganglion cells, suggesting that further characterization on the population of retinal cells is important, and modulation of cytokine expression in other cellular populations should be considered for neuroprotective treatments (45).

Treatment modalities that directly supply oxygen or glucose/glycogen have been shown to be beneficial. Pharmacological treatments targeted to inhibit the following events during retinal ischemia or glutamate excitotoxicity were found to preserve the retinal functional and prevent cell loss. Glutamate ionotropic (NMDA type and non-NMDA type) receptor antagonists, antioxidants, calcium channel blockers, and sodium channel modulators are all possibilities. Roth and co-workers demonstrated that ischemic preconditioning is protective to subsequent lethal ischemia in the retina. Further studies on the endogenous protective pathway to prevent loss of retinal ganglion cells are warranted in conjunction with our studies on endogenous neuroprotection against excitotoxicity, glaucoma, and other damaging insults (48, 51–53).

6. Future Opportunities

Due to the complexity and variability of the retinal blood supply, disturbance on the retinal vasculature causes different ischemic cascades with similar, but not exact, disease outcomes. Therefore, induction procedures should be carefully chosen so that the animal model can mimic a known human disease more precisely, and provide more useful data for further study. Future goals for animal models should not only be based on common or similar outcomes (for example, retinal ganglion cell loss), but should also share similar etiology, which becomes more relevant to human disease. For example, local administration of ET-1 to the optic nerve head seems to be more suitable for glaucoma research if the optic nerve head is hypothesized to be the site of injury or the area exposed to an ischemic insult.

Recently, procedures of retinal ischemia have been successfully performed in smaller-sized animals with minor modifications (54–56). Performing retinal ischemia procedures in genetically manipulated mice may significantly help investigate the underlying mechanisms of retinal ischemia with the particular gene of interest. In the near future, regenerative or replacement therapies such as neurotrophic factors, supplement, and stem cells, will hopefully replace the loss of retinal ganglion cells and other inner retinal neurons in retinal ischemic disorders. However, whether the replaced retinal ganglion cells make the proper connections to the neighboring cells or the axons of regenerating retinal ganglion cells grow back to the right target in the brain will be another vast challenge.

References

1. Osborne NN, Casson RJ, Wood JP, Chidlow G, Graham M, Melena J (2004) Retinal ischemia: mechanisms of damage and potential therapeutic strategies. Prog Retin Eye Res 23:91–147
2. Hayreh SS (2005) Prevalent misconceptions about acute retinal vascular occlusive disorders. Prog Retin Eye Res 24:493–519
3. Hayreh SS (1999) Retinal and optic nerve head ischemic disorders and atherosclerosis: role of serotonin. Prog Retin Eye Res 18:191–221
4. Schmidt DP, Schulte-Mönting J, Schumacher M (2002) Prognosis of central retinal artery occlusion: local intraarterial fibrinolysis versus conservative treatment. AJNR Am J Neuroradiol 23:1301–1307
5. Noble J, Weizblit N, Baerlocher MO, Eng KT (2008) Intra-arterial thrombolysis for central retinal artery occlusion: a systematic review. Br J Ophthalmol 92:588–593
6. Biousse V, Calvetti O, Bruce BB, Newman NJ (2007) Thrombolysis for central retinal artery occlusion. J Neuroophthalmol 27:215–230
7. Mitchell P, Smith W, Chang A (1996) Prevalence and associations of retinal vein occlusion in Australia. The Blue Mountains Eye Study. Arch Ophthalmol 114:1243–1247
8. The Central Vein Occlusion Study Group (1997) Natural history and clinical management of central retinal vein occlusion. Arch Ophthalmol 115:486–491
9. Green WR, Chan CC, Hutchins GM, Terry JM (1981) Central retinal vein occlusion: a prospective histopathologic study of 29 eyes in 28 cases. Trans Am Ophthalmol Soc 79:371–422
10. Hayreh SS, Rojas P, Podhajsky P, Montague P, Woolson RF (1983) Ocular neovascularization with retinal vascular occlusion-III. Incidence of ocular neovascularization with retinal vein occlusion. Ophthalmology 90:488–506
11. Sjølie AK, Stephenson J, Aldington S et al (1997) Retinopathy and vision loss in insulin-dependent diabetes in Europe. The EURODIAB IDDM complications study. Ophthalmology 104:252–260
12. Kaur C, Foulds WS, Ling EA (2008) Blood-retinal barrier in hypoxic ischaemic conditions: basic concepts, clinical features and management. Prog Retin Eye Res 27:622–647
13. Penn JS, Madan A, Caldwell RB, Bartoli M, Caldwell RW, Hartnett ME (2008) Vascular endothelial growth factor in eye disease. Prog Retin Eye Res 27:331–371
14. Klein R, Klein BE (1997) Diabetic eye disease. Lancet 350:197–204
15. Flammer J, Mozaffarieh M (2007) What is the present pathogenetic concept of glaucomatous optic neuropathy? Surv Ophthalmol 52(Suppl. 2):S162–S173
16. Osborne NN, Melena J, Chidlow G, Wood JP (2001) A hypothesis to explain ganglion cell death caused by vascular insults at the optic nerve head: possible implication for the treatment of glaucoma. Br J Ophthalmol 85: 1252–1259
17. Pang IH, Clark AF (2007) Rodent models for glaucoma retinopathy and optic neuropathy. J Glaucoma 16:483–505
18. Hayreh SS, Weingeist TA (1980) Experimental occlusion of the central artery of the retina. IV: retinal tolerance time to acute ischaemia. Br J Ophthalmol 64:818–825
19. Sims NR (1992) Energy metabolism and selective neuronal vulnerability following global cerebral ischemia. Neurochem Res 17:923–931
20. Phan TG, Wright PM, Markus R, Howells DW, Davis SM, Donnan GA (2002) Salvaging the ischaemic penumbra: more than just reperfusion? Clin Exp Pharmacol Physiol 29:1–10
21. Büchi ER, Suivaizdis I, Fu J (1991) Pressure-induced retinal ischemia in rats: an experimental model for quantitative study. Ophthalmologica 203:138–147
22. Büchi ER (1992) Cell death in the rat retina after a pressure-induced ischaemia-reperfusion insult: an electron microscopic study. I. Ganglion cell layer and inner nuclear layer. Exp Eye Res 55:605–613
23. Büchi ER (1992) Cell death in rat retina after pressure-induced ischaemia-reperfusion insult: electron microscopic study. II. Outer nuclear layer. Jpn J Ophthalmol 36:62–68
24. Safa R, Osborne NN (2000) Retinas from albino rats are more susceptible to ischaemic damage than age-matched pigmented animals. Brain Res 862:36–42
25. Morrison JC, Johnson EC, Cepurna WO, Funk RH (1999) Microvasculature of the rat optic nerve head. Invest Ophthalmol Vis Sci 40:1702–1709
26. Sugiyama K, Gu ZB, Kawase C, Yamamoto T, Kitazawa Y (1999) Optic nerve and peripapillary choroidal microvasculature of the rat eye. Invest Ophthalmol Vis Sci 40:3084–3090
27. Hangai M, Miyamoto K, Hiroi K et al (1999) Roles of constitutive nitric oxide synthase in postischemic rat retina. Invest Ophthalmol Vis Sci 40:450–458

28. Kuroiwa S, Katai N, Shibuki H et al (1998) Expression of cell cycle-related genes in dying cells in retinal ischemic injury. Invest Ophthalmol Vis Sci 39:610–617
29. Vidal-Sanz M, Lafuente M, Sobrado-Calvo P et al (2000) Death and neuroprotection of retinal ganglion cells after different types of injury. Neurotox Res 2:215–227
30. Vidal-Sanz M, Lafuente MP, Mayor S, de Imperial JM, Villegas-Pérez MP (2001) Retinal ganglion cell death induced by retinal ischemia. Neuroprotective effects of two alpha-2 agonists. Surv Ophthalmol 45(Suppl. 3):S261–S277
31. Stevens WD, Fortin T, Pappas BA (2002) Retinal and optic nerve degeneration after chronic carotid ligation: time course and role of light exposure. Stroke 33:1107–1112
32. Block F, Grommes C, Kosinski C, Schmidt W, Schwarz M (1997) Retinal ischemia induced by the intraluminal suture method in rats. Neurosci Lett 232:45–48
33. Mosinger JL, Price MT, Bai HY, Xiao H, Wozniak DF, Olney JW (1991) Blockade of both NMDA and non-NMDA receptors is required for optimal protection against ischemic neuronal degeneration in the in vivo adult mammalian retina. Exp Neurol 113:10–17
34. Büchi ER, Lam TT, Suvaizdis I, Tso MO (1994) Injuries induced by diffuse photodynamic action in retina and choroid of albino rats. Morphologic study of an experimental model. Retina 14:370–378
35. Daugeliene L, Niwa M, Hara A et al (2000) Transient ischemic injury in the rat retina caused by thrombotic occlusion-thrombolytic reperfusion. Invest Ophthalmol Vis Sci 41:2743–2747
36. MacCumber MW, Jampel HD, Snyder SH (1991) Ocular effects of the endothelins. Abundant peptides in the eye. Arch Ophthalmol 109:705–709
37. Murata M, Nakagawa M, Takahashi S (1998) Selective expression of endothelin 1 mRNA in rat retina. Ophthalmologica 212:331–333
38. Bursell SE, Clermont AC, Oren B, King GL (1995) The in vivo effect of endothelins on retinal circulation in nondiabetic and diabetic rats. Invest Ophthalmol Vis Sci 36:596–607
39. Cioffi GA, Orgül S, Onda E, Bacon DR, Van Buskirk EM (1995) An in vivo model of chronic optic nerve ischemia: the dose-dependent effects of endothelin-1 on the optic nerve microvasculature. Curr Eye Res 14:1147–1153
40. Chauhan BC, LeVatte TL, Jollimore CA et al (2004) Model of endothelin-1-induced chronic optic neuropathy in rat. Invest Ophthalmol Vis Sci 45:144–152
41. Masuzawa K, Jesmin S, Maeda S et al (2006) A model of retinal ischemia-reperfusion injury in rats by subconjunctival injection of endothelin-1. Exp Biol Med (Maywood) 231: 1085–1089
42. Hayreh SS, Zimmerman MB, Kimura A, Sanon A (2004) Central retinal artery occlusion. Retinal survival time. Exp Eye Res 78:723–736
43. Osborne NN, Larsen A, Barnett NL (1995) Influence of excitatory amino acids and ischemia on rat retinal choline acetyltransferase-containing cells. Invest Ophthalmol Vis Sci 36:1692–1700
44. Rosenbaum DM, Rosenbaum PS, Singh M et al (2001) Functional and morphologic comparison of two methods to produce transient retinal ischemia in the rat. J Neuroophthalmol 21:62–68
45. Sanchez RN, Chan CK, Garg S et al (2003) Interleukin-6 in retinal ischemia reperfusion injury in rats. Invest Ophthalmol Vis Sci 44:4006–4011
46. Wang J, Jiang S, Kwong JM, Sanchez RN, Sadun AA, Lam TT (2006) Nuclear factor-kappaB p65 and upregulation of interleukin-6 in retinal ischemia/reperfusion injury in rats. Brain Res 1081:211–218
47. Chun MH, Kim IB, Ju WK et al (1999) Horizontal cells of the rat retina are resistant to degenerative processes induced by ischemia-reperfusion. Neurosci Lett 260:125–128
48. Kwong JM, Lam TT, Caprioli J (2003) Hyperthermic pre-conditioning protects retinal neurons from N-methyl-D-aspartate (NMDA)-induced apoptosis in rat. Brain Res 970:119–130
49. Lam TT, Abler AS, Tso MO (1999) Apoptosis and caspases after ischemia-reperfusion injury in rat retina. Invest Ophthalmol Vis Sci 40:967–975
50. Lutty GA, McLeod DS (2003) Retinal vascular development and oxygen-induced retinopathy: a role for adenosine. Prog Retin Eye Res 22:95–111
51. Park KH, Cozier F, Ong OC, Caprioli J (2001) Induction of heat shock protein 72 protects retinal ganglion cells in a rat glaucoma model. Invest Ophthalmol Vis Sci 42:1522–1530
52. Ahn J, Piri N, Caprioli J, Munemasa Y, Kim SH, Kwong JM (2008) Expression of heat shock transcription factors and heat shock protein 72 in rat retina after intravitreal injection of low dose N-methyl-D-aspartate. Neurosci Lett 433:11–16

53. Caprioli J, Kitano S, Morgan JE (1996) Hyperthermia and hypoxia increase tolerance of retinal ganglion cells to anoxia and excitotoxicity. Invest Ophthalmol Vis Sci 37: 2376–2381
54. Lelong DC, Bieche I, Perez E et al (2007) Novel mouse model of monocular amaurosis fugax. Stroke 38:3237–3244
55. Berger S, Savitz SI, Nijhawan S et al (2008) Deleterious role of TNF-alpha in retinal ischemia-reperfusion injury. Invest Ophthalmol Vis Sci 49:3605–3610
56. Li SY, Fu ZJ, Ma H et al (2009) Effect of lutein on retinal neurons and oxidative stress in a model of acute retinal ischemia/reperfusion. Invest Ophthalmol Vis Sci 50:836–843

Chapter 11

Retinal Inflammation: Uveitis/Uveoretinitis

Reiko Horai and Rachel R. Caspi

Abstract

Retinal inflammatory disease of a putative autoimmune origin, known as autoimmune uveitis, affects 150,000 persons per year in the developed world and is a potentially blinding disease. The eye can be the only affected organ or uveitis can be part of a systemic syndrome. Animal models of uveitis induced by immunization with retinal antigens or through genetic engineering are used to study basic mechanisms, genetic control and therapeutic approaches. Although thymic expression of retinal antigens eliminates most autoreactive lymphocytes and positively selects natural regulatory T cells, peripheral tolerance to retina is inefficient due to the relative sequestration of retinal antigens. Therefore, residual autoreactive lymphocytes persist and can be activated by accidental encounter with self or a cross-reactive antigen. When exposed to retinal or cross-reactive antigens in the context of innate danger signals they mature into Th1 or Th17 effector cells that find their way into the eye. Upon recognition of specific antigen within the eye, they orchestrate a destructive inflammation by recruiting inflammatory leukocytes from the circulation. Regulatory T cells are also induced as part of the disease process and ultimately control inflammation and permit healing to take place. Novel immunotherapeutic approaches built upon the growing knowledge of basic mechanisms to target critical checkpoints in disease pathogenesis and restore immune homeostasis.

Key words: Uveitis, Uveoretinitis, Experimental autoimmune uveitis, EAU, T lymphocytes, Tolerance, Autoimmune disease

1. Retinal Inflammation in Human

Retinal inflammation includes retinopathy, retinitis (retinal vasculitis), and uveitis/uveoretinitis. Retinopathy has been described in Chaps. 6 and 7. In this chapter, we will focus primarily on uveitis/uveoretinitis and its induced as well as spontaneous animal models. Immune-mediated components were also recently shown to play a role in diseases that were once considered purely degenerative, such as age-related macular degeneration

Iok-Hou Pang and Abbot F. Clark (eds.), *Animal Models for Retinal Diseases*, Neuromethods, vol. 46
DOI 10.1007/978-1-60761-541-5_11, © Springer Science+Business Media, LLC 2010

(AMD) and glaucoma. These diseases and their models are described elsewhere in this volume.

Uveitis of a putative autoimmune nature is the one of the leading causes of blindness in the US and the developed world. It is estimated to affect about 150,000 Americans annually and to cause about 10% of severe visual handicap (1). When inflammation is predominately located in the vitreous, retina, and choroid, it is classified as posterior uveitis. The posterior uveitides are particularly likely to affect the photoreceptor cells and damage vision. Autoimmune uveitis can be part of a systemic autoimmune syndrome such as Behçet's disease, systemic sarcoidosis (SS) and Vogt–Koyanagi–Harada disease (VKH), or the eye can be the only target such as in birdshot retinochoroidopathy (BR) and sympathetic ophthalmia (SO) (2).

Uveitis patients often exhibit lymphocyte and antibody responses to antigens (Ags) present in the eye and show strong associations with particular HLA haplotypes (3, 4), supporting the autoimmune nature of the disease. The etiologic cause(s) of most types of uveitis are unknown. One of the theories proposes that in uveitic disease that cannot be linked to a trauma, lymphocytes capable of recognizing retinal Ags are primed in the periphery by a cross-reactive microbial stimulus (antigenic mimicry). Although it is difficult to determine whether the specific responses observed in the patients whose disease has been active for some time are primary, or have developed as a result of tissue breakdown, they are believed to help fuel progression of the disease.

1.1. Behçet's Disease

Behçet's disease is an autoimmune disease that results from damage to blood vessels. The cause of Behçet's disease is unknown. The affected tissues include the eye, mouth, genitals, and skin. Inflammation in the eye includes uveitis, retinitis, and iritis. HLA–B51 is a risk factor for this disease (5, 6).

1.2. Systemic Sarcoidosis

Sarcoidosis is an inflammatory disease that commonly targets lungs and lymph nodes, but also affects many other organs including eyes. About 25% of patients have ocular symptoms similar to uveitis.

1.3. Vogt–Koyanagi–Harada Syndrome

Vogt–Koyanagi–Harada syndrome (VKH) is a systemic autoimmune disorder involving multiple organs, including the ocular, auditory and nervous systems as well as the skin. Severe bilateral granulomatous uveitis associated with subretinal fluid accumulation is the hallmark of ocular VKH. There is an association with HLA-DR4 (7).

1.4. Birdshot Retinochoroidopathy

Birdshot retinochoroidopathy (BR) is a bilateral inflammatory disease affecting the choroid and the retina. It belongs to a group of disorders known as the idiopathic multifocal white dot syndromes, which themselves are one of the causes of chronic posterior uveitis. Its cause is not entirely clear but it is thought to be related to an inherited immune dysfunction and there is a strong association with HLA-A29 (6, 8).

1.5. Sympathetic Ophthalmia

Sympathetic ophthalmia (SO) is a rare granulomatous uveitis of both eyes. Following trauma or surgery to one eye, the other, uninjured eye, develops destructive inflammation that can lead to complete blindness. The main mechanism is thought to be an autoimmune reaction against melanin-containing cells in the uvea following exposure to Ags released from the injured eye (6).

1.6. Immune-Mediated Components in Degenerative Eye Disease

In recent years it has become increasingly clear that certain diseases affecting the retina, including AMD and glaucoma, have associated immunological components that are part of the mechanisms that predispose to disease or that drive its pathogenesis. This includes components of the complement cascade, chemokines, and their receptors as well as proteins that may serve as molecular chaperones (reviewed in (9)). Although in genetic studies the association of particular allelic forms of these molecules with predisposition to disease and with disease severity can be substantial, our understanding of the underlying mechanisms is still quite limited. The coming years should see important developments in this area. Because retinal degenerative diseases do exhibit a typical inflammation characterized by leukocytic infiltration and exudation, they will not be discussed further in this chapter.

2. Animal Models for Human Uveitis

2.1. Induced Models

2.1.1. The "Classical" EAU Model

The EAU model was originally established in the guinea pig using homologous uveal tissue and then was adapted to the rat using retinal extracts. To induce EAU, the immunizing Ag must be emulsified in complete Freund's adjuvant (CFA), a mineral oil supplemented with heat-killed mycobacteria, and in many cases an injection of pertussis toxin (PTX) is given as an additional adjuvant at the time of immunization (4, 10). This mimics the putative uveitogenic stimulus that is thought to trigger uveitis in humans, which is generally believed to involve an exposure to a retinal or mimic Ag combined with an infectious event providing inflammatory "danger" signals.

The EAU model in the rat induced with retinal arrestin (retinal soluble antigen, S–Ag) or with the interphotoreceptor retinoid-binding protein (IRBP), and the mouse EAU model induced with IRBP, have been by far the most widely used and the best characterized (10). Other uveitogenic proteins (studied mostly in the Lewis rat) include retina-derived rhodopsin, recoverin, and phosducin and retinal pigment epithelium-derived RPE65.

In addition, melanin preparations as well as purified tyrosinase-related proteins 1 and 2 (that participate in biosynthesis of melanin) are uveitogenic and result in anterior uveitis and choroiditis, targeting those ocular tissues that contain melanin. Despite differences in target Ags used to elicit disease, the immunological mechanisms driving the disease are similar. The various models of experimental uveitis have recently been reviewed (4).

Susceptibility to disease is strain-dependent. In mice and in rats, both major histocompatibility complex (MHC) and non-MHC genes have effects on the disease susceptibility. The Lewis rat is the most commonly used susceptible rat strain for EAU studies. In the mouse, the most susceptible strain identified so far is B10.RIII. The C57BL/6 strain is only moderately susceptible, but this is a very useful strain to analyze the basic mechanisms of pathogenesis, as various gene-manipulated (knockout or transgenic) lines are available on this background. B10.A mice have a susceptibility that is intermediate between C57BL/6 and B10.RIII. Each of these strains has been shown to recognize one or more defined epitopes of IRBP that can be synthesized in the laboratory and used to induce disease in place of the whole protein (Table 1). In addition, new Ags can be expressed in the retina through genetic engineering to serve as target Ags for uveitis. Foreign proteins such as hen egg lysozyme (HEL) or β-galactosidase (β-gal) have been transgenically expressed in the retina or lens under control of the appropriate tissue-specific promoters to serve as a neo-self Ag (11–14). Alternatively, retinal cells can be transduced to express a foreign protein such as influenza hemagglutinin (HA) using a viral vector (15).

The onset of disease is observed at day 9–12 after immunization with the retinal or neo-retinal Ag in CFA and peak of the disease occurs between day 16–21. The clinical course of the disease is typically monophasic, but a recurrent disease that is more

Table 1
Susceptibility of different mouse strains to IRBP-induced EAU

Strain	H-2	Susceptibility	Pathogenic epitopes (position)	References
B10.RIII	r	Very high	SGIPYIISYLHPGNTILHVD (161–180) HPGNTILHVDTIYNRPSNTT (171–190) SLGWATLVGEITAGNLLHTR (541–560)	(59, 60)
B10.A	a (I-A^k)	High	ADKDVVVLTSSRTGGV (201–216)	(61)
C57BL/6	b	Moderate	GPTHLFQPSLVLDMAKVLLD (1–20) LRHNPGGPSSAVPLLLSYFQ (461–480) LAQGAYRTAVDLESLASQLT (651–670)	(31, 59)

Additional pathogenic and/or immunological epitopes were reported in Cortes et al. for B10.RIII and C57BL/6 strains (59)

reminiscent of human uveitis was described in the B10.A mouse strain (16). Disease scores can be assessed by fundus examination (17, 18) or histopathological appearance using a semi-quantitative scoring on a scale of 0 (no disease) to 4 (severe/maximum disease) (10) (Figs. 1 and 2, Table 2). These scoring criteria are also applicable to other variants of the EAU model that will be described ahead.

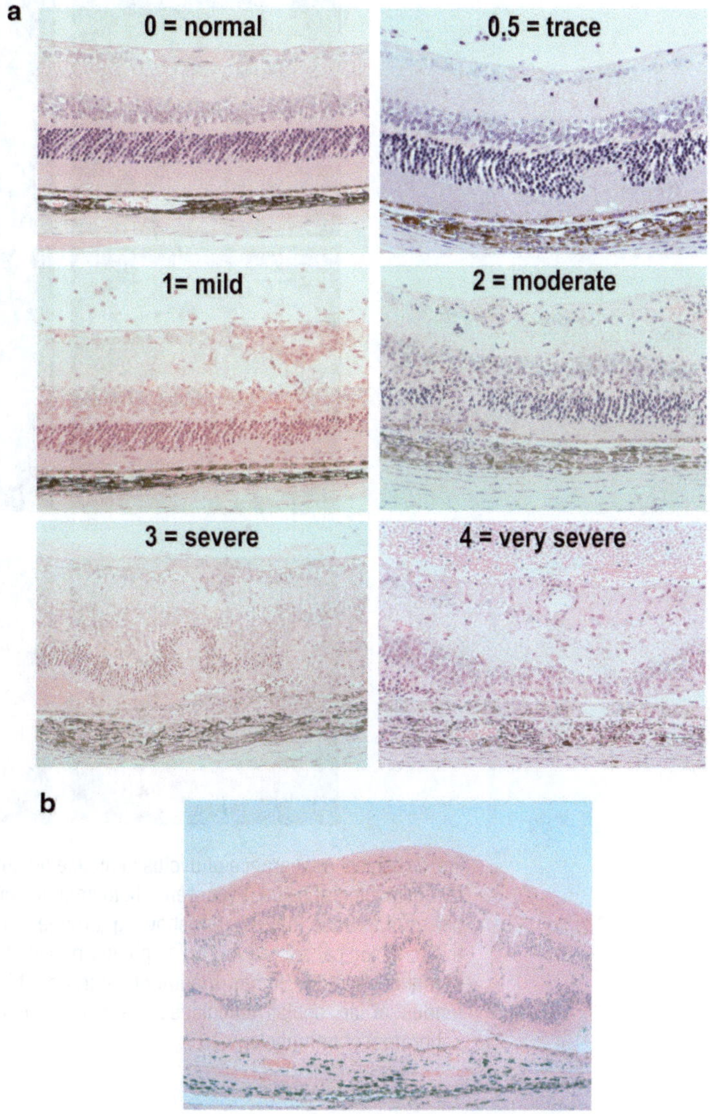

Fig. 1. Histopathology of mouse EAU compared with human uveitis. Eyes were collected from B10.RIII mice 21 days after uveitogenic immunization with IRBP, representing a range of disease scores (**a**). Uveitis (ocular sarcoidosis) in human (**b**). Note the similarity in the pathology. (Photographs provided by Dr. Chi-Chao Chan, Laboratory of Immunology, NEI; figure adapted from (59))

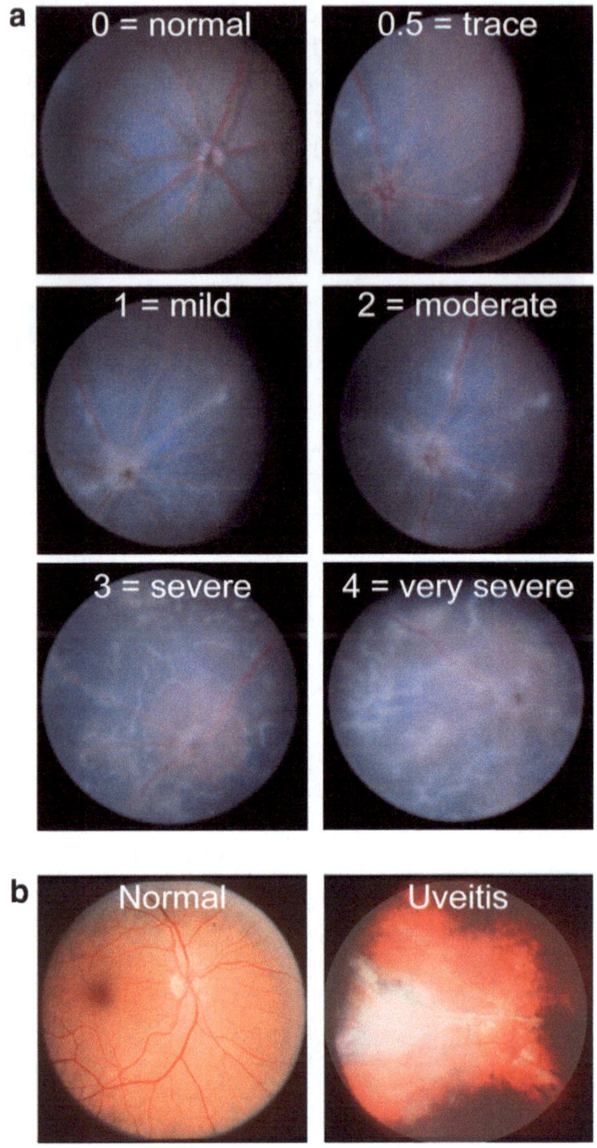

Fig. 2. Clinical appearance of uveitis in mouse ad human by fundus examination. Mouse fundus photographs (**a**). Eyes were photographed with a fundus camera during the acute phase of disease (day 14–21) showing a range of disease severity scores that parallel the pathological scores in Fig. 1 (photos provided by Dr. Yishay Falick, Laboratory of Immunology, NEI). Human fundus photographs (**b**) depict normal human eye and the fundus of an ocular sarcoidosis patient (photos provided by Dr. Chi-Chao Chan, Laboratory of Immunology, NEI)

2.1.2. "Humanized" EAU Models

Human uveitis is associated with specific HLA class I or class II types. To better understand these associations "humanized" EAU model have been developed. The HLA-A29 Tg mice develop spontaneous uveitis late in life and are described ahead, under

Table 2
Scoring of EAU in the mouse by histopathology and fundoscopy[a]

Score	Criteria for histopathology	Criteria for fundoscopy
0	No disease; normal retinal architecture	No change
0.5 (trace)	Mild inflammatory cell infiltration, no tissue damage	Few very small peripheral focal lesions, minimal vasculitis
1	Infiltrations in the uvea, vitreous, and retina; retinal folds and vasculitis, one small granuloma	Mild vasculitis, <5 small focal lesions, ≤1 linear lesion
2	Moderate infiltration in the uvea, vitreous and retina, retinal folds and focal shallow detachments, focal photoreceptor cell damage, small-size granulomas, retinal folds and vasculitis, focal loss of photorecptors	Multiple (>5) chorioretinal lesions and/or infiltrations, few (<5) linear lesions, severe vasculitis
3	Moderate to severe infiltration in the uvea, vitreous and retina, extensive retinal folding with large detachments, subretinal neovascularization, moderate photoreceptor cell damage, medium-size granulomatous lesions	Pattern of linear lesions, large confluent lesions, subretinal neovasculization, retinal hemorrhages, papilledema
4	Severe infiltration, diffuse retinal detachment, subretinal neovascularization and hemorrhages, extensive photoreceptor cell damage, large granulomatous lesions	Large retinal detachment, retinal atrophy

[a]Each disease score includes the criteria of the preceding score. (Table modified from ref. (10))

spontaneous EAU models. EAU can also be induced by immunization with retinal Ag in HLA class II Tg mice, in which the mouse class II molecules have been deleted and replaced with the human counterparts. HLA-DR3 (DRB1*0301), -DR4 (DRB1*0401), -DQ6 (DQB1*0601), or -DQ8 (DQB1*0302) Tg mice (19) present and respond to antigenic epitopes that would be recognized by humans bearing these class II molecules. Unlike the parental wild-type mice that are resistant to S-Ag-induced EAU, the HLA-Tg mice develop disease when immunized with S–Ag (20). More importantly, HLA DR3 Tg mice respond immunologically and develop typical EAU when immunized with S-Ag peptide N (LPLLANNRERRGIALDGKIKHE) that is known to elicit lymphocyte responses in uveitis patients ((21) and Mattapallil, MJ et al., unpublished). Because MHC class II molecules are involved in Ag recognition, the HLA class II Tg mice may help identify the critical regions of retinal Ags that involved in human disease and may be important for devising Ag-specific therapies for human uveitis.

2.1.3. Adoptive Transfer Model

Immune T cells from EAU-induced animals can pass on the disease to naïve, genetically compatible recipient animals by a process known as adoptive transfer. The donor T cells are activated with the immunizing ocular Ag in vitro and are infused in recipient animals. The recipients of these cells develop a destructive disease rapidly, usually within a week. This adoptive transfer model allows to avoid the use of adjuvants in the recipients and is useful to analyze the effector mechanism(s) of the disease, resembling the clinical situation (4).

The adoptive transfer model can also be used in mice made to express foreign neo-Ags in the eye by transgenic technology or by viral transduction. Uveitis is induced in retinal β-gal Tg mice as well as in lens HEL-Tg mice infused with activated uveitogenic T cells prepared from mice immunized with β-gal or HEL, respectively (12, 22). In a modified version of this adoptive transfer model, that permits particularly sophisticated approaches to the study of uveitogenic T cells, mice expressing a retinal neo-Ag are infused with T cells from Tg mice expressing a "monoclonal" T cell receptor (TCR) that recognizes this Ag (13, 15).

2.1.4. Ag-Pulsed Dendritic Cell EAU Model

Dendritic cells (DC) are professional Ag-presenting cells capable of stimulating naïve T cells, and are likely to be the main Ag-presenting cells in the early stages of EAU induction. A new model of EAU was developed by injection of matured splenic DC loaded with the major uveitogenic peptide of IRBP into B10.RIII mice (23). The duration of the disease is shorter, the pathology appears to be less severe, and the inflammatory infiltrate has a different composition than in the "classical" CFA–EAU model. Importantly, EAU elicited with Ag-pulsed DC is not only clinically distinct from CFA–EAU, but also is driven by unique effector mechanisms (see ahead). This model may represent some types of uveitis that were not adequately represented by the CFA–EAU model, and may offer new insights to the heterogenous nature of human uveitis.

2.2. Spontaneous Mouse Models of Uveitis

2.2.1. HLA-A29 Tg Mice

As mentioned above, BR is strongly associated with class I MHC, HLA-A29. A Tg mouse that expresses HLA-A29 spontaneously develops posterior uveitis resembling BR, VKH and SO. The development of spontaneous uveitis is observed in aged animals (>12 months old) with a frequency of about 80% (24, 25). This model is an important tool to study BR and its correlation with HLA-A29; however, the late onset of the disease makes it a difficult experimental model. The target retinal Ag in this model has not been identified, but is hypothesized to be derived from the HLA-A29 molecule itself, possibly in the form of a fragment immunologically cross-reactive with a component of the retina.

2.2.2. Spontaneous Models with Central Tolerance Deficiencies in the Thymus

Many tissue-specific Ags, including retinal Ags such as S-Ag and IRBP, are expressed ectopically in the thymus under control of the transcription factor AIRE (AutoImmune REgulator). In this way, self Ags that could potentially be targets of autoimmunity are presented to the developing T cells. Those T cells that express receptors having high affinity to self Ags (which would cause autoimmunity) are eliminated or anergized, while those with low affinity to self are positively selected and exit the thymus into the periphery, where they mediate host defense. Targeted disruption of the *aire* gene causes multiorgan autoimmunity, including anti-retinal antibodies, cellular responses to IRBP and a spontaneous EAU-like uveitis directed at IRBP (26, 27). Interestingly, although in AIRE-deficient mice thymic selection of T cells to S-Ag and other retinal Ags controlled by AIRE is also compromised, IRBP is the only Ag recognized by these mice as pathogenic (27).

The oldest spontaneous EAU model is in athymic ("nude") mice grafted with a neonatal rat thymus. The MHC-incompatible thymic tissue fails to efficiently eliminate the high-affinity autoreactive T cells. The spontaneous autoimmune disease that results resemble AIRE deficiency and mice exhibit an immune response directed against photoreceptors with specificity to IRBP, but not S–Ag (28).

2.2.3. Spontaneous Uveitis in Double Transgenic Retinal Neo-self Ag and Specific TCR Tg Mice

Transgenic mice expressing HEL in the eye as a neo-self Ag mice were generated under the control of the lens-specific promoter, αA-crystallin (29) or the retina-specific IRBP or rhodopsin promoters (13, 14). Spontaneous uveitis occurred when these HEL Tg mice were crossed with mice expressing a transgenic TCR that recognizes the HEL antigen (3A9 TCR Tg mice) by 6 weeks of age (13, 14, 29). In contrast, retinal β-gal crossed to β-gal TCR Tg mice do not seem to develop spontaneous uveitis (30). These neo-Ag x Ag-specific TCR double Tg mice are useful models to address Ag-specific questions in pathogenesis and regulation of uveitis. However, because the expression of these transgenes may be influenced by integration effects, causing an altered tissue distribution than native retinal Ag, findings in these mice must be interpreted with caution. These concerns are partly being addressed by development of IRBP-specific TCR Tg mice, described ahead.

In the aggregate, development of spontaneous uveitis in mice with increased proportion of retinal Ag-specific T cells, whether due to deficient elimination of autoreactive cells in the thymus or to transgenic expression of a retina-specific TCR, suggest that frequency of retinal Ag-specific cells can constitute an important factor in setting the threshold of EAU susceptibility.

3. Comparison of Different Animal Models

Due to ethical issues and other limitations affecting human studies, animal models of uveitis are an invaluable tool to understand basic mechanisms of disease. Because none of the animal models by itself can reproduce the full complexity of human uveitis, it is important to develop and use a variety of models to represent different aspects and diverse clinical and immunological manifestations of human disease. Each has its strengths and limitations, and data interpretation must be guided by what can, and what cannot, be concluded from each model.

The EAU model was initially developed in animals with relatively large eyes (e.g., rabbit, rat, and primate) (6), reflecting the interest in pathology and in therapeutics that dominated the field in earlier days. These models continue to be important in the pharmaceutical industry and preclinical testing, but the growing availability of well-defined inbred and genetically manipulated strains quickly catapulted the mouse model into the first place for basic immunological and genetic studies. Our current understanding of the mechanisms of uveitis is largely based on the "classical" EAU model induced by immunization with a retinal Ag in adjuvant. However, in order to induce EAU with high incidence the adjuvant enhances innate immune reactions and affects the type of response that is induced. Therefore, the data should be interpreted in comparison with adjuvant-free model(s), which are the spontaneous uveitis models and the adoptive transfer model. The latter has the added advantage that represents a pure effector situation, uncomplicated by emergence of newly stimulated, host-derived effector cells.

The spontaneous models avoid the use of active immunization and strong adjuvants, and in that regard are more reminiscent of the situation in humans, but they all rely on an artificially high frequency of Ag-specific uveitogenic T cells. That in itself, however, points to precursor frequency as one of the factors that may affect susceptibility. Interestingly, the spontaneous models with deficient central tolerance (AIRE-deficient and nude mice implanted with rat thymus) share the common underlying mechanism that the predominant retinal Ag is IRBP. The reason for this antigenic preference is unknown, but brings to mind the apparent "preference" for S–Ag in humans, which seems to be the most frequent retinal Ag recognized by lymphocytes of uveitis patients. HLA-Tg models, including the spontaneous model of uveitis in HLA-A29 Tg mice, are arguably the most relevant clinically, as they promise to help identify the molecules and their fragments that may be involved in driving human disease. Since humans express more than one HLA molecule, future studies should also include double-HLA-Tg mice, to study interactions between HLA molecules in determining Ag recognition, susceptibility, and resistance.

A summary of the comparison between models is shown in Table 3.

Table 3
Mouse models of human uveitis

Type of model (method of induction)	Examples of models	References	Comments
Active immunization with an ocular Ag emulsified in complete Freund's adjuvant (CFA–EAU)	EAU induced in B10.RIII or C57BL/6 mice with IRBP EAU in HLA-Tg mice with S-Ag Other Ags, or their peptide fragments, can be used	(10) (20)	Mycobacteria in CFA provide innate signals that polarize autoimmune lymphocytes towards the Th1 and Th17 lineages. PTX as additional adjuvant is needed in less susceptible strains
Infusion of Ag-pulsed mature dendritic cells (DC–EAU)	EAU variant induced in B10. RIII mice by injection of splenic DC elicited with Flt3L, matured in vitro and pulsed with IRBP p161–180	(23)	Requires two injections of DC and PTX. Less severe than CFA–EAU and appears mostly dependent on Th1 cells
Infusion of uveitogenic T cells (adoptive transfer)	Uveitis in mice transgenic for HEL or β-gal on an eye-specific promoter that are adoptively transferred with uveitogenic T cells. A variant is in vivo retrovirally induced expression of influenza HA in the retina and adoptive transfer of activated T cells specific for HA	(12, 13, 15)	Uveitogenic T cells needed to be activated
Deficient central tolerance	Spontaneous uveitis in AIRE deficient mice directed at IRBP, spontaneous uveitis in nude mice implanted with embryonic rat thymus	(26–28)	Higher frequency and higher affinity of cells bearing TCRs to IRBP
Double Tg mice expressing a neo-self Ag in the retina and the specific TCR	Spontaneous EAU-like uveitis in ocular HEL Tg mice × HEL TCR Tg mice	(13, 14, 29)	This may at least in part depend on the frequency and affinity of the transgenic TCR.
Mice transgenic for a TCR that recognizes a retinal Ag	Spontaneous EAU-like uveitis in IRBP-specific TCR Tg mice	Horai R et al., unpublished	High frequency of self TCR Tg T cells

4. Contributions of the Models in Understanding or Treatment of the Disease

4.1. Understanding Basic Mechanisms of Pathogenesis

Autoimmune inflammatory disease can be viewed as a failure of self tolerance. The process of central tolerance by selection on retinal Ags expressed in the thymus not only culls potentially pathogenic T cells, but also positively selects natural regulatory T cells (31, 32). Enhanced susceptibility to EAU of mice implanted with an IRBP deficient thymus (31), as well as spontaneous uveitis in AIRE-deficient mice and athymic mice transferred with rat thymus (26, 28), clearly demonstrated the importance of central tolerance in reducing susceptibility to EAU. Importantly, human thymic tissue from immunologically normal patients expresses variable yet detectable levels of tissue Ags, indicating that thymic control of the retina-specific T cell repertoire may also play a role in human uveitis (33).

Peripheral tolerance normally controls autoreactive T cells that escape thymic negative selection. This requires contact with Ags expressed in the healthy tissue, but retinal Ags are relatively sequestered from the immune system by the blood–retinal barrier, hindering peripheral tolerance (reviewed in (34)). Data obtained in retinal HEL-Tg mice with a normal T cell repertoire indicate that the immune system fails to recognize HEL in the retina, although the same Ag is recognized when expressed in the pancreas (14). This supports the notion that sequestration hinders peripheral tolerance. In keeping with this, forced expression of retinal Ags outside the eye results in improved peripheral tolerance and resistance to uveitis (35, 36). Recent data suggest that retinal expression of transgenic β-gal may generate some level of peripheral tolerance, but it appears too low to alter susceptibility of these mice to EAU induced with β-gal (35, 37) These studies lead to the conclusion that peripheral tolerance to Ags residing in the intact eye is relatively limited and supports the rationale for Ag-specific therapies aimed at enhancing peripheral tolerance.

In inflammatory and autoimmune disease, several T cell effector phenotypes have been defined, known as Th1 (IFN-γ producing), Th2 (IL-4, IL-5, and IL-13 producing) and Th17 (IL-17 producing). Because these signature cytokines are involved in the activity of these cells, knowledge of the effector responses driving uveitis has direct implications for therapy. Notably, both Th1 and Th17 responses have been associated with clinical uveitis (5, 38–40). Human uveitis is clinically heterogeneous even though patients often respond to the same retinal Ag(s). Because proving etiology in humans is difficult, animal models can step in as a tool to define causative relationships. Clues to the heterogeneous nature of human uveitis can be obtained from models of EAU. In the mouse EAU model, both Th1 and Th17 cells are induced in parallel. Th17 effector cells appear to play a critical role in EAU induced with IRBP immunization in CFA, but the EAU model induced with DC appears to be dependent on IFN-γ and the Th1 response (23, 41). Adoptive transfer experiments with Th17 and Th1 cells under conditions where the reciprocal signature cytokine was eliminated further determined that ocular autoimmunity can be driven by either a Th17 or a Th1 effector response. The dominant effector phenotype in a particular model appears to be influenced by the conditions under which Ag is first recognized by the immune system. Presentation of Ag in the context of bacterial products (mycobacteria are included in CFA) and by diverse APC present in the LN appear to promote a dominant Th17 response, whereas Ag presented by DC matured in vitro with LPS and anti-CD40 promotes a Th1 response. Importantly, despite having been induced with the same Ag in genetically identical mice, the CFA- and the DC-induced EAU models show differences both immunologically and clinically. It is thus conceivable that these EAU models shed light on the complex biology and the heterogeneous nature of human uveitis, and provide a more comprehensive representation of uveitic diseases of immune origin.

The regulatory T cells (Tregs) regulate induction and progression of immune responses. The contribution of Tregs in controlling human uveitic disease is not well-understood, but patients with uveitis associated with VKH disease appear to have reduced Treg function (42). In contrast to human uveitis that tends to be chronic with relapses and remissions, animal models of EAU are more monophasic and ultimately seem to resolve. Weak Treg activity may contribute to recurrent vs. monophasic uveitis in the rat EAU model (43, 44). Regulatory mechanisms, such as expansion of naturally occurring Tregs (nTregs) and generation of induced Tregs (iTregs) (32) have been studied in EAU models, but although humans are known to harbor T cells with regulatory activity, their role remains to be elucidated in uveitis patients.

The critical checkpoints in pathogenesis of EAU, as defined by studies in many laboratories over many years, are shown in Fig. 3.

4.2. Genetic Associations

Predisposition to uveitis follows a polygenic inheritance pattern, where no single gene is by itself necessary or sufficient to determine susceptibility. Studies in MHC-congenic mice indicated that susceptibility of EAU is controlled by both MHC and non-MHC genes (45). MHC genes are involved in Ag presentation, therefore MHC control is likely to be exerted at least in part through recognition of antigenic epitopes. To further identify pathogenic epitopes in the human uveitis and to study genetic mechanism of susceptibility to uveitis, the "humanized" HLA-transgenic models including HLA-A29 Tg mice and HLA class II Tg mice will serve as important tools. Non-MHC control is more complex, and can involve diverse mechanisms. Earlier studies had indicated

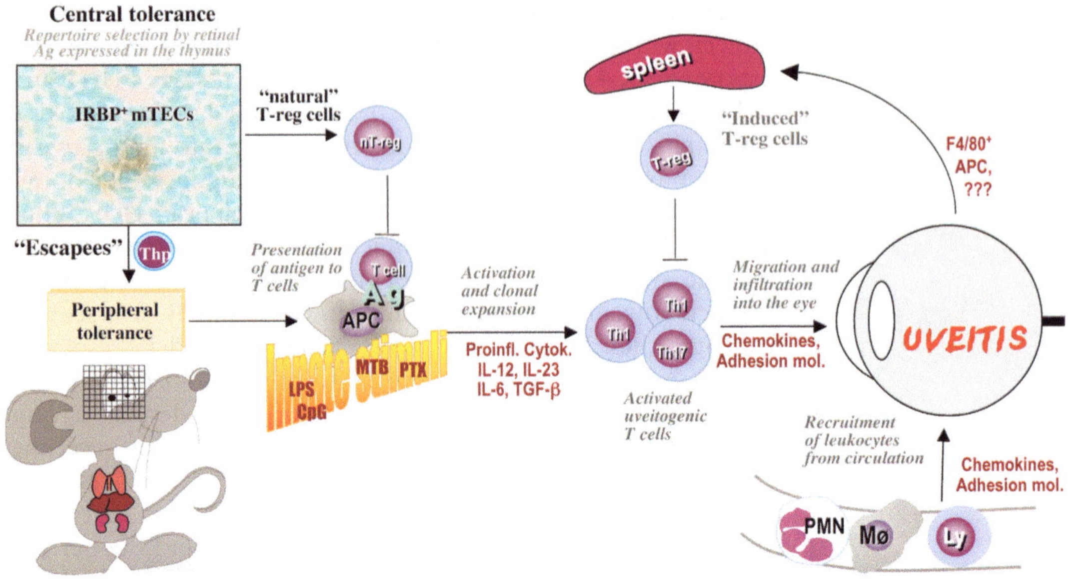

Fig. 3. Immunopathogenesis of uveitis, based on data from animal models of EAU. Autoreactive T cells that failed to be deleted in the thymus have limited access to retinal Ags due to their relative sequestration from the immune system, which hinders peripheral tolerance. Despite presence of natural Tregs, such non-tolerant thymic escapees may get primed by exposure to retinal or cross-reactive microbial Ag, and in the appropriate milieu differentiate to Th1 or Th17 effector lymphocytes that reach the eye, recognize their Ag in situ and recruit inflammatory leukocytes from the circulation. T regulatory cells that are induced during and as a result of the disease process ultimately control the pathogenic effector cells and downregulate inflammation. (Modified from (34))

that strains of rats and of mice genetically predisposed to mount a high Th1 response are more susceptible to EAU than Th2-biased strains (46, 47). More recent studies, performed after the Th17 effector lineage was defined and shown to participate in inflammatory disease, have indicated that at least the B10.RIII strain, which is particularly susceptible to EAU, is also a high Th17 responder (Luger D et al., unpublished). Much more work is needed to delineate the genetic loci involved in susceptibility and resistance, and to correlate them with functional effects.

4.3. Treatment of the Disease

Animal models of uveitis have traditionally served as a template for new therapeutic approaches. Successful modulation of the experimental disease has often been predictive of the efficacy of an investigational treatment in a clinical setting. Established therapies for uveitis are still largely based on general nonspecific immunosuppression (corticosteroids, antimetabolites, alkylating agents). Because of the severe side effects of these treatments, newer approaches aim to build on the increased understanding of basic disease mechanisms (Fig. 11.3), so as to target the pathogenic processes more specifically. The central involvement of T cells in pathogenesis makes them a natural target for therapy.

Cyclosporine, a T cell-targeting drug now in general use for ocular inflammation, was shown to have a therapeutic effect in the rat EAU model before going to clinical trials (48), as have the second-generation T cell targeting agents tacrolimus and sirolimus. Other studies have examined IL-2 receptor-directed therapy with monoclonal antibodies (daclizumab) as an approach to target activated T cells. This therapy is now in advanced clinical trials (49–51).

Ag-nonspecific therapies such as the ones above target common functions of activated T cells and, therefore, have the potential to negatively impact resistance to infections and cancer. Approaches targeting only the Ag-specific cells, based on tolerance-inducing regimens have the potential to overcome these limitations, but require knowledge of the inciting Ags. Experimental regimens have included forced expression of retinal Ags in the periphery (52, 53), infusion of tolerogenic Ag-pulsed DC and induction oral tolerance by feeding Ag. The last approach has shown promise in a clinical trial (2). Ag-specific therapies silence or delete the effector cells directly, and/or induce regulatory cells that can be expanded ex vivo (53). Notably, however, Ag-specific approaches pose a risk inherent in exposing the patient to an Ag to which there is preexisting immunity. The pros and the cons in the form of unintended side effects of the various immunotherapeutic approaches have recently been reviewed (54).

As a small and relatively closed organ, the eye lends itself to local therapies that have the potential to avoid systemic side effects. Intravitreal injections and intravitreal implants are already in use for local therapies. Biological products as well can be delivered into the eye. Locally produced IL-10 has been shown to be beneficial in animal models (55). This opens the possibility for intraocular injection of other anti-inflammatory molecules or T-regulatory cells, which can be isolated and expanded in vitro and are already known to control uveitis in animal models upon systemic injection (53). It will be important to deliver minimally invasive and highly efficient local drug delivery systems. A promising approach would be transduction of ocular tissues with therapeutic molecules to be produced in situ (56) or with molecules interfering with expression of inflammatory mediators.

5. New Developments and Future Opportunities

5.1. Further Insights into Ag-Specific Responses and Cells

Improved experimental tools are being developed that promise to refine our understanding of disease mechanisms and open new avenues into therapy. We have recently engineered a Tg mouse on the EAU-susceptible B10.RIII background that expresses a TCR specific to the major uveitogenic epitope of IRBP. The IRBP TCR

Tg mice generated IRBP-specific effector Th1, Th17, and T-regulatory cells in vivo and rapidly developed spontaneous uveitis, starting around 5 weeks of age and reaching 80% incidence by 8 weeks (Horai R et al., unpublished). These mice and the IRBP-specific T cells derived from them will make possible the study of pathogenic and regulatory T cells specific to a native retinal Ag, obviating the pitfalls inherent in neo-self Ag systems.

To advance development of Ag-specific therapies and diagnostics, it is important to have the ability to detect Ag-specific cells reliably and sensitively. In the mouse model, the uveitogenic IRBP epitope-specific reagent was developed in our laboratory (57). This reagent specifically binds to T cells expressing IRBP161-180-specific T cell receptors. We are currently developing similar reagents in the HLA-Tg model to detect uveitogenic fragments of S–Ag presented by human class II molecules (Mattapallil MJ et al., unpublished). Diagnostic use of HLA-Ag tetramers is being intensively explored in type 1 diabetes (58). Detection of retinal Ag-specific T cells using retinal Ag-specific reagents may have diagnostic or prediction potential in patients who suffer from or are at risk for developing uveitis.

Acknowledgments

We thank Dr. Chi-Chao Chan for providing EAU scoring criteria as well as histology pictures of EAU and human uveitis. We also thank Dr. Yishay Falick for providing mouse fundus pictures.

References

1. Gritz DC, Wong IG (2004) Incidence and prevalence of uveitis in Northern California; the Northern California Epidemiology of Uveitis Study. Ophthalmology 111(3):491–500 discussion
2. Nussenblatt RB, Whitcup SM (2004) Uveitis: fundamentals and clinical practice, 3rd edn. Mosby (Elsevier), Philadelphia, PA
3. Pennesi G, Caspi RR (2002) Genetic control of susceptibility in clinical and experimental uveitis. Int Rev Immunol 21(2–3):67–88
4. Caspi RR (2006) Animal models of autoimmune and immune-mediated uveitis. Drug Discov Today Dis Models 3:3–10 (http://dxdoiorg/101016/jddmec200605006)
5. Arayssi T, Hamdan A (2004) New insights into the pathogenesis and therapy of Behcet's disease. Curr Opin Pharmacol 4(2):183–188
6. Gery I, Nussenblatt RB, Chan CC, Caspi RR (2002) Autoimmune diseases of the eye. The molecular pathology of autoimmune diseases, 2nd edn. Taylor and Francis, New York, NY, pp 978–998
7. Damico FM, Kiss S, Young LH (2005) Vogt–Koyanagi–Harada disease. Semin Ophthalmol 20(3):183–190
8. Monnet D, Brezin AP (2006) Birdshot chorioretinopathy. Curr Opin Ophthalmol 17(6):545–550
9. Ding X, Patel M, Chan CC (2009) Molecular pathology of age-related macular degeneration. Prog Retin Eye Res 28(1):1–18
10. Agarwal RK, Caspi RR (2004) Rodent models of experimental autoimmune uveitis. Methods Mol Med 102:395–419
11. Lai JC, Fukushima A, Wawrousek EF et al (1998) Immunotolerance against a foreign antigen transgenically expressed in the lens. Invest Ophthalmol Vis Sci 39(11):2049–2057

12. Gregerson DS, Torseth JW, McPherson SW, Roberts JP, Shinohara T, Zack DJ (1999) Retinal expression of a neo-self antigen, beta-galactosidase, is not tolerogenic and creates a target for autoimmune uveoretinitis. J Immunol 163(2):1073–1080
13. Ham DI, Kim SJ, Chen J et al (2004) Central immunotolerance in transgenic mice expressing a foreign antigen under control of the rhodopsin promoter. Invest Ophthalmol Vis Sci 45(3):857–862
14. Lambe T, Leung JC, Ferry H et al (2007) Limited peripheral T cell anergy predisposes to retinal autoimmunity. J Immunol 178(7):4276–4283
15. Terrada C, Fisson S, De Kozak Y et al (2006) Regulatory T cells control uveoretinitis induced by pathogenic Th1 cells reacting to a specific retinal neoantigen. J Immunol 176(12):7171–7179
16. Caspi RR, Chan CC, Leake WC, Higuchi M, Wiggert B, Chader GJ (1990) Experimental autoimmune uveoretinitis in mice. Induction by a single eliciting event and dependence on quantitative parameters of immunization. J Autoimmun 3(3):237–246
17. Xu H, Koch P, Chen M, Lau A, Reid DM, Forrester JV (2008) A clinical grading system for retinal inflammation in the chronic model of experimental autoimmune uveoretinitis using digital fundus images. Exp Eye Res 87(4):319–326
18. Copland DA, Wertheim MS, Armitage WJ, Nicholson LB, Raveney BJ, Dick AD (2008) The clinical time-course of experimental autoimmune uveoretinitis using topical endoscopic fundal imaging with histologic and cellular infiltrate correlation. Invest Ophthalmol Vis Sci 49(12):5458–5465
19. Taneja V, David CS (1999) HLA class II transgenic mice as models of human diseases. Immunol Rev 169:67–79
20. Pennesi G, Mattapallil MJ, Sun SH et al (2003) A humanized model of experimental autoimmune uveitis in HLA class II transgenic mice. J Clin Invest 111(8):1171–1180
21. de Smet MD, Bitar G, Roberge FG, Gery I, Nussenblatt RB (1993) Human S-antigen: presence of multiple immunogenic and immunopathogenic sites in the Lewis rat. J Autoimmun 6(5):587–599
22. Lai JC, Lobanoff MC, Fukushima A et al (1999) Uveitis induced by lymphocytes sensitized against a transgenically expressed lens protein. Invest Ophthalmol Vis Sci 40(11):2735–2739
23. Tang J, Zhu W, Silver PB, Su SB, Chan CC, Caspi RR (2007) Autoimmune uveitis elicited with antigen-pulsed dendritic cells has a distinct clinical signature and is driven by unique effector mechanisms: initial encounter with autoantigen defines disease phenotype. J Immunol 178(9):5578–5587
24. Szpak Y, Vieville JC, Tabary T et al (2001) Spontaneous retinopathy in HLA-A29 transgenic mice. Proc Natl Acad Sci U S A 98(5):2572–2576
25. de Kozak Y, Camelo S, Pla M (2008) Pathological aspects of spontaneous uveitis and retinopathy in HLA-A29 transgenic mice and in animal models of retinal autoimmunity: relevance to human pathologies. Ophthalmic Res 40(3–4):175–180
26. Anderson MS, Venanzi ES, Klein L et al (2002) Projection of an immunological self shadow within the thymus by the aire protein. Science 298(5597):1395–1401
27. DeVoss J, Hou Y, Johannes K et al (2006) Spontaneous autoimmunity prevented by thymic expression of a single self-antigen. J Exp Med 203(12):2727–2735
28. Ichikawa T, Taguchi O, Takahashi T et al (1991) Spontaneous development of autoimmune uveoretinitis in nude mice following reconstitution with embryonic rat thymus. Clin Exp Immunol 86(1):112–117
29. Zhang M, Vacchio MS, Vistica BP et al (2003) T cell tolerance to a neo-self antigen expressed by thymic epithelial cells: the soluble form is more effective than the membrane-bound form. J Immunol 170(8):3954–3962
30. McPherson SW, Heuss ND, Gregerson DS (2009) Lymphopenia-induced proliferation is a potent activator for CD4+ T cell-mediated autoimmune disease in the retina. J Immunol 182(2):969–979
31. Avichezer D, Grajewski RS, Chan CC et al (2003) An immunologically privileged retinal antigen elicits tolerance: major role for central selection mechanisms. J Exp Med 198(11): 1665–1676
32. Grajewski RS, Silver PB, Agarwal RK et al (2006) Endogenous IRBP can be dispensable for generation of natural CD4+CD25+ regulatory T cells that protect from IRBP-induced retinal autoimmunity. J Exp Med 203(4):851–856
33. Takase H, Yu CR, Mahdi RM et al (2005) Thymic expression of peripheral tissue antigens in humans: a remarkable variability among individuals. Int Immunol 17(8): 1131–1140
34. Caspi RR (2006) Ocular autoimmunity: the price of privilege? Immunol Rev 213:23–35
35. McPherson SW, Roberts JP, Gregerson DS (1999) Systemic expression of rat soluble

retinal antigen induces resistance to experimental autoimmune uveoretinitis. J Immunol 163(8):4269–4276
36. Xu H, Wawrousek EF, Redmond TM et al (2000) Transgenic expression of an immunologically privileged retinal antigen extraocularly enhances self tolerance and abrogates susceptibility to autoimmune uveitis. Eur J Immunol 30(1):272–278
37. Gregerson DS, Heuss ND, Lehmann U, McPherson SW (2008) Evidence for extrathymic generation of regulatory T cells specific for a retinal antigen. Ophthalmic Res 40(3–4):154–159
38. Amadi-Obi A, Yu CR, Liu X et al (2007) TH17 cells contribute to uveitis and scleritis and are expanded by IL-2 and inhibited by IL-27/STAT1. Nat Med 13(6):711–718
39. Chi W, Yang P, Li B et al (2007) IL-23 promotes CD4+ T cells to produce IL-17 in Vogt–Koyanagi–Harada disease. J Allergy Clin Immunol 119(5):1218–1224
40. Chi W, Zhu X, Yang P et al (2008) Upregulated IL-23 and IL-17 in behcet patients with active uveitis. Invest Ophthalmol Vis Sci 49(7):3058–3064
41. Luger D, Silver PB, Tang J et al (2008) Either a Th17 or a Th1 effector response can drive autoimmunity: conditions of disease induction affect dominant effector category. J Exp Med 205(4):799–810
42. Chen L, Yang P, Zhou H et al (2008) Decreased frequency and diminished function of CD4+CD25high regulatory T cells are associated with active uveitis in patients with Vogt–Koyanagi–Harada syndrome. Invest Ophthalmol Vis Sci 49(8):3475–3482
43. Ke Y, Jiang G, Sun D, Kaplan HJ, Shao H (2008) Ocular regulatory T cells distinguish monophasic from recurrent autoimmune uveitis. Invest Ophthalmol Vis Sci 49(9):3999–4007
44. Han G, Shao H, Peng Y et al (2007) Suppressor role of rat CD8+CD45RClow T cells in experimental autoimmune uveitis (EAU). J Neuroimmunol 183(1–2):81–88
45. Caspi RR, Grubbs BG, Chan CC, Chader GJ, Wiggert B (1992) Genetic control of susceptibility to experimental autoimmune uveoretinitis in the mouse model. Concomitant regulation by MHC and non-MHC genes. J Immunol 148(8):2384–2389
46. Caspi RR, Silver PB, Chan CC et al (1996) Genetic susceptibility to experimental autoimmune uveoretinitis in the rat is associated with an elevated Th1 response. J Immunol 157(6):2668–2675
47. Sun B, Sun SH, Chan CC, Wiggert B, Caspi RR (1999) Autoimmunity to a pathogenic retinal antigen begins as a balanced cytokine response that polarizes towards type 1 in a disease-susceptible and towards type 2 in a disease-resistant genotype. Int Immunol 11(8):1307–1312
48. Nussenblatt RB, Dinning WJ, Fujikawa LS, Chan CC, Palestine AG (1985) Local cyclosporine therapy for experimental autoimmune uveitis in rats. Arch Ophthalmol 103(10):1559–1562
49. Nussenblatt RB, Fortin E, Schiffman R et al (1999) Treatment of noninfectious intermediate and posterior uveitis with the humanized anti-Tac mAb: a phase I/II clinical trial. Proc Natl Acad Sci U S A 96(13):7462–7466
50. Nussenblatt RB (2002) Bench to bedside: new approaches to the immunotherapy of uveitic disease. Int Rev Immunol 21(2–3):273–289
51. Yeh S, Wroblewski K, Buggage R et al (2008) High-dose humanized anti-IL-2 receptor alpha antibody (daclizumab) for the treatment of active, non-infectious uveitis. J Autoimmun 31(2):91–97
52. Agarwal RK, Kang Y, Zambidis E, Scott DW, Chan CC, Caspi RR (2000) Retroviral gene therapy with an immunoglobulin-antigen fusion construct protects from experimental autoimmune uveitis. J Clin Invest 106(2):245–252
53. Silver PB, Agarwal RK, Su SB et al (2007) Hydrodynamic vaccination with DNA encoding an immunologically privileged retinal antigen protects from autoimmunity through induction of regulatory T cells. J Immunol 179(8):5146–5158
54. Caspi RR (2008) Immunotherapy of autoimmunity and cancer: the penalty for success. Nat Rev Immunol 8(12):970–976
55. Smith JR, Verwaerde C, Rolling F et al (2005) Tetracycline-inducible viral interleukin-10 intraocular gene transfer, using adeno-associated virus in experimental autoimmune uveoretinitis. Hum Gene Ther 16(9):1037–1046
56. Bejjani RA, Andrieu C, Bloquel C, Berdugo M, BenEzra D, Behar-Cohen F (2007) Electrically assisted ocular gene therapy. Surv Ophthalmol 52(2):196–208
57. Karabekian Z, Lytton SD, Silver PB, Sergeev YV, Schneck JP, Caspi RR (2005) Antigen/MHC class II/Ig dimers for study of uveitogenic T cells: IRBP p161–180 presented by both IA and IE molecules. Invest Ophthalmol Vis Sci 46(10):3769–3776

58. Buckner JH, Holzer U, Novak EJ, Reijonen H, Kwok WW, Nepom GT (2002) Defining antigen-specific responses with human MHC class II tetramers. J Allergy Clin Immunol 110(2):199–208
59. Cortes LM, Mattapallil MJ, Silver PB et al (2008) Repertoire analysis and new pathogenic epitopes of IRBP in C57BL/6 (H-2b) and B10.RIII (H-2r) mice. Invest Ophthalmol Vis Sci 49(5):1946–1956
60. Silver PB, Rizzo LV, Chan CC, Donoso LA, Wiggert B, Caspi RR (1995) Identification of a major pathogenic epitope in the human IRBP molecule recognized by mice of the H-2r haplotype. Invest Ophthalmol Vis Sci 36(5):946–954
61. Namba K, Ogasawara K, Kitaichi N et al (1998) Identification of a peptide inducing experimental autoimmune uveoretinitis (EAU) in H-2Ak-carrying mice. Clin Exp Immunol 111(2):442–449

Index

MIX
Papier aus verantwortungsvollen Quellen
Paper from responsible sources
FSC® C105338

If you have any concerns about our products,
you can contact us on
ProductSafety@springernature.com

In case Publisher is established outside the EU,
the EU authorized representative is:
Springer Nature Customer Service Center GmbH
Europaplatz 3, 69115 Heidelberg, Germany

Printed by Libri Plureos GmbH
in Hamburg, Germany